Disease Control in Crops

Disease Control in Crops

Biological and Environmentally Friendly Approaches

Edited by

Dale Walters
Crop and Soil Systems Research Group
Scottish Agricultural College
Edinburgh, UK

WILEY-BLACKWELL
A John Wiley & Sons, Ltd., Publication

This edition first published 2009

Blackwell Publishing was acquired by John Wiley & Sons in February 2007. Blackwell's publishing programme has been merged with Wiley's global Scientific, Technical, and Medical business to form Wiley-Blackwell.

Registered office
John Wiley & Sons Ltd, The Atrium, Southern Gate, Chichester, West Sussex, PO19 8SQ, United Kingdom

Editorial offices
9600 Garsington Road, Oxford, OX4 2DQ, United Kingdom
2121 State Avenue, Ames, Iowa 50014-8300, USA

For details of our global editorial offices, for customer services and for information about how to apply for permission to reuse the copyright material in this book please see our website at www.wiley.com/wiley-blackwell.

A catalogue record for this book is available from the British Library and the Library of Congress.

ISBN: 978-1-4051-6947-9

Contents

List of contributors

Dr Botond Balogh	Plant Pathology Department, University of Florida 1453 Fifield Hall, Gainesville, FL 32611, USA E-mail: bbalogh@ufl.edu
	Department of Plant Pathology and Ecology The Connecticut Agricultural Experiment Station 123 Huntington Street, New Haven, CT 06504, USA
Dr Ian Bingham	Crop & Soil Systems Research Group Scottish Agricultural College, King's Buildings West Mains Road, Edinburgh EH9 3JG, UK E-mail: ian.bingham@sac.ac.uk
Dr Elke Bloem	Institute for Crop and Soil Science Julius Kühn-Institut, Federal Research Centre for Cultivated Plants, Bundesallee 50 D-38116 Braunschweig, Germany E-mail: elke.bloem@jki.bund.de
Dr Abraham Gamliel	Laboratory for Pest Management Research, Institute of Agricultural Engineering, ARO Volcani Center, Bet Dagan, 50250 Israel E-mail: agamliel@volcani.agri.gov.il
Dr Silvia Haneklaus	Institute for Crop and Soil Science Julius Kühn-Institut, Federal Research Centre for Cultivated Plants, Bundesallee 50 D-38116 Braunschweig, Germany E-mail: silvia.haneklaus@jki.bund.de
Professor Jeffrey Jones	Plant Pathology Department, University of Florida Fifield Hall, P.O. Box 110680 Gainesville, FL 32611, USA E-mail: jbjones@ufl.edu
Professor Jaacov Katan	Department of Plant Pathology and Microbiology Faculty of Agricultural, Food and Environmental Sciences, The Hebrew University, Rehovot 76100 Israel E-mail: katan@agri.huji.ac.il

Dr John Kirkegaard CSIRO Plant Industry, GPO Box 1600, Canberra
Australia 2601
E-mail: john.kirkegaard@csiro.au

Dr Audrey Litterick SAC Environmental, Scottish Agricultural College
Craibstone, Aberdeen AB21 9YA, UK
E-mail: audrey.litterick@sac.co.uk

Dr Mark P. McQuilken Life Sciences Teaching Group, Scottish Agricultural College, Ayr Campus, Auchincruive Estate
Auchincruive, Ayr KA6 5HW, UK
E-mail: mark.mcquilken@sac.ac.uk

Dr Timur Momol Plant Pathology Department, University of Florida,
1052 McCarty Hall,
Gainesville, FL 32611, USA
E-mail: tmomol@ufl.edu

Dr Adrian Newton Scottish Crop Research Institute, Invergowrie,
Dundee DD2 5DA, UK
E-mail: Adrian.newton@scri.ac.uk

Dr Aleksa Obradovic Plant Pathology Department, Faculty of Agriculture
University of Belgrade, Nemanjina 6, 11080
Belgrade-Zemun, Serbia
E-mail: aleksao@agrifaculty.bg.ac.yu

Dr Tony Reglinski Plant and Food Research, Ruakura Research Centre,
East Street, Private Bag 3123, Hamilton
New Zealand
E-mail: treglinski@hortresearch.co.nz

Professor Ewald Schnug Institute for Crop and Soil Science,
Julius Kühn-Institut, Federal Research Centre
for Cultivated Plants, Bundesallee 50, D-38116
Braunschweig, Germany
E-mail: popschnug@jki.bund.de

Dr Hugh Wallwork South Australian Research and Development Institute
GPO Box 397, Adelaide, SA 5001, Australia
E-mail: hugh.wallwork@sa.gov.au

Professor Dale Walters Crop & Soil Systems Research Group
Scottish Agricultural College, King's Buildings
West Mains Road, Edinburgh EH9 3JG, UK
E-mail: dale.walters@sac.ac.uk

Professor John M. Whipps Warwick HRI, University of Warwick, Wellesbourne
Warwick CV35 9EF, UK
E-mail: john.whipps@warwick.ac.uk

Dr Martin Wood Earthcare Environmental Ltd
PO Box 8036, Reading RG6 9ED, UK
E-mail: martin@earthcare.co.uk

Preface

Although more food is produced than required to feed the entire world population, inadequate distribution systems mean that in various parts of the world, many people go hungry. Nevertheless, increasing global population growth, urbanisation and rising *per capita* incomes will require continuing increases in agricultural productivity.

However, simply maintaining food production requires crops to be adequately protected against disease. Devastating pathogen epidemics are not a thing of the past and the last decade has seen, for example, major epidemics of cassava mosaic virus in West Africa, rhizomania disease of sugar beet in the United Kingdom and Europe, and soybean rust in Asia, South America and more recently the United States. Tackling this pathogen onslaught requires a range of weapons, including fungicides and host resistance, both of which continue to make an enormous contribution to plant disease control. But pathogens are highly variable and adaptable, and failures in both fungicide sensitivity and host plant resistance still occur with monotonous regularity. There is a need, therefore, to regularly strengthen the disease control armoury. The need to have a diversity of control measures, coupled with increasing concern with environmental protection, has led to renewed interest in biologically based and environmentally friendly approaches to disease control in crops. The chapters in this book cover a wide range of such approaches, starting with cultural control, the foundation for successful disease management in crops. Some of the approaches dealt with in this book are already in practice in various parts of the world, while others are several years away from practice. Approaches range from tolerance to plant disease through to use of bacteriophages in disease control.

It is my belief that effective, lasting disease control requires a sound understanding of crop ecology, both above and below ground. Using control measures in the absence of such understanding is the equivalent of firing blank ammunition at the enemy – at the very least, it is asking for trouble.

Dale Walters
June 2008

Chapter 1
Introduction

Dale Walters

Scottish Agricultural College, Edinburgh, UK

1.1 The importance of plant disease

Plant disease has been a continual problem for mankind ever since the beginnings of agriculture some 10 000 years ago. These problems persist, despite the advent of fungicides and resistant varieties, due to the genetic adaptability of the pathogens which cause plant disease. Although crop losses at farm and local level can have serious implications for growers, plant disease can inflict much more serious damage on a larger scale. The often-cited potato blight epidemic of the 1840s in Europe is an example of how devastating plant disease can be. This disease, caused by the Oomycete pathogen *Phytophthora infestans,* decimated crops across Europe, and in Ireland, led to the death of some one million people and the emigration of several million more (Large, 1940; Strange, 2003). Incredibly, today, more than 170 years later, potato blight still poses a major problem for potato growers across the globe. The devastation caused by potato blight is but one example of the damage plant diseases can inflict on mankind. Thus, in the Great Bengal Famine of 1943, the fungal pathogen *Cochliobolus miyabeanus* devastated rice crops and led to the starvation and death of an estimated two million people (Padmanabhan, 1973), while in Ceylon (now Sri Lanka), the arrival of the rust fungus *Hemileia vastatrix* in 1875 heralded the beginning of the end of coffee growing on that island. Here, coffee rust had virtually wiped out coffee plantations by 1889, forcing the islanders to switch to growing tea (Schumann, 1991).

1.1.1 Crop losses due to disease

It might be expected that given the historical importance of plant disease in arable agriculture, accurate estimates of crop losses as a result of disease would be readily available. In reality, the availability of quantitative data on the effects of pathogens on crop losses is very limited (Oerke, 2006). This should not be surprising, since as pointed out by Oerke (2006), the generation of such data is laborious and time-consuming and to complicate matters, crop losses will vary between seasons because of variations in pathogen incidence and disease severity.

Crop loss as a result of disease can be expressed in absolute terms (e.g. $kg\,ha^{-1}$) or relative terms (e.g. % loss), while the loss rate can be expressed as the proportion of attainable yield, although the proportion of the actual yield is sometimes used

Table 1.1 Estimated loss potential and actual losses due to pathogens (fungi and bacteria) in six major crops worldwide in 2001–2003 (adapted from Oerke, 2006, © Cambridge University Press, reproduced with permission).

	Crop losses (%) due to pathogens	
Crop	**Potential**	**Actual**
Wheat	15.6 (12–20)	10.2 (5–14)
Rice	13.5 (10–15)	10.8 (7–16)
Maize	9.4 (8–13)	8.5 (4–14)
Potatoes	21.2 (20–23)	14.5 (7–24)
Soybeans	11.0 (7–16)	8.9 (3–16)
Cotton	8.5 (7–10)	7.2 (5–13)

(Oerke, 2006). Loss rates can be looked at in terms of potential loss and actual loss. Potential loss compares yields in a system without any form of crop protection treatment, with yields from a system with a similar intensity of crop production, but receiving crop protection treatments. Actual losses are those sustained despite the use of crop protection (Oerke, 2006). The efficacy of crop protection can be calculated as the percentage of potential losses prevented. Table 1.1 illustrates the potential and actual losses as a result of disease for a range of crops. Potential losses range from 8.5% for cotton to 21.2% for potatoes, while actual losses range from 7.2% for cotton to 14.5% for potatoes (Table 1.1). These figures indicate the importance of crop protection in reducing potential losses in all of these crops. It is important to note here that disease-induced crop losses and the efficacy of crop protection practices will vary with geographical area, as a result of differences in cropping intensity, climatic conditions and cropping systems (Oerke, 2006).

A comparison of the actual losses due to disease in wheat and maize in the period 1964–2003 shows increases in crop losses from 9.1% to 12.6% in wheat and from 9.4% to 11.2% in maize (Table 1.2). In contrast, although the actual losses due to disease in cotton increased from 9.1% to 10.5% from 1964 to 1990, actual losses dropped to 7.9% by 2003 (Table 1.2). According to Oerke (2006), the differences in the estimates of actual crop losses over this period are likely to be the result of several factors, including:

- The use of varieties with high yield potential, but high susceptibility to pathogens
- Increased fertiliser use, further increasing susceptibility to some pathogens
- Large-scale cropping of genetically uniform plants, providing ideal conditions for rapid pathogen spread
- Expansion of crops into less suitable regions with higher incidence of other pathogens; here, less well-adapted, but high-yielding varieties replace well-adapted local varieties
- The import and spread of pathogens into regions without the natural restrictions (e.g., climate, natural enemies) of the region of origin.

Globally, agricultural production has grown faster than the human population over the past few decades (Hazell & Wood, 2008). In most parts of the world, this has been

Table 1.2 Estimates of actual losses due to diseases in worldwide production of wheat, maize and cotton for the years 1964/65, 1988–90 and 2001–03 (adapted from Oerke, 2006, © Cambridge University Press, reproduced with permission).

Period	Actual loss (%)
Wheat	
1964/65[a]	9.1
1988–90[b]	12.4
2001–03	12.6
Maize	
1964/65[a]	9.4
1988–90[b]	10.8
2001–03	11.2
Cotton	
1964/65[a]	9.1
1988–90[b]	10.5
2001–03	7.9

[a]From Cramer (1967).
[b]From Oerke *et al.* (1994).

the result, not of increased area of cropped land, but of increased inputs, including pesticides (Hazell & Wood, 2008). In the period from 1963 to 2002, cereal yields increased by 114% globally, although the annual rate of growth fell from 3.14% in the period 1963–1976, to 0.84% in the period 1989–2002 (Hazell & Wood, 2008). In the period from 1960 to 2004, pesticide sales worldwide increased more than 10-fold to some $30 billion (Oerke, 2006). However, despite this increased pesticide use, crop losses as a result of pests, diseases and weeds have not fallen significantly in the past 40 years.

1.2 Problems associated with controlling plant disease

Plant disease can be controlled using a variety of approaches. The first line of defence is the exclusion of the pathogen through plant quarantine and, for example, the use of pathogen-free propagating material. The next line of defence is to exclude, eliminate or reduce pathogen inoculum. This can be achieved in various ways, including cultural control (see Chapter 2), use of host plant resistance (see Chapter 6) and chemical control. Cultural methods provide the foundation for disease control in crops. However, in many parts of the world, diverse ecosystems have been replaced with simple agro-ecosystems, which are more vulnerable to pathogen attack. The planting of large areas of genetically uniform crops facilitates pathogen spread and in conjunction with widespread use of race-specific resistance, leads to the appearance of new strains of the pathogen able to infect and colonise previously resistant crop varieties. Cereal powdery mildews are a case in point. These fungi can evolve rapidly to overcome host resistance without apparent loss of fitness (Bronson & Ellingboe, 1986; Brown, 2003). In some interesting work, the barley powdery mildew fungus *Blumeria graminis* f. sp. *hordei* was found to possess a repertoire of avirulence (*AVR*) genes, which function as effectors and contribute to virulence (Ridout *et al.*, 2006). The authors suggested that multiple copies of related but distinct AVR effector paralogues might enable populations of *B. graminis* f. sp. *hordei* to rapidly overcome host resistance genes while maintaining virulence (Ridout *et al.*, 2006).

Fungicides are an integral part of crop production in most parts of the world. Without fungicides, crop losses in the region of 10% for wheat and rice, for example, would be considerably higher (Oerke, 2006). In soybean, yield losses due to the rust *Phakopsora pachyrhizi* are reported to be in the region of 80%, with projected losses in the Mato Grosso and Bahia areas of Brazil alone amounting to 2.2 million tonnes ($487 m). These losses were incurred despite the use of fungicides at a cost of $544 m (Yorinori *et al.*, 2005). One of the problems with repeated, high frequency use of fungicides is that fungi can become insensitive to them. The development of fungicide resistance in pathogen populations has been a major problem since the use of single-site mode of action fungicides became widespread. For example, the benzimidazole group of fungicides was introduced in the 1970s and disruptive selection for resistance occurred quickly in a number of pathogens leading to a loss of fungicide efficacy in the field. The mid-1990s saw the introduction of the Quinone outside Inhibitor (QoI) group of fungicides, which includes the strobilurins. These fungicides quickly took a large share of the cereal fungicide market because of their effectiveness against important cereal pathogens. Although the risk of resistance developing was predicted to be moderate, the first signs of resistance were reported within two years in the wheat powdery mildew pathogen, *Blumeria graminis* f. sp. *tritici*, in northern Germany. By 2000, resistance to this group of fungicides in populations of wheat powdery mildew was widespread throughout north west Europe and in 2002, isolates of *Mycosphaerella graminicola* (anamorph *Septoria tritici*, the cause of septoria leaf spot) resistant to strobilurin fungicides were detected in five European countries (Leadbeater, 2005). Resistance to QoI fungicides has also been detected in other cereal pathogens, including *Pyrenophora tritici-repentis* on wheat and *Drechslera teres* on barley (Leadbeater, 2005). Moreover, in a recent study, 4200 isolates of *Alternaria solani* were collected in the 5-year period from 2002 to 2006 from 11 potato-producing states in the United States. Of these isolates, 96% exhibited reduced sensitivity to QoI fungicides and/or had the F129L mutation in the cytochrome *b* gene (Pasche & Gudmestad, 2008).

In the past 50 years, pesticide use has greatly increased the quantity and improved the quality of food for the increasing world population. However, as pesticide use increased, so did concern about their adverse effects on non-target organisms, including humans. Pesticide poisoning of non-target organisms has been identified as a cause of mortality in fish, reproductive failure in birds and illness in humans (Rao *et al.*, 1993). Increasing public concern about the accumulation of pesticides in the environment and the impact on non-target organisms has led to the introduction of rigorous regulatory processes (Holm *et al.*, 2005; Stark, 2008). Nevertheless, there is continued concern over the impacts of pesticides on wildlife, including invertebrate populations, wild plants and farmland birds (Best & Burn, 2005). These concerns have led to reviews of active substances used in plant protection, with the resulting withdrawal of an increasing number of crop protection products from the market (Richardson, 2005). This has created problems and in some situations, effective control measures are no longer available to meet all the challenges posed by pathogens, pests and weeds (Richardson, 2005).

1.3 Conclusions

The continued ability of pathogens to overcome host resistance genes and to develop resistance to fungicides seriously erodes our ability to provide effective, lasting disease control on important crops. These problems, combined with the withdrawal of active substances from the market and increasing public concern with the effects of pesticides on the environment creates a huge challenge for plant pathology in the future. Plant disease control has an important role to play in our efforts to feed the ten billion (Evans, 1998). However, providing effective and lasting disease control, without harming the environment, will require more than a sensible approach to the use of host resistance and fungicides. It also requires a well-developed arsenal that includes a range of innovative approaches. In some situations, innovative control methods might be used to complement existing approaches. In other cases, for example for those diseases for which no adequate control measures exist, innovative control options might provide the only solution. The following chapters examine the use of a range of biological and environmentally friendly approaches to control plant diseases. It starts with cultural control, which, although not necessarily innovative, provides the foundation for disease control in crops.

1.4 Acknowledgements

Scottish Agricultural College receives financial support from the Scottish Government Rural and Environment Research and Analysis Directorate (RERAD).

1.5 References

Best J, Burn A, 2005. 'Polluter pays' or 'public good': tackling indirect effects of pesticides. *Proceedings of the BCPC International Congress – Crop Science and Technology 2005,* Vol. 2, 1097–1102.

Bronson CR, Ellingboe AH, 1986. The influence of four unnecessary genes for virulence on the fitness of *Erysiphe graminis* f. sp. *tritici. Phytopathology* **76,** 154–158.

Brown JKM, 2003. Little else but parasites. *Science* **299,** 1680–1681.

Cramer HH, 1967. Plant protection and world crop production. *Bayer Pflanzenschutz-Nachrichten* **20,** 1–524.

Evans LT, 1998. *Feeding the Ten Billion: Plants and Population Growth.* Cambridge: Cambridge University Press.

Hazell P, Wood S, 2008. Drivers of change in global agriculture. *Philosophical Transactions of the Royal Society B* **363,** 495–515.

Holm RE, Baron JJ, Kunkel D, 2005. The IR-4 programme and its cooperation with the crop protection industry to provide new pest control solutions to US speciality crop growers. *Proceedings of the BCPC International Congress – Crop Science and Technology 2005,* Vol. 1, 239–250.

Large EC, 1940. *The Advance of the Fungi.* London: Jonathan Cape.

Leadbeater AJ, 2005. Ensuring the long-term effectiveness of fungicides – an industry perspective. *Proceedings of the BCPC International Congress – Crop Science and Technology 2005,* Vol. 1, 257–266.

Oerke E-C, 2006. Crop losses to pests. *Journal of Agricultural Science* **144,** 31–43.

Oerke E-C, Dehne H-W, Schonbeck F, Weber A, 1994. *Crop Production and Crop Protection – Estimated Losses in Major Food and Cash Crops.* Amsterdam: Elsevier Science.

Padmanabhan SY, 1973. The great Bengal famine. *Annual Review of Phytopathology* **11,** 11–26.

Pasche JS, Gudmestad NC, 2008. Prevalence, competitive fitness and impact of the F129L mutation in *Alternari solani* from the United States. *Crop Protection* **27,** 427–435.

Richardson DM, 2005. The registration process, its effect on active substance availability, and initiatives to reduce the impact on minor crops at both UK and EU level. *Proceedings of the BCPC International Congress – Crop Science and Technology 2005,* Vol. 1, 231–238.

Ridout CJ, Skamnioti P, Porritt O, Sacristan S, Jones JDG, Brown JKM, 2006. Multiple avirulence paralogues in cereal powdery mildew fungi may contribute to parasite fitness and defeat of plant resistance. *The Plant Cell* **18,** 2402–2414.

Schumann GL, 1991. *Plant Diseases: their Biology and Social Impact.* St. Paul, MN: APS Press.

Stark G, 2008. The EU thematic strategy for pesticides and the UK national pesticides strategy. *Proceedings Crop Protection in Northern Britain 2008,* 23–26.

Strange RN, 2003. *Introduction to Plant Pathology.* Chichester, UK: Wiley.

Yorinori JT, Frederick RD, Costamilan LM, Bertagnolli PF, Hartman GE, Godoy CV, Nunes J, 2005. Epidemics of soybean rust (*Phakopsora pachyrhizi*) in Brazil and Paraguay from 2001 to 2003. *Plant Disease* **89,** 675–677.

Chapter 2

Managing crop disease through cultural practices

Dale Walters

Scottish Agricultural College, Edinburgh, UK

2.1 Introduction

Methods used to control plant disease vary depending on the host plant, the type of pathogen, the interaction between the two, as well as a range of other factors including environmental conditions. Most control methods are aimed at protecting crops against pathogens rather than curing them once they have been infected. Cultural control falls into this category, since it aims to prevent contact with the pathogen, to create environmental conditions unfavourable to the pathogen or at least to avoid favourable conditions, or to reduce the amount of pathogen inoculum available to infect crop plants. Methods used for cultural control include host eradication, crop rotation, sanitation, irrigation, tillage and improving crop growth conditions, for example through appropriate fertiliser use. Cultural control provides the foundation for disease control in crops and yet its importance is often overlooked. This chapter aims to provide an overview of the various cultural methods used to control crop disease and to highlight the importance of these methods in improving crop health and productivity.

2.2 Reducing the amount of pathogen inoculum

2.2.1 Host eradication

Host eradication, or roguing, refers to the removal and disposal of whole infected plants. This method is used routinely in nurseries, greenhouses and fields to prevent the spread of pathogens, since it eliminates the infected plants that act as a source of inoculum. In potato cultivation, pathogens can overwinter in infected tubers left in the field and give rise to infected plants (known as volunteers) in the spring. These volunteers can act as sources of inoculum and their removal from the field and subsequent destruction will reduce levels of pathogen inoculum. Eradication has also been used on a somewhat larger scale to stop the spread of destructive pathogens. Continual vigilance is required, however, since the pathogen may reappear. For example, Sharka disease caused by the *Plum pox virus* (PPV) was first detected in Switzerland in 1967. By the end of the 1970s, PPV was thought to be successfully eradicated through a combination of survey work and destruction and disposal of infected trees. However, PPV was detected again in 2004, leading to a new campaign of survey and eradication (Ramel *et al.*, 2006).

If a pathogen requires two hosts to complete its life cycle, control is possible by eradication of the less important host. The wheat stem rust fungus, *Puccinia graminis* f. sp. *tritici*, is a case in point. It requires two hosts, wheat and barberry, to complete its life cycle and until the 1950s was the most important pathogen of wheat in the United States (Leonard, 2001). Since the 1950s, however, stem rust has declined in importance in the United States, due in part to the successful eradication of its alternate host, common barberry (Campbell & Long, 2001).

2.2.2 Sanitation

Sanitation refers to eliminating or reducing the amount of inoculum present by various means, including removal of infected plant parts and plant debris. Destroying crop residues is an important practice, but how it is performed depends upon the type of crop and the type of pathogen. Burying crop debris (see Section 2.3.1) can destroy certain pathogens, particularly if the residues are ploughed in deeply enough, while burning crop residue is a common practice for cereal crops in some parts of the world and will destroy many pathogens. However, burning has some drawbacks, particularly, loss of nutrients and increased soil erosion.

Removal of infected plant parts by pruning has been recommended for the control of fungal pathogens of perennial crops, for example, black Sigatoka disease during the establishment phase of plantains (Emebiri & Obiefuna, 1992), while pruning of infected plant parts and removal and destruction of plant debris form an integral part of the management of *Botrytis* in greenhouses (Hausbeck & Moorman, 1996). Field sanitation is also recommended for control of late blight in potatoes (Sherf & Macnab, 1986; Cohen, 1987). Removal of plant debris by burning was shown to reduce severity of tan spot (*Pyrenophora tritici-repentis*) in wheat and to increase yields (Carignano *et al.*, 2008), while burning of chickpea stubble minimised stubble-borne inoculum of *Ascochyta rabei* (Gan *et al.*, 2006). However, some workers consider that burning has limited utility for plant disease control, since elevated soil temperatures are unlikely to be uniformly intense enough at the soil surface and throughout the upper soil profile where pathogen survival structures are found (Felton *et al.*, 1987). Thus, burning crop residue in Saskatchewan was shown to increase the incidence of plants infected with the common root rot pathogen, *Cochliobolus sativus* (Ledingham *et al.*, 1960) and although burning wheat residue in Brazil reduced the population of *C. sativus*, disease severity was not reduced (Reis & Abrao, 1983; Reis *et al.*, 1990).

2.2.3 Crop rotation

Crop rotation is an ancient cultural practice and a form of crop rotation is described in *Leviticus* 25:3–5, whereby fields were not to be sown and vineyards not to be pruned once every seven years, as a means of providing complete rest for the land (Howard, 1996). Indeed, the benefits of crop rotation include maintenance of soil structure and organic matter, and a reduction in soil erosion that is often associated with continuous row crops (Janvier *et al.*, 2007). The main purpose of rotating crops in conventional arable rotations is to reduce the incidence of diseases, pests or weeds that are difficult to control with pesticides and for this reason, short rotations of two to three crops are usually employed. In the United States, for example, the majority of the maize crop is grown on a two to three year rotation, while in the United Kingdom, barley and wheat usually form the main part of the rotation, with breaks of oilseed rape, beans, peas or potatoes (Ball *et al.*, 2005).

Continuous cropping with the same susceptible host plant will result in the establishment of a soil population of pathogenic microbes. Crop rotation avoids this and is often associated with a reduction in crop diseases caused by soil-borne pathogens (Janvier *et al.*, 2007). Using non-host or less susceptible crop plants for the rotation can lead to a decline in the specific populations of plant pathogens in the soil and is best suited for biotrophs, since they require the presence of the specific living host for survival, or pathogens with low saprophytic ability (Bailey & Duczek, 1996; Peters *et al.*, 2003). Crop rotation is less suitable for controlling root-inhabiting pathogens that survive saprophytically or can exist for long periods in soil, for example, pathogens with tough survival structures such as *Rhizoctonia solani, Sclerotinia sclerotiorum* and *Pythium* spp. (Sumner, 1982; Umaerus *et al.*, 1989). *S. sclerotiorum* is among the most non-specific and successful plant pathogens and can attack a wide range of crops including oilseed rape (canola), sunflower, flax, pea, bean, clover and potato (Morrall & Dueck, 1982). The number of viable sclerotia of *S. sclerotiorum* in soil was unchanged after three consecutive barley crops following canola (Williams & Stelfox, 1980), suggesting that the four-year rotation between susceptible crops recommended to control sclerotinia stem rot might be ineffective (Kharbanda & Tewari, 1996). In contrast, rotation involving at least three years between canola crops is usually associated with a reduction in severity of blackleg caused by *Leptosphaeria maculans* (Petrie, 1986). Although *L. maculans* can survive in stubble for more than five years, its viability decreases with age and most damaging infections arise from inoculum produced on two- to three-year-old stubble (Kharbanda & Tewari, 1996). Indeed, in South Australia, *L. maculans* was not detected in soils three or more years after a canola crop (Sosnowski *et al.*, 2006) and in this region a two-year rotation between successive canola crops is encouraged (Gladders *et al.*, 2006). With the barley leaf scald pathogen *Rhynchosporium secalis,* the amount of primary inoculum available for initiating epidemics can be decreased by rotation (Shipton *et al.*, 1974), whereas short rotations and reduced tillage which leave infected debris on the soil surface could lead to severe epidemics of *R. secalis* in crops exposed to more primary inoculum (Zhan *et al.*, 2008).

2.3 Reducing pathogen spread within the crop

The rate of pathogen spread within a crop can be reduced by altering the environment within the crop, for example, by manipulating crop density, humidity and moisture levels.

2.3.1 Tillage

Tillage has indirect effects on pathogen spread and can also be used to reduce pathogen inoculum in the soil. Conventional tillage uses primary and secondary cultivation to prepare a seed-bed for planting and results in considerable soil disturbance, while reduced tillage uses a single cultivation, or even no cultivation (no-tillage, zero tillage, direct drilling), and as a result leads to minimal soil disturbance. Minimum tillage and no-tillage practices can be grouped together under the generic term conservation tillage (Sturz *et al.*, 1997).

Tillage can bury pathogens deeper in the soil where they are less likely to become a problem. It can alter soil texture, aeration, temperature, moisture and density, and can also influence nutrient release in the soil with benefits to the crop (Ball *et al.*, 2005). Tillage also leads to clear fluctuations in microbial activity and biomass in the soil (van Bruggen *et al.*, 2006). Reduced tillage or no-tillage is often associated with higher microbial biomass and activity in upper soil layers compared to regular tillage (ploughing) (van Diepeningen *et al.*, 2005). This concentration of crop debris in the top layers of the soil can promote the over-wintering and survival of numerous pathogens and has prompted concern that increased disease and decreased yields will be the inevitable result of using conservation tillage practices. Although this has proved to be the case under some conditions, there have also been reports of decreases in the incidence of soil-borne diseases (Sturz *et al.*, 1997). As suggested by Sturz *et al.* (1997), such contradictory reports may reflect differences in root development and soil microbial biomass and activity under different regimes. Thus, conservation tillage practices can lead to pathogen inoculum concentrations several orders of magnitude greater than those found under conventional tillage (Khan, 1975; McFadden & Sutton, 1975) and, as a result, plant roots growing in the upper soil layers might be more prone to pathogen infection (Sturz *et al.*, 1997). In contrast however, increased microbial biomass and activity in the top soil layers can give rise to greater root density and root activity (Lynch & Panting, 1980; Carter & Rennie, 1984), which may offset the damaging effects of disease on yield, and might also provide a highly competitive soil environment with resulting disease-suppressive effects (Chen *et al.*, 1988).

In the United States, in the 1990s, losses of wheat and barley as a result of infection by *Fusarium graminearum* (the cause of ear blight, head blight or scab) were nearly $3 billion (Windels, 2000). These losses were blamed, in part, on the use of conservation tillage, allowing pathogen inoculum to survive on crop residues, although the evidence for increased disease severity under minimum tillage has not always been clear cut (Bateman *et al.*, 2007). Thus, minimum tillage was identified as a risk factor for *F. graminearum* infection in wheat in mid-western USA, if the preceding crop was wheat or maize (Dill-Macky & Jones, 2000). In Germany, the risk of *F. graminearum* infection in wheat was not clear cut following no tillage, if maize, and not wheat, was the previous crop (Yi *et al.*, 2001). In the United Kingdom, there is evidence that minimum tillage and maize cropping increase the risk of infection of wheat ears by *F. graminearum*, although the risk depends on the effects of weather conditions on, for example, infection and inoculum accumulation (Bateman *et al.*, 2007).

In some recent work, severity of tan spot in wheat was found to increase under no-tillage conditions, but was reduced following reduced tillage (Carignano *et al.*, 2008). To control blackleg (*Leptosphaeria maculans*) on canola (oilseed rape), it is recommended that crop debris is buried in the autumn and a non-host crop be direct seeded the following spring to avoid re-exposing the buried residue (Gladders & Musa, 1980; Kolte, 1985). Recent research suggests that inoculum production by *L. maculans* decreased with increasing duration of stubble burial in the field over 10 months, before stopping completely

(Naseri *et al.*, 2008). This effect may be due to the mycobiota associated with the buried stubble and these workers suggest that it might be possible to manipulate the population of saprophytic microbiota present on oilseed rape stubble to facilitate the decline of *L. maculans* (Naseri *et al.*, 2008).

2.3.2 Sowing practices

Altering sowing practices such as time of sowing, sowing depth and crop density can help to protect crop plants from pathogens they are susceptible to at particular stages of their development.

2.3.2.1 Time of sowing

Altering the time of sowing to avoid high levels of pathogen inoculum or conditions conducive for development of a particular disease can lead to reduced severity of several crop diseases. For example, in the United Kingdom, sowing winter oilseed rape in August rather than September exposes the earlier sown crop to inoculum from stubble of the previous crop, resulting in more severe *Alternaria* infection on pods. In contrast, the risk of infection is reduced in the later sown crop because the stubble is buried by tillage (Humpherson-Jones, 1992). Late sowing may also be recommended for autumn-sown barley crops, in order to decrease exposure of newly emerging seedlings to inoculum of *R. secalis* produced on previous barley crops in the area (Zhan *et al.*, 2008). In Turkey, rainfall during spring increases the risk of infection of chickpea by the ascochyta blight pathogen, *A. rabei,* while severe drought conditions from late April onwards can lead to reduced crop yields (Dusunceli *et al.*, 2007). Therefore, determining the best time to sow chickpeas requires a balance between the resistance rating of the chickpea cultivar and the weather conditions. Thus, it is recommended that susceptible varieties are sown later, in late March to early April, since this will avoid the precipitation required for *A. rabei* infection and development, while resistant varieties can be sown early (Dusunceli *et al.*, 2007).

2.3.2.2 Depth of sowing

Sowing depth can influence the risk of infection, since the pre-emergence stage of the seedling, which is usually more susceptible to pathogen infection, is longer when seeds are sown deeper. In *Brassica rapa*, for example, rapid emergence of seedlings reduces pre-emergence damping -off because the period of contact between the emerging seedlings and *R. solani* in the soil is reduced (Nuttall, 1982). Thus, significantly higher seedling emergence was reported for several cultivars of *B. rapa* sown at a depth of 1.5 cm compared to 3.0 cm (Nuttall, 1982).

2.3.2.3 Crop density

Crop density can exert considerable influence over disease incidence due to the ease with which pathogen inoculum can be transferred between closely spaced plants and alterations in crop microclimate. In densely planted crops, temperatures are more uniform, humidity is increased and leaves are wet for longer time, all of which provides favourable conditions for pathogen infection and subsequent development. Crop density can be manipulated in various ways, for example, sowing, pruning and fertilisation.

Reducing the sowing density of barley can decrease severity of *R. secalis* epidemics (Hoad & Wilson, 2006), probably by decreasing the density of the crop canopy, thereby altering microclimate and ensuring inadequate leaf wetness for germination of *R. secalis* conidia (Davis & Fitt, 1994). A similar effect can also be achieved in barley by reducing nitrogen applications (Hoad & Wilson, 2006), mediated possibly by altered microclimate, although there might also be effects of reduced nitrogen on the pathogen (Zhan *et al.*, 2008).

2.4 Soil amendments and mulching

2.4.1 Mulching

Mulches are used to conserve organic matter and moisture and to reduce soil erosion. A variety of materials can be used as mulches, including straw, manure, plastics and paper. Mulching can lead to water retention and nutrient enrichment in the soil and can decrease soil temperature, all of which can influence pathogen infection and disease development in plants. Although mulching can reduce the spread of splash-dispersed pathogens, by altering the environment, it could lead to increased severity of some diseases. Further, if crop residues are used in mulching, disease incidence could increase, since the residues could be used as a food source by a range of pathogens.

Working on capsicum, Stirling & Eden (2008) found that damage from Pythium root rot was more severe with plastic mulches than mulches of plant residue, probably because the organic mulch reduced soil temperatures by roughly 12°C. However, mulching increased losses from cutworms and increased the severity of infection by *Xanthomonas campestris* pv. *vesicatoria* and moreover, the organic mulch reduced fruit yield, mainly due to nutrient leaching from the soil. These workers suggest that mulches have the potential to reduce losses from soil-borne pathogens in vegetable crops, providing crop nutrition is managed adequately (Stirling & Eden, 2008). UV-reflective mulch was found to be much more effective than black polythene mulch in reducing colonisation of tomato by thrips and subsequent infections by tomato spotted wilt virus (Momol *et al.*, 2004). In some years, virus incidence was reduced further by use of the mulch and application of the plant activator acibenzolar-*S*-methyl (Momol *et al.*, 2004). In fact, the use of reflective mulches to delay the onset of infestations of whitefly and associated viruses is well documented (e.g. Summers & Stapleton, 2002; Summers *et al.*, 2004).

2.4.2 Fertilisers

Adequate mineral nutrition is central to crop production. However, it can also exert considerable influence on disease development (Datnoff *et al.*, 2007a; Walters & Bingham, 2007). Fertiliser application can increase or decrease development of diseases caused by different pathogens, and the mechanisms responsible are complex, including effects of nutrients on plant growth, plant resistance mechanisms and direct effects on the pathogen (Walters & Bingham, 2007). The effects of mineral nutrition on plant disease and the mechanisms responsible for those effects have been dealt with comprehensively elsewhere (Datnoff *et al.*, 2007a; Walters & Bingham, 2007). The sections below will deal briefly with the influence of nitrogen, phosphorus, potassium, calcium and silicon on plant disease. The effects of sulphur on plant disease, in particular the phenomenon of sulphur-induced resistance, is dealt with in Chapter 11.

2.4.2.1 Nitrogen

Nitrogen fertiliser applied above the recommended rates can result in increased disease incidence and lesion area. This has been demonstrated for biotrophic fungal pathogens such as powdery mildews and rusts (e.g. Mascagni *et al.*, 1997; Hoffland *et al.*, 2000) and necrotrophic pathogens such as *Magnaporthe grisea,* the cause of rice blast (Talukder *et al.,* 2005). It is commonly thought that application of nitrogen fertiliser can increase disease severity via effects on crop canopy development. Thus, large canopies with high shoot densities may be more conducive to spore transfer and pathogen infection than sparse canopies. For example, nitrogen has been shown to increase the severity of *Fusarium* head blight in wheat, and it has been suggested that this might be the result of a nitrogen-induced increase in canopy size, leading to an altered microclimate (Lemmens *et al.,* 2004). In contrast, work on yellow rust on winter wheat suggested that the impact of nitrogen on disease was the result of effects of nitrogenous substances in wheat leaves on pathogen growth, rather than effects on canopy growth and microclimate (Neumann *et al.,* 2004).

However, nitrogen fertilisation is not always associated with increased disease. Several studies have reported no effect of nitrogen on disease severity (e.g. Buschbell & Hoffmann, 1992; Olesen *et al.*, 2000), while Hoffland *et al.* (2000) found that the effect of nitrogen depended on the type of pathogen. Thus, nitrogen increased susceptibility of tomato to the powdery mildew pathogen *Oidium lycopersicum* and the bacterial pathogen *Pseudomonas syringae* pv. *tomato,* while it had no effect on susceptibility to the vascular wilt pathogen *Fusarium oxysporum* f. sp. *lycopersici* (Hoffland *et al.*, 2000). In contrast, tomato plants were more susceptible to *Botrytis cinerea* when grown under low nitrogen conditions (Hoffland *et al.,* 1999). These results do not support the view that nutrient-limited plants are better defended (Bryant *et al.*, 1983; Herms & Mattson, 1992). Indeed, nitrogen limitation has been found to severely compromise the ability of *Arabidopsis thaliana* to express induced resistance to pathogen infection (Dietrich *et al.*, 2004, 2005). It is clear therefore that generalising about the effects of nitrogen on plant disease is unwise and practically, although manipulation or assessment of crop nitrogen status might be used as part of disease control strategies, the approach adopted will depend on the crop and the pathogens from which it is most at risk (Walters & Bingham, 2007).

2.4.2.2 Phosphorus

In an analysis of some 2440 studies of the effects of fertiliser on more than 400 diseases and pests, Perrenoud (1990) found that, in general, phosphorus fertilisation tended to improve plant health, with reductions in disease recorded in 65% of cases. Nevertheless, phosphorus fertilisation increased disease and pest problems in 28% of the cases examined

(Perrenoud, 1990). As with nitrogen, the effects of phosphorus on plant disease may be the result of direct effects on the pathogen, host plant metabolism, leading to effects on pathogen food supply, and effects on plant defences (Walters & Bingham, 2007). Indeed, foliar application of phosphate salts has been shown to induce resistance to pathogens in a range of crop plants, including cucumber (Mucharromah & Kuc, 1991), broad bean (Walters & Murray, 1992), grapevine (Reuveni & Reuveni, 1995), maize (Reuveni *et al.*, 1994) and rice (Mandahar *et al.*, 1998).

Clearly, an adequate phosphorus supply is important for crop growth and in turn, may well help to reduce disease. However, the regime of phosphorus fertiliser used will depend on a range of factors, including the crop and the pathogens likely to be important. Reuveni & Reuveni (1998) suggested that foliar-applied phosphate might be used as part of an integrated disease control programme, although grower adoption of such an approach will depend on the existence of other effective disease control measures and the economics of disease control in the particular crop.

2.4.2.3 Potassium

In an analysis of 181 papers reporting effects of potassium on plant disease, Prabhu *et al.* (2007) found that 120 (66%) reported reductions in disease, while 49 (27%) reported an increase in disease. Although this suggests that in many cases, potassium is associated with disease reductions, Prabhu *et al.* (2007) point out that inadequate consideration has been given to the effects of associated anions, nutrient balance and nutrient status to allow the definitive role of potassium to be determined. For example, it has been suggested that in some cases, the effects of potassium, applied as potassium chloride fertiliser, might be due to the chloride ion rather than potassium (Fixen *et al.*, 1986). Further, chloride fertilisation has been shown to suppress disease in cereal crops (Engel *et al.*, 1994).

As indicated in the section above, there has been much interest in the application of fertilisers to crop foliage, including the effects of foliar fertiliser application for crop disease control (Reuveni & Reuveni, 1998; Ehret *et al.*, 2002). Foliar-applied potassium chloride has been shown to control *Blumeria graminis* and *Septoria tritici* on wheat in field studies (Cook *et al.*, 1993; Mann *et al.*, 2004), probably due to osmotic effects on the fungal pathogens, disrupting pathogen development and subsequent infection (Kettlewell *et al.*, 2000; Mann *et al.*, 2004).

Application of potassium to deficient soils usually increases plant resistance to diseases (Prabhu *et al.*, 2007). This might be partly related to the effect of potassium in increasing epidermal cell wall thickness or disease escape as a result of vigorous crop growth (Prabhu *et al.*, 2007), although the mechanisms by which potassium affects plant disease are not well understood.

2.4.2.4 Calcium

There are many reports that application of calcium to soils, foliage and fruit reduces the incidence and severity of a range of diseases of crops, including cereals, vegetable crops, legumes, fruit trees, as well as post-harvest diseases of tubers and fruits (Rahman & Punja, 2007). For example, calcium has been shown to inhibit anthracnose (caused by

Colletotrichum gloeosporiodes or *C. acutatum*) in apples (Biggs, 1999) and to decrease post-harvest disease development on strawberry (Cheour *et al.*, 1990), while treatment of tomato with calcium carbonate reduced fusarium crown rot disease (Woltz *et al.*, 1992). In contrast, Nam *et al.* (2006) could find no effect of calcium on anthracnose on strawberry. Because calcium increases resistance of plant cell membranes and cell walls to microbial enzymes, increasing calcium concentrations in storage organs could lead to enhanced resistance to pathogens (Conway & Sams, 1984; Biggs & Peterson, 1990). However, the form in which the calcium is applied can influence the mechanism by which calcium affects disease. For example, the addition of lime can affect disease by altering pH, while calcium salts (e.g. propionate) can be directly inhibitory to pathogens (Rahman & Punja, 2007). Making general recommendations for the use of calcium in plant disease control would be unwise due to the range of crops and pathogens affected by calcium application. Instead, the appropriate amount and form of calcium to be applied needs to be determined for individual crop–pathogen interactions. The dwindling availability of fungicides, together with increasing public concern for the environment means that the use of calcium to control plant disease, especially post-harvest, is attracting increased attention.

2.4.2.5 Silicon

Although the effects of silicon in reducing disease severity have been known since 1940 (Wagner, 1940), it was not until the 1980s that more detailed work was carried out in this area. In this work, cucumbers grown in nutrient solutions supplemented with silicon were found to have significantly less powdery mildew infection than plants not receiving silicon supplementation (Miyake & Takahashi, 1983; Adatia & Besford, 1986). Indeed, silicon has been shown to suppress both foliar and soil-borne pathogens in curcubits (Belanger *et al.*, 1995) and to reduce susceptibility of rice to various pathogens (Datnoff *et al.*, 2007b). Wheat grown in soil amended with silicon showed reduced infection by several pathogens, including *B. graminis* f. sp. *tritici*, *S. tritici* and *Oculimacula yallundae* (Rodgers-Gray & Shaw, 2000, 2004).

It has been suggested that the effects of silicon in providing disease control are due to the creation of a mechanical barrier to penetration (Kim *et al.*, 2002). However, this has been disputed by studies which could find no evidence for the creation of a physical barrier following silicon treatment in wheat inoculated with powdery mildew and bitter gourd and tomato inoculated with *Pythium aphanidermatum* (Samuels *et al.*, 1991; Heine *et al.*, 2007). Rather, several studies have suggested that silicon activates defences in plants. For example, in wheat inoculated with *B. graminis* f. sp. *tritici*, epidermal cells of silicon-treated plants were shown to react to attempted infection with specific defences, including papilla formation and callose production (Belanger *et al.*, 2003). In the rice–*M. grisea* pathosystem, silicon-mediated resistance was found to be associated with accumulation of antimicrobial compounds at infection sites, including diterpenoid phytoalexins (Rodrigues *et al.*, 2004). In fact, phytoalexin accumulation occurs in silicon-mediated resistance in both dicots and monocots and since phytoalexins are highly specific to plant species, it has been suggested that silicon might be acting on mechanisms shared by all plant species, for example, those resulting in activation of plant stress genes (Fauteux *et al.*, 2005).

2.4.2.6 Crop nutrition and plant disease

The sections above on nitrogen, phosphorus, potassium, calcium and silicon show that, in spite of some inconsistencies, crop nutrition clearly influences disease incidence and severity in a range of pathosystems. Many of the studies suggest that managing crop nutrition through appropriate fertiliser practice could be a useful aid to control plant disease. However, because nutrition will also affect crop yield and quality, a balance needs to be struck between maximising yield and quality and minimising disease. Therefore, the extent to which fertiliser regimes can be modified to enhance disease control will depend on the relative effects of crop nutrition on disease development, the response of crop yield and quality to disease, and the crop's potential yield and quality in the absence of disease.

2.4.3 Organic amendments

Organic amendments cover a range of inputs, including animal manure, solid wastes and composts. These amendments are often used to improve soil quality, usually by contributing to general suppressiveness through enhanced microbial biomass and activity (Janvier *et al.*, 2007). Organic amendments are rich in labile carbon fractions which are an energy source for microorganisms and moreover, they can themselves contain antagonistic microbes. A substantial body of data indicates that organic materials can reduce incidence of diseases caused by a range of plant pathogens (see Bailey & Lazarovits, 2003). Since composts are dealt with in Chapter 5, this section will look at the effects of high nitrogen amendments and manures on plant disease.

2.4.3.1 Animal manures

The impact of animal manures on disease incidence and severity is much less predictable than that of composts. Thus, fresh chicken manure was shown to reduce survival of *Phytophthora cinnamomi* and disease incidence on seedlings of *Lupinus albus,* while cow, sheep and horse manure did not consistently suppress populations of *P. cinnamomi* or disease symptoms (Aryantha *et al.,* 2000). Interestingly, in this work, only chicken manure stimulated populations of endospore-forming bacteria, a factor that was strongly associated with seedling survival. Animal manures have been implicated in increasing the incidence of common scab on potato (Bailey & Lazarovits, 2003). However, Conn & Lazarovits (1999) found that a single application of liquid swine manure reduced the incidence of wilt and common scab in potato and reduced numbers of plant parasitic nematodes for three years after the treatment. Microsclerotia of the wilt pathogen *Verticillium dahliae* were killed by exposure to liquid swine manure (Conn & Lazarovits, 2000), apparently due to the presence of volatile fatty acid mixtures in the manure (Tenuta *et al.,* 2002). Subsequent work demonstrated that in acidic soils, liquid swine manure killed microsclerotia of *V. dahliae* by volatile fatty acids and/or nitrous acid toxicity, while in alkaline soils, microsclerotia were killed by ammonia toxicity (Conn *et al.,* 2005). The authors suggested that for these mechanisms to be operational and effective in practice, the chemical composition of the manure, rate of application and soil characteristics need to be determined in each case (Conn *et al.,* 2005).

In some recent work, Messiha *et al.* (2007) examined the incidence and severity of brown rot in different soil types. They found that cow manure amendment significantly reduced disease incidence in organic Dutch sandy soils, although populations of the bacterial pathogen *Ralstonia solanacearum* were not affected. In Egyptian sandy soils, however, population density of the bacterium was reduced, probably as a result of microbial competition (Messiha *et al.*, 2007). This work indicates that the mechanism of disease suppression of soil-borne plant pathogens may vary greatly depending on the soil type.

2.4.3.2 High nitrogen amendments

There are numerous reports of the efficacy of a range of high nitrogen amendments in controlling a variety of pathogens (see Bailey & Lazarovits, 2003). For example, meat and bone meal and soy meal significantly reduced the incidence of verticillium wilt, common scab and populations of plant parasitic nematodes in potato field trials (Conn & Lazarovits, 1999; Lazarovits *et al.*, 1999). Amendments such as soy meal and blood meal are degraded in the soil leading to the release of ammonia, which is toxic to many organisms, including the resting structures of plant pathogens (Bailey & Lazarovits, 2003). A low level of organic carbon in the soil was found to be critical for the accumulation of ammonia, while high levels of soil organic matter prevents ammonia accumulation (Tenuta & Lazarovits, 1999).

2.4.4 Irrigation

Although an adequate water supply is vital to crop production, irrigation can play a detrimental rather than a beneficial role in managing plant diseases. For example, irrigation water can spread pathogen propagules and under dry conditions can prevent desiccation of such propagules, thereby effectively increasing the level of inoculum in soil. Watering from overhead prolongs leaf wetness, thereby providing favourable conditions for germination and infection by fungal spores. Overhead watering also increases the risk of splash-dispersal of spores, thus increasing pathogen spread. However, irrigation can be used to reduce the level of pathogen inoculum. Thus, the activity of microbes that destroy fungal sclerotia can be increased by alternate wetting and drying of the soil. Generally, drip or trickle irrigation, which delivers water directly to the root zone at a rate insufficient to lead to pathogen spread, is least likely to encourage disease development.

In a study of the control of the downy mildew pathogen *Peronospora sparsa* on blackberry, O'Neill *et al.* (2002) found that the use of sub-irrigated sand beds resulted in very low disease incidence, whereas the use of overhead irrigation led to the disease developing in 97% of plants. In southern Israel, less frequent and reduced irrigation was shown to lower the incidence of *Monosporascus cannonballus* on melon and to postpone plant collapse, although yields were reduced (Pivonia *et al.*, 2004). Gitaitis *et al.* (2004) examined the effects of a number of treatments, including irrigation, on centre rot of onion caused by the bacterium *Pantoea ananatis.* They found no effect of drip or overhead sprinkler irrigation on the incidence or severity of centre rot (Gitaitis *et al.*, 2004). *Phytophthora capsici* is a serious soil-borne pathogen of pepper (*Capsicum annuum* L.) and causes significant crop losses worldwide. The pathogen has been shown to spread under high soil moisture conditions. However, Sanogo (2006) could find no evidence that soil water saturation predisposed pepper to infection by *P. capsici.*

Appropriate treatment of irrigation water can be used to reduce pathogen inoculum, thereby reducing spread of the pathogen. *Phytophthora* root rot is a problem in container-grown hardy nursery stock and the pathogen can be spread by irrigation water. In some recent work, Pettitt *et al.* (2007) examined the efficacy of a non-woven capillary matting fabric (Tex-R® Pro), coated with a latex polymer-based formulation of cupric hydroxide (Spin Out®), in controlling *Phytophthora* root rot in container-grown *Chamaecyparis lawsoniana.* The fabric was used as bed covers and was also cut into discs which were used to cover the tops of the plant containers or were inserted to cover the holes at the bottoms of the containers. Bed covers and disc inserts were found to significantly reduce disease spread, while pot toppers were not effective. In addition to reducing spread of the pathogen in irrigation water, survival of zoospores and zoospore cysts was also significantly reduced by this fabric (Pettitt *et al.,* 2007).

2.4.5 Flooding

Flooding can be used in crop protection, since it reduces weeds as well as numbers of fungal propagules, nematodes and insects in the soil. However, flooding can also spread pathogens and indeed, its success in disease management is variable, depending on the pathogens present in the soil. Thus, Teo *et al.* (1989) found that 65% of sclerotia of *S. sclerotiorum* were destroyed after two years in the field at high moisture (5.9–26.2%) compared with 45% at low soil moisture (0.0–1.7%). As a result, they concluded that incorporating an appropriate irrigation schedule into the crop rotation system might reduce inoculum of *S. sclerotiorum* (Teo *et al.,* 1989). Unfortunately, frequent irrigation could increase alternaria blackspot and root rot (Teo *et al.,* 1988; Saharan, 1992). Interestingly, in a study of the population dynamics of *M. cannonballus,* Beltran *et al.* (2005) found that although ascospore numbers declined in fields that were in fallow and flooded for three years, soil-borne inoculum was viable and capable of infecting muskmelon. Clearly, *M. cannonballus* is well adapted to survive in soils which maintain a high water table or under flooding (Beltran *et al.,* 2005). Eradication of some pathogens can be achieved effectively with flooding, but it is expensive, adversely affects soil structure and its effect in controlling disease is temporary (Kharbanda & Tewari, 1996).

2.5 Suppressive soils

A range of root-inhabiting pathogens, for example, *Pythium* spp., some *Phytophthora* spp. and some *Fusarium* spp., survive saprophytically on soil organic matter or exist for long periods in the soil in the absence of the host plant, making them difficult to control. Interestingly, some soils have the capacity to suppress such pathogens, with the result that crops grown in such soils exhibit less disease, even if other environmental conditions are favourable (van Bruggen, 1995). Baker & Cook (1974) described suppressive soils as those in which disease severity or incidence remains low, despite the presence of a pathogen, a susceptible host plant, and environmental conditions favouring pathogen infection and subsequent disease development. Soil suppressiveness can be the result of different mechanisms (Baker & Cook, 1974): (a) inability of the pathogen to establish or persist, (b) the pathogen establishes, but causes little or no damage, (c) the pathogen establishes and causes disease, but the disease becomes less important, despite the presence of the pathogen in the soil. Höper & Alabouvette (1996) distinguished between pathogen suppression and disease suppression, the former referring to the ability of the soil to limit the inoculum density of the pathogen and its saprophytic activity, and the latter to the capacity of the soil to restrict disease development even though the host, quantity of pathogen inoculum and the environment appear favourable. Therefore, the capacity of a soil for disease suppression will be determined by its effects on the processes of colonisation and infection by the pathogen, and the subsequent development of disease symptoms. Importantly, pathogen suppression and disease suppression might not necessarily be coupled, and some soils might be pathogen suppressive but not disease suppressive and vice versa (Höper & Alabouvette, 1996).

It is thought that the suppressiveness of a soil is determined mainly by its microbial properties, especially since the suppressive effect can be destroyed by sterilisation (Peters *et al.,* 2003). These microbial properties include the presence of rhizosphere and root endophytic bacteria, which disrupt pathogen infection by various means, including production of antibiotics, siderophores, nutrient competition and induction of systemic resistance (Peters *et al.,* 2003; Sturz & Christie, 2003). Other possible mechanisms include predation of fungal hyphae by soil microfauna and competition from arbuscular mycorrhizal fungi (Workneh & van Bruggen, 1994; Knudsen *et al.,* 1995; Azcon-Aguilar & Barea, 1996).

It has been suggested that given the effects of soil structure on soil micro-heterogeneity and microbial activity, suppressiveness is likely to be dependent on soil structure (Ball *et al.,* 2005). For example, a poorly structured soil will restrict the activity and movement of soil organisms, with consequences for predator–prey relationships. This, in turn, will have consequences on pathogen survival, the spread and survival of introduced microorganisms (e.g. biocontrol agents) and infection by fungal pathogens (Rattray *et al.*, 1993; Young & Ritz, 2000; Otten *et al.*, 2001).

van Bruggen *et al.* (2006) argue that healthy soils are more suppressive to soil-borne plant pathogens than biologically impoverished soils. These authors define a healthy soil as a stable soil system with high levels of biological diversity and activity, internal nutrient cycling, and resilience to disturbance. The implication is that microbial fluctuations following a disturbance would dampen more quickly in a healthy than a biologically impoverished soil (van Bruggen *et al.,* 2006). The authors suggest that regular addition of soil organic matter might increase background levels of microbial activity, increase nutrient cycling, lower the concentrations of easily available nutrient sources, increase microbial diversity and enhance natural disease suppression (van Bruggen *et al.,* 2006).

2.6 Intercropping

The simultaneous planting of more than one crop in the same area is called intercropping and is an important feature of cropping systems in the tropics. Intercropping has been reported to provide protection against pathogens in component crops (e.g. Boudreau & Mundt, 1992; Fininsa, 1996), although effectiveness can vary depending on location and crop variety (Boudreau, 1993; Boudreau & Mundt, 1994; Bulson *et al.*, 1997) and there can be effects on yield. Thus, under-sowing leeks with clover reduced the incidence of rust (*Puccinia allii*) and although quality was improved, crop growth was reduced (Theunissen & Schelling, 1996). Similarly, intercropping tomato with soybean or sesame, in combination with sanitation, limited late blight (*Phytophthora infestans*) development, but taller intercrops reduced tomato growth and production (Tumwine *et al.*, 2002). Interestingly, Kinane & Lyngkjaer (2002) found that in barley–legume intercrops, disease incidence was reduced, irrespective of location, although not always significantly. For example, net blotch (*Pyrenophora teres*) on barley was reduced whenever it was intercropped with grain legumes, while on pea, ascochyta blight (*Ascochyta pisi*) was reduced (Kinane & Lyngkjaer, 2002). In this work, although brown rust (*P. hordei*) on barley was reduced when intercropped with legumes, these reductions were not significant. In some recent work on barley–grain legume intercropping, disease reductions were observed in all intercrop combinations, compared to the sole crop (Hauggaard-Nielsen *et al.*, 2008).

Mechanisms proposed to account for reductions in disease in intercropped systems include alterations in microclimate, competition and induced resistance. Fininsa & Yuen (2002) examined the effects of intercropping bean (*Phaseolus vulgaris*) with maize and/or sorghum in four cropping systems (sole cropping, row, mixed and broadcast intercropping) on common bacterial blight caused by *Xanthomonas campestris* pv. *phaseoli*. They found that intercropping delayed epidemic onset, lowered disease incidence and severity and reduced the disease progress rate, although the magnitude of these effects varied depending on the cropping system. Fininsa & Yuen (2002) suggested that in mixed and broadcast intercropping, where the plants are under the associate crop and not in separate rows, there are likely to be competition and dispersal interference effects. In contrast, in row intercropping competition, microclimate changes and interference effects would be less important (Fininsa & Yuen, 2002). According to these workers, the final effect on disease would depend on the canopy and root structure and tillering capacity of the associate crop.

Intercropping has been shown to enhance and stabilise yields, reduce weeds and plant diseases and improve resource use. However, there is a need for increased understanding of the ecological mechanisms associated with planned spatial diversity, in order to enhance the benefits achieved from intercropping.

2.7 Conclusions

Cultural control can be an effective and sustainable approach to the management of plant disease. Indeed, continuing problems with fungicide resistance and breakdown of host plant resistance, together with increasing concern for the environment means that there is renewed interest in cultural practices for the management of crop diseases. However, the choice and use of cultural practices will depend on the crop and the pathogen, although it might be possible to integrate the management of more than one disease by combining several appropriate cultural practices. In order to maximise the potential of cultural practices in disease control, a sound understanding of the mechanisms by which they exert their effects is required.

2.8 Acknowledgements

Scottish Agricultural College receives financial support from the Scottish Government Rural and Environment Research and Analysis Directorate (RERAD).

2.9 References

Adatia MH, Besford RT, 1986. The effects of silicon on cucumber plants grown in recirculating nutrient solution. *Annals of Botany* **58**, 343–351.

Aryantha IP, Cross R, Guest DI, 2000. Suppression of *Phytophthora cinnamomi* in potting mixes amended with uncomposted and composted animal manures. *Phytopathology* **90,** 775–782.

Azcon-Aguilar C, Barea JM, 1996. Arbuscular mycorrhizas and biological control of soil-borne pathogens – an overview of the mechanisms involved. *Mycorrhiza* **6,** 457–464.

Bailey DJ, Duczek LJ, 1996. Managing cereal diseases under reduced tillage. *Canadian Journal of Plant Pathology* **18,** 159–167.

Bailey KL, Lazarovits G, 2003. Suppressing soil-borne diseases with residue management and organic amendments. *Soil & Tillage Research* **72**, 169–180.

Baker KF, Cook RJ, 1974. *Biological Control of Plant Pathogens*. San Francisco: APS Press.

Ball BC, Bingham I, Rees RM, Watson CA, Litterick A, 2005. The role of crop rotations in determining soil structure and crop growth conditions. *Canadian Journal of Soil Science* **85,** 557–577.

Bateman GL, Gutteridge RJ, Gherbawy Y, Thomsett MA, Nicholson P, 2007. Infection of stem bases and grains of winter wheat by *Fusarium culmorum* and *F. graminearum* and effects of tillage method and maize-stalk residues. *Plant Pathology* **56,** 604–615.

Belanger RR, Bowen P, Ehret DL, Menzies JG, 1995. Soluble silicon: Its role in crop and disease management of greenhouse crops. *Plant Disease* **79,** 329–336.

Belanger RR, Benhamou N, Menzies JG, 2003. Cytological evidence of an active role of silicon in wheat resistance to powdery mildew (*Blumeria graminis* f. sp. *tritici*). *Phytopathology* **93,** 402–412.

Beltran R, Vicent A, Sales R, Garcia-Jimenez J, Armengol J, 2005. Population dynamics of *Monosporascus cannonballus* ascospores in marsh soils in eastern Spain. *European Journal of Plant Pathology* **113,** 357–365.

Biggs AR, 1999. Effects of calcium salts on apple bitter rot caused by two *Colletotrichum* spp. *Plant Disease* **83,** 1001–1005.

Biggs AR, Peterson CA, 1990. Effect of chemical applications to peach bark wounds on accumulation of lignin and suberin and susceptibility to *Leucostoma persoonii. Phytopathology* **80,** 861–865.

Boudreau MA, 1993. Effect of intercropping with maize on the severity of angular leaf spot of beans in Kenya. *Plant Pathology* **42,** 16–25.

Boudreau MA, Mundt CC, 1992. Mechanisms of alteration in bean rust epidemiology due to intercropping with maize. *Phytopathology* **82,** 1051–1060.

Boudreau MA, Mundt CC, 1994. Mechanisms of alteration in bean rust development due to intercropping, in computer simulated epidemics. *Ecological Applications* **4,** 729–740.

Bryant JP, Chapin III FS, Klein DR, 1983. Carbon/nutrient balance of boreal plants in relation to vertebrate herbivory. *Oikos* **40,** 357–368.

Bulson HAJ, Snaydon RW, Stopes CE, 1997. Effects of plant density on intercropped wheat and field beans in an organic farming system. *Journal of Agricultural Science* **128,** 59–71.

Buschbell T, Hoffmann GM, 1992. The effects of different nitrogen regimes on the epidemiologic development of pathogens on winter-wheat and their control. *Journal of Plant Diseases and Protection* **99,** 381–403.

Campbell CL, Long DL, 2001. The campaign to eradicate the common barberry in the United States. In: Peterson P, ed. *Stem Rust of Wheat: From Ancient Enemy to Modern Foe.* St. Paul, MN: APS Press, 16–50.

Carignano M, Staggenborg SA, Shroyer JP, 2008. Management practices to minimize tan spot in a continuous wheat rotation. *Agronomy Journal* **100,** 145–153.

Carter MR, Rennie DA, 1984. Nitrogen transformations under zero and shallow tillage. *Soil Science Society of America Journal* **48,** 1077–1081.

Chen W, Hoitink HAJ, Madden LV, 1988. Micrbial activity and biomass in container media for predicting suppressiveness to damping-off caused by *Pythium ultimum. Phytopathology* **78,** 1447–1450.

Cheour F, Willemot C, Arul J, Desjardins Y, Makhlouf J, Charest PM, Gosselin A, 1990. Foliar application of calcium chloride delays postharvest ripening of strawberry. *Journal of the American Society for Horticultural Science* **115,** 789–792.

Cohen Y. 1987. Late blight of potatoes: Epidemiology and control. *Hassadeh* **68,** 268–271.

Conn KL, Lazarovits G, 1999. Impact of animal manures on verticillium wilt, potato scab, and soil microbial populations. *Canadian Journal of Plant Pathology* **21,** 81–92.

Conn KL, Lazarovits G, 2000. Soil factors influencing the efficacy of liquid swine manure added to soil to kill *Verticillium dahliae. Canadian Journal of Plant Pathology* **22,** 400–406.

Conn KL, Tenuta M, Lazarovits G, 2005. Liquid swine manure can kill *Verticillium dahliae* microsclerotia in soil by volatile fatty acid, nitrous acid and ammonia toxicity. *Phytopathology* **95,** 28–35.

Conway WS, Sams CE, 1984. Possible mechanisms by which postharvest calcium treatment reduces decay in apples. *Phytopathology* **74,** 208–210.

Cook JW, Kettlewell PS, Parry DW, 1993. Control of *Erysiphe graminis* and *Septoria tritici* on wheat with foliar-applied potassium chloride. *Journal of the Science of Food and Agriculture* **63,** 126.

Datnoff LE, Elmer WH, Huber DM, 2007a. *Mineral Nutrition and Plant Disease.* St. Paul, MN: APS Press.

Datnoff LE, Rodrigues FA, Seebold KW, 2007b. Silicon and plant disease. In: Datnoff LE, Elmer WH, Huber DM, eds. *Mineral Nutrition and Plant Disease.* St. Paul, MN:APS Press, 233–246.

Davis H, Fitt BDL, 1994. Effects of temperature and leaf wetness on the latent period of *Rhynchosporium secalis* (leaf blotch) on leaves of winter barley. *Journal of Phytopathology* **140,** 269–279.

Dietrich R, Ploss K, Heil M, 2004. Constitutive and induced resistance to pathogens in *Arabidopsis thaliana* depends on nitrogen supply. *Plant, Cell and Environment* **27,** 896–906.

Dietrich R, Ploss K, Heil M, 2005. Growth responses and fitness costs after induction of pathogen resistance depend on environmental conditions. *Plant, Cell and Environment* **28,** 211–222.

Dill-Macky R, Jones RK, 2000. The effect of previous crop residues and tillage on fusarium head blight of wheat. *Plant Disease* **84,** 71–76.

Dusunceli F, Meyveci K, Cetin L, Avci M, Surek D, Albustan S, Mert Z, Akan K, Karacam M, Strange RN, 2007. Determination of agronomic practices for the management of blight of chickpea caused by *Aschochyta rabei* in Turkey: 1. Appropriate sowing date. *European Journal of Plant Pathology* **119,** 449–456.

Ehret DL, Menzies JG, Bogdanoff C, Utkhede RS, Frey B, 2002. Foliar Applications of fertiliser salts inhibit powdery mildew on tomato. *Canadian Journal of Plant Pathology* **24,** 437–444.

Emebiri LC, Obiefuna JC, 1992. Effect of leaf removal and intercropping on the incidence and severity of black Sigatoka disease at the establishment phase of plantains (*Musa* spp. AAB). *Agriculture, Ecosystems and Envirnoment* **39,** 213–219.

Engel RE, Eckhoff J, Berg RK, 1994. Grain yield, kernel weight and diseases responses of winter wheat cultivars to chloride fertilization. *Agronomy Journal,* **86,** 891–896.

Fauteux F, Remus-Borel W, Menzies JG, Belanger RR, 2005. Silicon and plant disease resistance against pathogenic fungi. *FEMS Microbiology Letters* **249,** 1–6.

Felton WL, Freebairn DM, Fettell NA, Thomas JB, 1987. Crop residue management. In: Cornish PS, Pratley JE, eds. *Tillage – New Directions in Australian Agriculture.* Melbourne: Inkata Press, 171–193.

Fininsa C, 1996. Effect of intercropping bean with maize on bean common bacterial blight and rust diseases. *International Journal of Pest Management* **42,** 51–54.

Fininsa C, Yuen J, 2002. Temporal progression of bean bacterial blight (*Xanthomonas campestris* pv. *phaseoli*) in sole and intercropping systems. *European Journal of Plant Pathology* **108,** 485–495.

Fixen PE, Gelderman RH, Gerwing J, Cholick FA, 1986. Response of spring wheat, barley and oats to chloride in potassium chloride fertilizers. *Agronomy Journal* **78,** 664–668.

Gan YT, Siddique KHM, MacLeod WJ, Jayakumar P, 2006. Management options for minimizing the damage by Aschochyta blight (*Aschochyta rabei*) in chickpea (*Cicer arietinum* L.). *Field Crops Research* **97,** 121–134.

Gitaitis RD, Walcott RR, Sanders HF, Zolobowska L, Diaz-Perez JC, 2004. Effects of mulch and irrigation system on sweet onion II. The epidemiology of center rot. *Journal of the American Society for Horticultural Science* **129,** 225–230.

Gladders P, Evans N, Marcroft S, Pinochet X, 2006. Dissemination of information about management strategies and changes in farming practices for the exploitation of resistance to *Leptosphaeria maculans* (phoma stem canker) in oilseed rape cultivars. *European Journal of Plant Pathology* **114,** 117–126.

Gladders P, Musa TM, 1980. Observations of the epidemiology of *Leptosphaeria maculans* stem canker in winter oilseed rape. *Plant Pathology* **29,** 28–37.

Hall FR, Menn JJ, 1999. *Biopesticides: Use and Delivery*. Totowa, NJ: Humana Press.

Hauggaard-Nielsen H, Jørnsgaard B, Kinane J, Jensen ES, 2008. Grain legume–cereal intercropping: The practical application of diversity, competition and facilitation in arable and organic cropping systems. *Renewable Agriculture and Food Systems* **23,** 3–12.

Hausbeck MK, Moorman GW, 1996. Managing *Botrytis* in greenhouse-grown flower crops. *Plant Disease* **80,** 1212–1219.

Heine G, Tikum G, Horst WJ, 2007. The effect of silicon on the infection by and spread of *Pythium aphanidermatum* in single roots of tomato and bitter gourd. *Journal of Experimental Botany* **58,** 569–577.

Herms DA, Mattson WJ, 1992. The dilemma of plants: To grow or to defend. *Quarterly Review of Biology* **67,** 283–335.

Hoad SP, Wilson GW, 2006. Influence of plant population density, nitrogen fertiliser rate and variety on *Rhynchosporium secalis* in winter barley. *Proceedings Crop Protection in Northern Britain* 2006, Dundee, UK, 185–190.

Hoffland E, van Beusichem ML, Jeger MJ, 1999. Nitrogen availability and susceptibility of tomato leaves to *Botrytis cinerea. Plant and Soil* **210,** 263–272.

Hoffland E, Jeger MJ, van Beusichem ML, 2000. Effect of nitrogen supply rate on disease resistance in tomato depends on the pathogen. *Plant and Soil* **218,** 239–247.

Höper H, Alabouvette C, 1996. Importance of physical and chemical soil properties in the suppressiveness of soils to plant diseases. *European Journal of Soil Biology* **32,** 41–58.

Howard RJ, 1996. Cultural control of plant diseases: A historical perspective. *Canadian Journal of Plant Pathology* **18,** 145–150.

Humpherson-Jones FM, 1992. Epidemiology and control of dart leaf spot of brassicas. In: Chelkowski J, Visconti A, eds. *Alternaria Biology, Plant Diseases and Metabolites*. Amsterdam, The Netherlands: Elsevier Science Publishers, 267–288.

Janvier C, Villeneuve F, Alabouvette C, Edel-Hermann V, Mateille T, Steinberg C, 2007. Soil health through soil disease suppression: Which strategy descriptors to indicators? *Soil Biology and Biochemistry* **39,** 1–23.

Kettlewell PS, Cook JW, Parry DW, 2000. Evidence for an osmotic mechanism in the control of powdery mildew disease of wheat by foliar-applied potassium chloride. *European Journal of Plant Pathology,* **106,** 297–300.

Khan AG, 1975. The effect of vesicular arbuscular mycorrhizal associations on growth of cereals II. Effects on wheat growth. *Annals of Applied Biology* **80,** 27–36.

Kharbanda PD, Tewari JP, 1996. Integrated management of canola diseases using cultural methods. *Canadian Journal of Plant Pathology* **18,** 168–175.

Kim SG, Kim KW, Park EW, Choi D, 2002. Silicon-induced cell wall fortification of rice leaves: A possible cellular mechanism of enhanced host resistance to blast. *Phytopathology* **92,** 1095–1103.

Kinane J, Lyngkjaer ML, 2002. Effect of barley–legume intercrop on disease frequency in an organic farming system. *Plant Protection Science* **38,** 227–231.

Knudsen IM, Elmholt S, Hockenhull J, Jensen DF, 1995. Distribution of saprophytic fungi antagonistic to *Fusarium culmorum* in two differently cultivated field soils, with special emphasis on the genus *Fusarium. Biological Agriculture and Horticulture* **12,** 61–76.

Kolte SJ, 1985. *Diseases of Annual Edible Oilseed Crops. Vol. II. Rapeseed–Mustard and Sesame Diseases.* Boca Raton, FL: CRC Press.

Lazarovits G, Conn KL, Potter J, 1999. Reduction of potato scab, verticillium wilt, and nematodes by soymeal and meat and bone meal in two Ontario potato fields. *Canadian Journal of Plant Pathology* **21,** 345–353.

Ledingham RJ, Sallans BJ, Wenhardt A, 1960. Influence of cultural practices on incidence of common root rot in wheat in Saskatchewan. *Canadian Journal of Plant Science* **40,** 310–316.

Lemmens M, Buerstmayr H, Krska R, Schumacher R, Grausgruber H, Rackenbauer P, 2004. The effect of inoculation treatment and long term application of moisture on Fusarium head blight symptoms and deoxynivalenol contamination in wheat grains. *European Journal of Plant Pathology* **110,** 299–308.

Leonard KJ, 2001. Stem rust – future enemy? In: Peterson P, ed. *Stem Rust of Wheat: From Ancient Enemy to Modern Foe.* St. Paul, MN: APS Press, 119–146.

Lynch JM, Panting LM, 1980. Cultivation and the soil biomass. *Soil Biology and Biochemistry* **12,** 29–33.

Mandahar HK, Lyngs Jorgensen HJ, Mathur SB, Smedegaard-Petersen V, 1998. Resistance to rice blast induced by ferric chloride, di-potassium hydrogen phosphate and salicylic acid. *Crop Protection* **17,** 323–329.

Mann RL, Kettlewell PS, Jenkinson P, 2004. Effect of foliar-applied potassium chloride on septoria leaf blotch of winter wheat. *Plant Pathology* **53,** 653–659.

Mascagni HJ Jr, Harrison SA, Russin JS, Desta HM, Colyer PD, Habetz RJ, Hallmark WB, Moore SH, Rabb JL, Hutchinson RL, Boquet DJ, 1997. Nitrogen and fungicide effects on winter wheat produced in the Louisiana Gulf Coast region. *Journal of Plant Nutrition* **20,** 1375–1390.

McFadden AG, Sutton JC, 1975. Relationships of populations of *Trichoderma* spp. in soil to disease in maize. *Canadian Journal of Plant Science* **55,** 579–586.

Messiha NAS, van Bruggen AHC, van Diepeningen AD, de Vos OJ, Termorshuizen AJ, Tjou-Tam-Sin NNA, 2007. Potato brown rot incidence and severity under different management and amendment regimes in different soil types. *European Journal of Plant Pathology* **119,** 367–381.

Miyake Y, Takahashi E, 1983. Effect of silicon on the growth of solution cultured cucumber plant. *Soil Science and Plant Nutrition* **29,** 71–83.

Momol MT, Olson SM, Funderburk JE, Stavisky J, Marois JJ, 2004. Integrated management of tomato spotted wilt on field-grown tomatoes. *Plant Disease* **88,** 882–890.

Morrall RAA, Dueck J, 1982. Epidemiology of Sclerotinia stem rot of rapeseed in Saskatchewan. *Canadian Journal of Plant Pathology* **4,** 161–168.

Mucharromah E, Kuc J, 1991. Oxalates and phosphates induce systemic resistance against disease caused by fungi, bacteria and viruses in cucumber. *Crop Protection* **10,** 265–270.

Nam MH, Jeong SK, Lee YS, Choi JM, Kim HG, 2006. Effects of nitrogen, phosphorus, potassium and calcium nutrition on strawberry anthracnose. *Plant Pathology* **55,** 246–249.

Naseri B, Davidson JA, Scott ES, 2008. Survival of *Leptosphaeria maculans* and associated mycobiota on oilseed rape stubble buried in soil. *Plant Pathology* **57,** 280–289.

Neumann S, Paveley ND, Beed FD, Sylvester-Bradley R, 2004. Nitrogen per unit area affects the upper asymptote of *Puccinia striiformis* f. sp. *tritici* epidemics in winter wheat. *Plant Pathology* **53,** 725–732.

Nuttall WF, 1982. The effect of seeding depth, soil moisture regime, and crust strength on emergence of rape cultivars. *Agronomy Journal* **74,** 1018–1022.

Olesen JE, Mortensen JV, Jorgensen LN, Andersen MN, 2000. Irrigation strategy, nitrogen application and fungicide control in winter wheat on a sandy soil. I.Yield, yield components and nitrogen uptake. *Journal of Agricultural Science, Cambridge* **134,** 1–11.

O'Neill TM, Pye D, Locke T, 2002. The effect of fungicides, irrigation and plant density on the development of *Peronospora sparsa,* the cause of downy mildew in rose and blackberry. *Annals of Applied Biology* **140,** 207–214.

Otten W, Hall D, Harris K, Ritz K, Young IM, Gilligan CA, 2001. Soil physics, fungal epidemiology and the spread of *Rhizoctonia solani. New Phytologist* **151,** 459–468.

Perrenoud S, 1990. *Potassium and Plant Health,* 2nd edition. Bern, Switzerland: International Potash Institute. Peters RD, Sturz AV, Carter MR, Sanderson JB, 2003. Developing disease suppressive soils through crop rotation and tillage management practices. *Soil & Tillage Research* **72,** 181–182.

Petrie GA, 1986. Consequences of survival of *Leptosphaeria maculans* (blackleg) in canola stubble residue through an entire crop rotation sequence. *Canadian Journal of Plant Pathology* **8,** 353.

Pettitt TR, Monaghan JM, Crawford MA, 2007. Assessment of the control of *Phytophthora* root rot disease spread by Spin Out®-treated fabrics in container-grown hardy nursery stock. *Crop Protection* **27,** 198–207.

Pivonia S, Cohen R, Cohen S, Kigel J, Levita R, Katan J, 2004. Effect of irrigation regimes on disease expression in melon plants infected with *Monosporascus cannonballus. European Journal of Plant Pathology* **110,** 155–161.

Prabhu AS, Fageria NK, Huber DM, Rodrigues FA, 2007. Potassium and plant disease. In: Datnoff LE, Elmer WH, Huber DM, eds. *Mineral Nutrition and Plant Disease.* St. Paul, MN: APS Press, 57–78.

Rahman M, Punja ZK, 2007. Calcium and plant disease. In: Datnoff LE, Elmer WH, Huber DM, eds. *Mineral Nutrition and Plant Disease.* St. Paul, MN: APS Press, 79–94.

Ramel ME, Gugerli P, Bunter M, 2006. Control and monitoring: Eradication of *Plum pox virus* in Switzerland. *EPPO Bulletin* **36,** 312–314.

Rattray EAS, Tyrrell JA, Prosser JI, Killham K, 1993. Effect of soil bulk density and temperature on wheat rhizosphere colonisation by *lux*-marked *Pseudomonas fluorescens. European Journal of Soil Biology* **29,** 73–82.

Reis EM, Abrao JJR, 1983. Effect of tillage and wheat residue management on the vertical distribution and inoculum density of *Cochliobolus sativus* in soil. *Plant Disease* **67,** 1088–1089.

Reis EM, Denardin JE, Kochhann RA, 1990. Fluctuations of *Cochliobolus sativus* propagules in the soil related to systems of soil tillage. *Fitopatologia Brasileira* **15,** 267–269.

Reuveni R, Agapov V, Reuveni M, 1994. Foliar spray of phosphates induces growth increase and systemic resistance to *Puccinia sorghi* in maize. *Plant Pathology* **43,** 245–250.

Reuveni M, Reuveni R, 1995. Efficacy of foliar application of phosphates in controlling powdery mildew fungus on field-grown winegrapes: Effects on cluster yield and peroxidase activity. *Journal of Phytopathology* **143,** 21–25.

Reuveni R, Reuveni M, 1998. Foliar fertilizer therapy – a concept in integrated pest management. *Crop Protection* **17,** 111–118.

Rodgers-Gray BS, Shaw MW, 2000. Substantial reductions in winter wheat diseases caused by addition of rice straw but not manure to soil. *Plant Pathology* **49,** 590–599.

Rodgers-Gray BS, Shaw MW, 2004. Effects of straw and silicon soil amendments on some foliar and stem base diseases in pot-grown winter wheat. *Plant Pathology* **53,** 733–740.

Rodrigues FA, McNally DJ, Datnoff LE, Jones JB, Labbe C, Benhamou N, Menzies JG, Belanger RR, 2004. Silicon enhances the accumulation of diterpenoid phytoalexins in rice: A potential mechanism for blast resistance. *Phytopathology* **94,** 177–183.

Saharan GS, 1992. Management of rapeseed and mustard diseases. In: Kumar D, Rai M, eds. *Advances in Oilseeds Research* Vol. 1. Jodhpur, India: Scientific Publishers, 151–188.

Samuels AL, Glass ADM, Ehret DL, Menzies JG, 1991. Mobility and deposition of silicon in cucumber plants. *Plant Cell and Environment* **14,** 485–492.

Sanogo S, 2006. Predispositional effect of soil water saturation on infection of Chile pepper by *Phytophthora capsici. HortScience* **41,** 172–175.

Sherf AF, Macnab AA, 1986. *Vegetable Diseases and their Control.* New Jersey: John Wiley and Sons.

Shipton WA, Boyd WJR, Ali SM, 1974. Scald of barley. *Review of Plant Pathology* **53,** 839–861.

Sosnowski MR, Scott ES, Ramsey MD, 2006. Survival of *Leptosphaeria maculans* in soil on residues of *Brassica napus* in South Australia. *Plant Pathology* **55,** 200–206.

Stirling GR, Eden LM, 2008. The impact of organic amendments, mulching and tillage on plant nutrition, pythium root rot, root-knot nematode and other pests and diseases of capsicum in a subtropical environment, and implications for the development of more sustainable vegetable farming systems. *Australasian Plant Pathology* **37,** 123–131.

Sturz AV, Christie BR, 2003. Beneficial microbial allelopathies in the root zone: The management of soil quality and plant disease with rhizobacteria. *Soil & Tillage Research* **72,** 107–123.

Sturz AV, Carter MR, Johnston HW, 1997. A review of plant disease, pathogen interactions and microbial antagonism under conservation tillage in temperate humid agriculture. *Soil & Tillage Research* **41,** 169–189.

Summers CG, Stapleton JJ, 2002. Use of UV reflective mulch to delay the colonisation and reduce the severity of *Bemisia argentifolii* (Homoptera: Aleyrodidae) infestations in curcubits. *Crop Protection* **21,** 921–928.

Summers CG, Mitchell JP, Stapleton JJ, 2004. Management of aphid-borne viruses and *Bemisia argentifolii* (Homoptera: Aleyrodidae) in zucchini squash by using UV reflective plastic and wheat straw mulches. *Environmental Entomology* **33,** 1447–1457.

Sumner DR, 1982. Crop rotation and plant productivity. In: Rechcigl J, ed. *Handbook of Agricultural Productivity* Vol. I. Boca Raton, FL: CRC Press, 273–315.

Talukder ZI, McDonald AJS, Price AH, 2005. Loci controlling partial resistance to rice blast do not show marked QTL X environment interaction when plant nitrogen status alters disease severity. *New Phytologist* **168,** 455–464.

Tenuta M, Lazarovits G, 1999. Mechanisms of action for control of soilborne pathogens by high nitrogen containing soil amendments. *Proceedings of the 6th Annual International Research Conference on Methyl Bromide Alternatives and Emission Reduction, San Diego, CA*, 27–28.

Tenuta M, Conn KL, Lazarovits G, 2002. Volatile fatty acids in liquid swine manure can kill microsclerotia of *Verticillium dahliae. Phytopathology* **92,** 548–552.

Teo BK, Morrall RAA, Verma PR, 1989. Influence of soil moisture, seeding date and canola cultivars (Tobin and Westar) on the germination and rotting of sclerotia of *Sclerotinia sclerotiorum. Canadian Journal of Plant Pathology* **11,** 393–399.

Teo BK, Yitbarek SM, Verma PR, Morrall RAA, 1988. Influence of soil moisture, seeding date, and *Rhizoctonia solani* isolates (AG 2–1 and AG 4) on disease incidence and yield in canola. *Canadian Journal of Plant Pathology* **10,** 151–158.

Theunissen J, Schelling G, 1996. Pest and disease management by intercropping: Suppression of thrips and rust in leek. *International Journal of Pest Management* **42,** 227–234.

Tumwine J, Frinking HD, Jeger MJ, 2002. Integrating cultural control methods for tomato late blight (*Phytophthora infestans*) in Uganda. *Annals of Applied Biology* **141,** 225–236.

Umaerus VR, Scholte K, Turkensteen LJ, 1989. Crop rotation and the occurrence of fungal diseases in potatoes. In: Vos J, Van Loon CD, Bollen GJ, eds. *Effects of Crop Rotation on Potato Productions in the Temperate Zones*. Dordrecht: Kluwer Academic Publications, 171–189.

van Bruggen AHC, 1995. Plant disease severity in high-input compared to reduced input and organic farming systems. *Plant Disease* **79,** 976–984.

van Bruggen AHC, Semenov AM, van Diepeningen AD, de Vos OJ, Blok WJ, 2006. Relation between soil health, wave-like fluctuations in microbial populations, and soil-borne plant disease management. *European Journal of Plant Pathology* **115,** 105–122.

van Diepeningen AD, de vos OJ, Zelenev VV, Semenov AM, van Bruggen AHC, 2005. DGGE fragments oscillate with or counter to fluctuations of cultivable bacteria along wheat roots. *Microbial Ecology* **50**, 506–517.

Wagner F, 1940. Die bedeutung der kieselsnure far das wachstum einiger kulturpflanzen, ihren nnhrstoffhaushalt und ihre anfnlligkeit gegen echte mehltaupilze. *Phytopathologische Zeitschrift* **12,** 427–479.

Walters DR, Bingham IJ, 2007. Influence of nutrition on disease development caused by fungal pathogens: Implications for plant disease control. *Annals of Applied Biology* **151,** 307–324.

Walters DR, Murray DC, 1992. Induction of systemic resistance to rust in *Vicia faba* by phosphate and EDTA: Effects of calcium. *Plant Pathology* **41**, 444–448.

Williams JR, Stelfox D, 1980. Influence of farming practices in Alberta on germination and apothecium production of sclerotia of *Sclerotinia sclerotiorum. Canadian Journal of Plant Pathology* **2,** 169–172.

Windels CE, 2000. Economics and social impacts of fusarium head blight: Changing farms and rural communities in the northern Great Plains. *Phytopathology* **90,** 17–21.

Woltz SS, Jones JP, Scott JW, 1992. Sodium chloride, nitrogen source and lime influence Fusarium crown rot severity in tomato. *HortScience* **27,** 1087–1088.

Workneh F, van Bruggen AHC, 1994. Suppression of corky root of tomatoes in soils from organic farms, associated with soil microbial activity and nitrogen status of soil and tomato tissue. *Phytopathology* **84,** 688–694.

Yi C, Kaul H-P, Kubler E, Schwadorf K, Aufhammer W, 2001. Head blight (*Fusarium graminearum*) and deoxynivalenol concentration in winter wheat as affected by pre-crop, soil tillage and nitrogen fertilization. *Zeitschrift fur Pflanzenkrankheiten und Pflanzenschutz* **108,** 217–230.

Young IM, Ritz K, 2000. Tillage, habitat space and function of soil microbes. *Soil & Tillage Research* **53,** 201–213.

Zhan J, Fitt BDL, Pinnschmidt HO, Oxley SJP, Newton AC, 2008. Resistance, epidemiology and sustainable management of *Rhynchosporium secalis* populations on barley. *Plant Pathology* **57,** 1–14.

Chapter 3
Biological control agents in plant disease control

John M. Whipps[1] *and Mark P. McQuilken*[2]

[1]Warwick HRI, University of Warwick, Wellesbourne, Warwick, UK
[2]The Scottish Agricultural College, Ayr, UK

3.1 Introduction

Over the last 50 years, disease control has relied heavily on the use of chemical fungicides, bacteriocides and soil fumigants. However, there are now many problems associated with their continued deployment including increasing pressure to reduce chemical use in the environment in general, development of pesticide resistance in many pathogens, and decreasing availability of active ingredients through stricter registration and difficulty in finding novel active compounds. Consequently, the search for alternative non-chemical methods of disease control continues to gain significance. Indeed, for organic growers where chemical control measures are not permitted, these considerations have been paramount for many years.

In the broadest sense, alternative, biologically based disease control measures have been used for many years. These include plant breeding for resistance, crop rotations, tillage systems and fertilizer practices that affect pathogens directly or alter microbial populations to inhibit pathogens, exploitation of disease suppressive soils and growing media, as well as environmental controls, particularly in the glasshouse. However, the greatest interest has been in the development of biological control agents (BCAs) used as microbial inoculants, mimicking the use of chemical pesticides. Many aspects of the understanding, development and use of BCAs have been reviewed extensively in the last 10 years (e.g. Whipps, 1997a, 2001, 2004; Whipps & Davies, 2000; Montesinos, 2003; Harman *et al.*, 2004; Kiss *et al.*, 2004; Kloepper *et al.*, 2004; Compant *et al.*, 2005a; Fravel, 2005; Haas & Défago, 2005; Jacobsen, 2006; Harman, 2006; Woo *et al.*, 2006; Bakker *et al.*, 2007; Lugtenberg & Leveau, 2007; Whipps & Gerhardson, 2007; Weller, 2007) to which reference should be made for detailed background information. This chapter uses selected examples to focus on some key areas of importance or where advances have taken place relatively recently. It also attempts to indicate future directions that will enhance the development and commercial marketing of BCAs.

3.2 Modes of action

Perhaps the greatest recent advances in biological disease control have been concerned with understanding modes of action. This reflects the huge developments in molecular biology of bacteria, fungi and plants providing the tools to dissect the many types of interactions that can occur, particularly through the use of mutants and genetically marked strains of microorganism as well as gene expression studies. General modes of action include competition, antibiosis, parasitism, induced resistance and plant-growth promotion along with highly specialised mechanisms such as that associated with hypovirulence. Commonly, biological disease control by a single BCA can involve a number of modes of action and no one mode of action is necessarily mutually exclusive to another. This holds true for BCAs active in the phyllosphere, spermosphere, rhizosphere and post-harvest environments.

3.2.1 Competition for space and nutrients

One of the classic demonstrations of competition for space, infection sites or nutrients as a mode of action concerns the control of fireblight caused by the pathogenic bacterium, *Erwinia amylovora* by the non-pathogenic bacterium *Pseudomonas fluorescens* A506 (Lindow & Leveau, 2002). By spraying flowers of apple and pear with *P. fluorescens* A506 just as they open, the BCA colonises the flowers, utilises the available nutrients, and prevents multiplication of small numbers of *E. amylovora* that might then encounter the flowers, thereby preventing infection by pre-emptive exclusion. Pre-emptive colonisation of necrotic leaf tissues by the fungus *Ulocladium atrum* to control the fungal pathogen, *Botrytis cinerea,* is another case of this type of mode of action which may involve both competition for infection sites and nutrients, resulting in reduced pathogen sporulation (Kessel *et al.*, 2005). Competition for infection sites and/or nutrients, in or on roots, has also been recorded. For example, between non-pathogenic and pathogenic strains of *Fusarium oxysporum* on tomato (Olivain & Alabouvette, 1999; Bolwerk *et al.*, 2005; Olivain *et al.*, 2006) and between non-pathogenic strains of *Rhizoctonia* and the pathogen *Rhizoctonia solani* on several plant species (Herr, 1995). *Pseudomonas* spp. have also been shown to act in part by competition for space and nutrients during the colonization of tomato roots by *Fusarium oxysporum* f. sp. *radicis-lycopersici* (Bolwerk *et al.*, 2003) and may involve secretion of a site-specific recombinase (Dekkers *et al.*, 2000). There is also evidence that ectomycorrhizal fungi, by way of their physical sheathing of the root, may also occupy pathogen infection sites, thereby preventing infection (Whipps, 2004).

Competition for nutrients alone, ranging from simple carbon- and nitrogen-containing compounds to complex plant residues, in a range of situations is also a commonly reported mode of action for both bacterial and fungal BCAs. For example, the yeasts *Cryptococcus laurentii* BSR-Y22 and *Sporobolomyces roseus* FS43–238 controlled *Botrytis cinerea* in apple wounds by competing for fructose, glucose and sucrose (Filonow, 1998) and *Candida guillermondii* competed for nitrates during control of *Penicillium expansum* on apple (Scherm *et al.*, 2003). Competition in soil for carbon in the form of glucose has also been shown to be involved in the suppression of pathogenic *F. oxysporum* by non-pathogenic strains of the same species (Larkin & Fravel, 1999). A related mode of action concerns the ability of some bacteria and fungi to metabolise organic compounds released from germinating seeds which normally stimulate pathogen propagules responsible for damping-off to germinate or stimulate zoospore attraction. This mechanism has been recorded for control of *Pythium ultimum* by *Enterobacter cloacae* on a number of plants (Kageyama & Nelson, 2003), *Pythium aphanidermatum* by *Burkholderia cepacia* on pea (Heungens & Parke, 2000) and for *P. ultimum* and/or *Rhizopus oryzae* by *Trichoderma* spp. on cotton (Howell, 2002). However, the most commonly cited example for competition, particularly in soil and the rhizosphere, involves that for iron, as often its low bioavailability particularly in high pH soils is a factor limiting growth of microorganisms. Consequently, most microorganisms produce iron-chelating compounds termed siderophores to competitively acquire ferric iron. Many siderophores produced by bacteria have a very high affinity for ferric iron, and their release sequesters the limited supply of iron making it unavailable to pathogenic fungi, thereby restricting their growth (Loper & Henkels, 1999). Thus, the pyoverdine siderophores produced by numerous *Pseudomonas* species have been shown to be involved in the control of both *Pythium* and *Fusarium* species (Loper & Buyer, 1991; Duijff *et al.*, 1993). Some bacterial BCAs can even utilise the siderophores produced by other bacteria enhancing their ability to colonise the rhizosphere and potentially their biocontrol activity (Loper & Henkels, 1999).

3.2.2 Production of antibiotics

Production of antibiotics and inhibitory metabolites by microorganisms has been well established as a mode of action. Microorganisms commonly produce such metabolites during the course of their growth and only if production at the site of biocontrol is confirmed, or activity implied by use of either non-producing or over-producing mutants, or reporter strains, can a role in biocontrol be assured. Against this background, compounds such as amphisin, 2,4-diacetylphloroglucinol, hydrogen cyanide, oomycin A, phenazine, pyoluteorin, pyrrolnitrin, tensin, tropolone and cyclic lipopolysaccharides produced by *Pseudomonas* spp. (Défago, 1993; Nielsen *et al.*, 2002; Raaijmakers *et al.*, 2002; de Souza *et al.*, 2003; Nielsen & Sørensen, 2003) and gramicidin S, oligomycin A, kanosamine, iturin, zwittermycin A and xanthobaccin produced by *Bacillus*, *Streptomyces* and *Stenotrophomonas* spp. (Milner *et al.*, 1995, 1996; Hashidoko *et al.*, 1999; Kim *et al.*, 1999; Nakayama *et al.*, 1999; Edwards & Seddon, 2001; Romero *et al.*, 2007) have been identified to have a role in disease biocontrol. The regulation of many of these bacterial antibiotics has been explored and involvement of regulatory genes, and sigma factors, and key signal molecules has been found. These include GacA/GacS or GrrA/Grrs, RpoD, RpoN, RpoS, prsA, and *N*-acyl homoserine lactone derivatives (Pierson *et al.*, 1998; Chancey *et al.* 1999; Bloemberg & Lugtenberg, 2001; Haas & Keel, 2003; Chin-A-Woeng *et al.*, 2005; Péchy-Tarr *et al.*, 2005) as well as positive autoregulation systems (Schnider-Keel *et al.*, 2000; Brodhagen *et al.*, 2004). Production is also influenced by nutrient availability, plant type and age, environmental conditions, microorganisms present including other BCAs and the pathogen itself, which all involve complex signalling pathways (Molina *et al.*, 2003; Duffy *et al.*, 2004; Maurhofer *et al.*, 2004; Morello *et al.*, 2004; Compant *et al.*, 2005b) and so the spectrum of antibiotic production by any of the strains of BCA may differ depending on the situation under consideration. Interestingly, interference with signalling processes controlling antibiotic production may actually be a novel mode of action in control of plant pathogens by some BCAs which deserves further study (Molina *et al.*, 2003; Uroz *et al.*, 2003; Dong *et al.*, 2004).

Gene clusters responsible for production of several antibiotics produced by bacteria have now been cloned and manipulated to examine the potential to enhance biocontrol activity in other strains (Hammer *et al.*, 1997; Nowak-Thompson *et al.*, 1999; Timms-Wilson *et al.*, 2000; Chin-A-Woeng *et al.*, 2001; Delany *et al.*, 2001; Huang *et al.*, 2004). The potential for detection of additional, and potentially novel, gene clusters responsible for antibiotic production and other modes of action is likely to ensue as complete genomic sequences such as that for the BCA strain *Pseudomonas fluorescens* Pf-5 (Paulsen *et al.*, 2005; Loper *et al.*, 2007) become available.

The situation is less clear cut with fungi where molecular manipulation methods are generally less advanced and more complex, especially for producing single insert gene mutants. Nevertheless, flocculosin produced by *Pseudozyma flocculosa* (Cheng *et al.*, 2003) and gliotoxin and peptaibols produced by *Trichoderma* spp. (Wilhite *et al.*, 1994; Wiest *et al.*, 2002) have been clearly shown to be involved in disease biocontrol. Interestingly, UV mutants of *Trichoderma virens* deficient in either antibiotic production, mycoparasitism or both, were equally effective at controlling damping-off in cotton as the parent strains, indicating that other mechanisms such as some form of induced resistance in the plant were more important for biocontrol in this system (Howell & Stipanovic, 1995; Howell *et al.*, 2000; Howell, 2002). Production of volatile groups of inhibitory compounds including alcohols, esters, ketones, acids and lipids by *Muscador albus* and *M. roseus* may be responsible for the control of seedling diseases of sugar beet and eggplant (Strobel *et al.*, 2001; Stinson *et al.*, 2003) and use of these BCAs as 'mycofumigants' may be an interesting avenue to pursue.

3.2.3 Parasitism and production of extracellular lytic enzymes

Parasitism and associated production of extracellular lytic enzymes has been thoroughly explored as a mode of action in biocontrol. This is a relatively simple phenomenon for bacteria where degradation of target cell walls is generally considered to reflect parasitism, and may range from simple attachment of bacterial cells to hyphae with minimal degradation, through biofilm formation to complete lysis and cell wall breakdown (Mitchell & Hurwitz, 1965; Nelson *et al.*, 1986; Bolwerk *et al.*, 2003). Not surprisingly, lists of extracellular enzymes produced by bacterial BCAs have been produced but only relatively rarely have unequivocal roles of enzymes produced by bacteria been demonstrated. These include a number of chitinases (Chernin *et al.*, 1995, 1997; Pleban *et al.*, 1997; Kamensky *et al.*, 2003), proteases (Dunne *et al.*, 1997) and β-1,3-glucanases (Fridlender *et al.*, 1993; Palumbo *et al.*, 2005). Regulation of protease and chitinase production in bacteria involves the two part regulatory systems GacA/GacS and GrrA/GrrS (Sacherer *et al.*, 1994; Corbell & Loper, 1995; Ovadis *et al.*, 2004) similar to the production of siderophores and antibiotics.

With fungal BCAs the process of parasitism of fungal plant pathogens, or mycoparasitism, is more complex than that for bacteria and a series of interlinked phases of hyphal–hyphal interactions have been recorded especially for *Trichoderma* spp. including: sensing, directed growth, contact and binding, sometimes involving production of appressoria, coiling or alignment of hyphae of the mycoparasite around the host, penetration and then degradation (Chet *et al.*, 1998; Steyaert *et al.*, 2003; Lu *et al.*, 2004). Numerous variations on the basic process exist. For example, the production of haustoria in the biotrophic phase of infection by the BCA *Verticillium biguttatum* during parasitism of *Rhizoctonia solani* hyphae (van den Boogert & Deacon, 1994), intracellular growth by *Ampelomyces quisqualis* in powdery mildew hyphae (Kiss *et al.*, 2004) and inter- or intracellular growth of BCA hyphae of many mycoparasites during infection of complex propagules such as sclerotia (Jeffries & Young, 1994; Rey *et al.*, 2005), but cell wall degrading enzymes always seem to be involved in the process. Molecular proof of the involvement of specific extracellular enzymes is difficult to obtain for some mycoparasites due to the presence of multiple copies of similar genes where knock-out of one gene leaves others functional, despite potentially having a role in the parasitism process (Carsolio *et al.*, 1999; Woo *et al.*, 1999; Grevesse *et al.*, 2003). Nevertheless, through some molecular expression studies, gene knock-out experiments and microscopical cytochemical labelling procedures, along with a history of correlative studies, involvement of chitinases, glucanases, proteases and cellulases seems clear (Benhamou & Chet, 1996; Baek *et al.*, 1999; Rotem *et al.*, 1999; Markovich & Kononova, 2003; Steyaert *et al.*, 2003; Pozo *et al.*, 2004; Rey *et al.*, 2005; Morissette *et al.*, 2006; Woo *et al.*, 2006; Friel *et al.*, 2007). Earlier, directed growth was thought to reflect growth along a chemical gradient of amino acids and sugars but recent evidence in *Trichoderma* spp. suggests that low level expression of cell wall degrading enzymes by the BCA leads to the release of low-molecular-weight oligosaccharides from the host cell wall that in turn are recognised by *Trichoderma*, resulting in stimulated growth, antibiotic and enzyme production, and mycoparasitism (Viterbo *et al.*, 2002; Brunner *et al.*, 2003; Zeilinger *et al.*, 2003; Woo *et al.*, 2006). On contact, the recognition and coiling appear to be lectin mediated (Barak *et al.*, 1985). Recent molecular studies suggest that the Gα subunit of heterotrimeric G proteins plays a role in the signal transduction to increase chitinase expression, antibiotic production and coiling during mycoparasitism (Rocha-Ramirez *et al.*, 2002; Mukerjee *et al.*, 2004; Reithner *et al.*, 2005). Mitogen-activated protein kinase (MAPK) cascades may also be involved in mycoparasite signal transduction in *Trichoderma* but may differ in response with isolate, host and environment (Mendoza-Mendoza *et al.*, 2003; Mukherjee *et al.*, 2003). Enhanced antagonistic effects may consequently occur in *Trichoderma* through combined action of cell wall degrading enzymes with different target polysaccharides or through combination with release of antibiotics (Schirmbock *et al.*, 1994). However, recent studies suggest that for some *Trichoderma* isolates, modes of action involving interactions with the plant may be more significant than previously thought (Harman *et al.*, 2004; Howell, 2006).

Production of extracellular enzymes that degrade virulence factors such as *N*-acyl homoserine lactones have been mentioned previously as a mode of action involved with regulation of antibiotic production (Molina *et al.*, 2003; Dong *et al.*, 2004) but this general concept is growing in importance. For example, *Pantoea dispersa* produces an esterase that detoxifies the toxin albicidin produced by *Xanthomonas albilineans* (Zhang & Birch, 1996, 1997) and *Trichoderma harzianum* produces proteases that degrade hydrolytic enzymes produced by *Botrytis cinerea* on bean leaves, preventing the pathogen from infecting its host (Kapat *et al.*, 1998; Elad & Kapat, 1999). A related concept is detoxification of virulence factors rather than their degradation. For instance, detoxification of albicidin by proteins produced by *Klebsiella oxytoca* (Walker *et al.*, 1988) and

Alcaligenes denitrificans (Basnayake & Birch, 1995), detoxification of fusaric acid from *Fusarium* species by proteins from *Ralstonia solonacearum* (Toyoda *et al.*, 1988) and production of the antifungal metabolite, cladosporol, by *Cladosporium tenuissima* that inhibits β-1,3-glucan biosynthesis, thereby preventing growth of *Cronartium flaccidum* and *Peridermum pini* (Moricca *et al.*, 2001).

3.2.4 Induced resistance

Induced resistance, defined as 'the process of active resistance dependent on the host plant's physical or chemical barriers, activated by biotic or abiotic agents (inducing agents)' (Kloepper *et al.*, 1992), has been a major area of recent research (see Chapter 4). Biological control involves microorganisms as the inducers and a combination of experiments involving spatial or temporal separation between inducer microorganisms and the pathogen has enabled this mode of action to be identified, even though inducers may exhibit other modes of actions in the same pathosystem. Use of mutants lacking expression of other modes of action has also helped to identify a role for induced resistance.

Bacteria, filamentous fungi and yeasts have all been shown to exhibit induced resistance in the phyllosphere, rhizosphere, spermosphere and post-harvest environments, with individual isolates or species showing activity against a number of viral, bacterial and fungal plant pathogens, either locally or systemically (Wilson *et al.*, 1994; Whipps, 2001; Kloepper *et al.*, 2004; Compant *et al.*, 2005a; Jacobsen, 2006; Woo *et al.*, 2006; Bakker *et al.*, 2007). Some well-established examples of microorganisms exhibiting induced resistance include bacteria such as *Bacillus* spp. (Bargabus *et al.*, 2002, 2003; Collins & Jacobsen, 2003; Kloepper *et al.*, 2004), *Pseudomonas* spp. (van Peer *et al.*, 1991; Bakker *et al.*, 2007; Weller, 2007) and *Lysobacter enzymogenes* (Kilic-Ekici & Yuen, 2003) and fungi such as *Trichoderma* spp. (Yedidia *et al.*, 2003; Shoresh *et al.*, 2005; Woo *et al.*, 2006), *Fusarium* spp. (Fravel *et al.*, 2003), mycorrhizal fungi (Whipps, 2004), binucleate *Rhizoctonia* isolates (Jabaji-Hare & Neate, 2005) and the fungus-like Straminopile *Pythium oligandrum* (Benhamou *et al.*, 1997).

The mechanisms involved in resistance induced by biocontrol microorganisms, largely determined from studies on bacteria, appear to differ subtly to those involved with abiotic inducers and necrotising pathogens. The former (termed induced systemic resistance (ISR)) generally involves ethylene and jasmonate signalling along with expression of the regulatory gene *NPR1* and the latter (termed systemic acquired resistance (SAR)) involves salicylate (SA) signalling with activation of pathogenesis-related (PR) proteins (Bakker *et al.*, 2003). However, these separations are not absolute. For example, there is cross-talk between the signalling pathways (van Loon & Glick, 2004), and some *Pseudomonas* spp. apparently can induce SA-dependent signalling by producing small amounts of SA in the rhizosphere (de Meyer *et al.*, 1999) although this behaviour has been challenged (Ran *et al.*, 2005b). In addition, some *Bacillus* spp. exhibit ISR dependent on SA, but independent of jasmonate and expression of *NPR1* (Ryu *et al.*, 2003; Kloepper *et al.*, 2004). Interestingly, induced resistance involving *Trichoderma* spp. seems to involve the jasmonate/ethylene signalling pathway (Shoresh *et al.*, 2005) similar to ISR in most bacteria. Consequently, to ease description the general term induced resistance is used here.

Numerous physiological changes have been associated with microbially induced resistance. These include (a) strengthening of epidermal and cortical walls and deposition of barriers beyond infection sites including callose, lignin and phenolics (Benhamou *et al.*, 1996; Duijff *et al.*, 1997; M'Piga *et al.*, 1997; Benhamou *et al.*, 1998, 2000; Yedidia *et al.*, 2003); (b) increased levels of enzymes such as chitinases, β-1,3-glucanases, β-1,4-glucosidase, PR-1 protein, glucosidases, peroxidases, phenylalanine ammonia lyase, hydroxyperoxidolyase and an associated oxidative burst (Tamietti *et al.*, 1993; Fuchs *et al.*, 1997; Duijff *et al.*, 1998; Recorbet *et al.*, 1998; Chen *et al.*, 2000; Bargabus *et al.*, 2002, 2003; Yedidia *et al.*, 2003; Harman *et al.*, 2004); (c) enhanced phytoalexin production (Ongena *et al.*, 1999; Howell *et al.*, 2000; Howell & Puckhaber, 2005); and (d) enhanced expression of stress-related genes and resistance to the oxidative burst (Timmusk & Wagner, 1999; Castoria *et al.*, 2003). Not all the biochemical changes are found in all inducer–plant combinations and responses may even differ between bacterial and fungal inducers on the same plant (Duijff *et al.*, 1998).

Not surprisingly, numerous elicitors of induced resistance are known, especially from bacteria. These include lipopolysaccharides (Leeman *et al.*, 1995; Meziane *et al.*, 2005), siderophores (Metraux *et al.*, 1990; Leeman *et al.*, 1996; Ran *et al.*, 2005a; Meziane *et al.*, 2005), flagella or the protein subunit flagellin (Gomez-Gomez & Boller, 2000; Zipfel *et al.*, 2004; Meziane *et al.*, 2005), bacterial volatiles (Ryu *et al.*, 2003, 2004), salicylate (de Meyer *et al.*, 1999), the cyclic peptide syringolin (Waspi *et al.*, 1998), antibiotics such 2,4-diacetylphloroglucinol (Iavicoli *et al.*, 2003) and a *N*-trialkylated benzylamine derivative (Ongena *et al.*, 2005). There are fewer, less well characterised elicitors from fungi. For instance, these include from *Trichoderma* spp. enzymes or peptides such as a 22 kDa xylanase and an 18 kDa serine proteinase (Hansen & Howell, 2004; Woo *et al.*, 2006), pectic and other oligosaccharides released from host cell walls (Jabaji-Hare *et al.*, 1999; Woo *et al.*, 2006) and avirulence proteins with homology to Avr 4 and Avr 9 genes from *Cladosporium fulvum* (Woo *et al.*, 2006) and from *Pythium oligandrum*, elicitin-like proteins (Picard *et al.*, 2000; Benhamou *et al.*, 2001; Takenaka *et al.*, 2006). This area of research is advancing rapidly.

The ability to colonise host tissues internally (as an endophyte) may be a key feature in some inducer microorganisms (Steijl *et al.*, 1999; Harman *et al.*, 2004). Population density of the inducer is also another important feature. For instance, dose–response studies have shown that ISR was only mediated by *Pseudomonas fluorescens* WCS374 in radish when population densities reached 10^5 colony-forming units per gram of root (Raaijmakers *et al.*, 1995) and similar effects of dose rate were found for induced resistance by various non-pathogenic *Fusarium* spp. but effects were also dependent on host cultivar or species (Hervás *et al.*, 1995; Larkin & Fravel, 1999).

3.2.5 Plant-growth promotion

Both fungal and bacterial BCAs can exhibit the phenomenon of plant-growth promotion. To some extent this may reflect the ability of BCAs to control well-known diseases caused by major pathogens or by reducing the deleterious effects of minor pathogens as discussed previously. However, many BCAs are also able to promote plant growth in the absence of any pathogens, thereby exhibiting additional physiological activities.

Historically, most attention has been placed on root-colonising bacteria that can enhance plant growth resulting in the term plant-growth-promoting rhizobacteria (PGPR; see also Chapter 4). Those PGPR BCA strains are typically *Bacillus* and *Pseudomonas* but numerous other PGPR species are known (Vessey, 2003; Zahir *et al.*, 2004). Mechanisms involved may include associative N_2 fixation, solubilisation of nutrients such as P, promotion of mycorrhizal and rhizobia function, regulating ethylene production in roots, release of phytohormones, and decreased heavy metal toxicity (Whipps, 2001; Vessey, 2003; Zahir *et al.*, 2004). Root and seed colonisation to various degrees are the key features for growth promotion with the significance of endophytic growth in this process gradually being recognised (Rosenblueth & Martinez-Romero, 2006).

Of the fungal plant-growth promoters, *Trichoderma* spp. are probably the most well known (Harman *et al.*, 2004; Harman, 2006). *Trichoderma* have the ability to solubilise P and numerous other micronutrients, enhance efficiency of N use, increase root development and root hair formation (Altomare *et al.*, 1999; Harman, 2006), and like most bacterial PGPs, exhibit ability to colonise roots, often becoming endophytic (Yedidia *et al.*, 2000; Harman, 2006).

3.2.6 Hypovirulence

A highly specialised mode of action concerns the use of hypovirulent isolates of fungal pathogens. Hypovirulent fungal isolates contain mycoviruses that intrinsically cause the fungus to be less fit. When hypovirulent isolates are introduced into plant tissues infected with a virulent pathogen isolate, the viruses can be transmitted via hyphal anastamoses, spreading the viral infection, and decreasing disease. The classic example of this process is that of hypovirulent isolates of *Cryphonectria parasitica*, containing unencapsidated double-stranded RNA viruses of the virus family Hypoviridae which have been used to control Chestnut blight in Europe (Heiniger & Rigling, 1994). Hypovirus infection is persistent and non-lytic, and is associated with inability to effectively penetrate the host plant, reduced sexual sporulation, female infertility, and reduced pigmentation. Genetically modified strains of *C. parasitica* transformed with a severe hypovirus are being explored for control of Chestnut blight in the United States (Dawe & Nuss, 2001).

3.3 Production, formulation and application

Production, formulation and application of BCAs have been investigated extensively with the aim of producing successful and cost-effective products (Burges, 1998; Hall & Menn, 1999). A major aim is to produce the greatest quantity of viable propagules with the best quality for formulation as cheaply as possible, preferably using inexpensive growing media such as industrial wastes. Production of bacteria and fungi can be done using large-scale liquid fermentation which often involves manipulating the culture medium to induce production of the desired propagules for formulation. Factors which are often manipulated include temperature, pH and osmotic potential, as well as nutritional factors such as carbon source and C:N ratio (Jackson, 1997). Recently, solid-state fermentation has been used for the production of fungal biomass. For example, conidia produced by solid-state fermentation are incorporated into the wettable granule formulation of the commercial *C. minitans* product, Contans WG (de Vrije *et al.*, 2001).

Unless inocula of BCAs are used immediately following production, cells or biomass are usually dried and formulated as products capable of storage, distribution and application (Fravel *et al.*, 1998; Fravel, 2005). Drying can be done by a range of different methods, including air- and freeze-drying, drying on silica gel and spray- and fluid bed-drying. These methods reduce the metabolic rate of the inoculum by removing the available water, which tends to preserve the inoculum with high viability depending on the BCA. Once the inoculum is dried, it is usually mixed with various components such as carriers, bulking agents, diluents and food bases. BCAs have been formulated as dusts, gels, emulsions, prills, pellets and granules for seed treatments, dips, wettable powders and sprays for application to aerial plant parts, and drenches for incorporation into soil and growing media (Fravel *et al.*, 1998). Most work on formulation closely involves agrochemical, biotechnology or seed-treatment companies and, unfortunately, tends not to be published. The final formulated product should be convenient to use, safe to handle and have an adequate shelf life with stability for at least 1 year. Other desirable characteristics of a formulation include compatibility with application machinery, and ease of integration into integrated pest and disease control systems.

Both quality assurance and technical support are important to ensure that the formulated product contains the appropriate active BCA without contamination, and is applied correctly to ensure efficacy. Quality assurance and extensive technical support have been instrumental in the success of Serenade, a product containing the bacterium *Bacillus subtilis*, and used to reduce post-harvest diseases of citrus and pome fruit (Janisiewicz & Korsten, 2002). The success of products based on *Bacillus* (Tables 3.1 and 3.2) is largely related to their ability to form spores and their ease of formulation and storage (Schisler *et al.*, 2004).

Large-scale field application of BCAs poses practical problems in terms of producing sufficient amounts required to reach the target plant pathogen, and achieve efficacy, as well as concerns over production costs. The target and timing of application depends on the BCA, the pathogen and also the crop. There has been extensive research directed at improving the application and performance of BCAs, and reducing the amounts required for control. One way of reducing the amount of BCA required to control both seed and soil-borne diseases is to apply the agent to seed rather than in-furrow, or as a soil or growing medium incorporation (McQuilken *et al.*, 1998). Application of BCAs to seed has the potential to deliver the agent 'in the right amount, at the right place and at the right time'. However, the process of applying BCAs to seeds presents a special set of technical considerations. For example, sufficient numbers of the BCA must survive the process, and be able to grow and colonise the environment of the germinating seed fast enough to provide control. The BCA must also be able to survive a period of low water activity as the seed has to be stored at low moisture levels. Colonisation of seeds by BCAs during germination can be improved by incorporating the agent during seed priming, a process used for the physiological enhancement of germination (McQuilken *et al.*, 1998). This has been done successfully for *Trichoderma* spp. applied through solid matrix priming (Harman, 1991), and recently in the United Kingdom through drum priming for several bacterial species, including *Bacillus subtilis* and *Pseudomonas fluorescens* (Wright *et al.*, 2003). The use of drum priming is a major advance in the application of BCAs to seeds and has significant commercial potential.

There has been considerable interest in the use of insects to apply and disseminate BCAs to aerial microbiomes. For example, honey and bumble bees have been used successfully to spread both bacterial and fungal BCAs to specific sites such as soft and pome fruit flowers to control diseases including grey mould and fireblight, caused by

Table 3.1 Examples of bacteria and fungi registered or commercially marketed as biological control agents for control of soil- and seed-borne plant pathogens.

Antagonist	Target pathogen(s)/Activity	Disease/Host	Product name and source
Bacteria			
Agrobacterium radiobacter	*Agrobacterium tumefaciens* causing root galls	Ornamentals and other plants sensitive to *A. tumefaciens* root galls	Norback 84-C, Galltrol-A, Nogall, Diegall, Dygall (Becker Underwood Pty Ltd., Australia; Bio-Care Technology Pty Ltd., Australia; New BiProducts Inc., USA; AgBiChem Inc., USA; Agbioresearch Ltd., New Zealand)
Bacillus cereus BP01	Plant-growth promotion	Cotton	Mepplus (MicroFlo Co. LLC, USA)
Bacillus pumilus GB34	*Fusarium* spp., *Rhizoctonia solani*	Soybean	YieldShield concentrate; GB34 Biological Fungicide (Gustafson LLC, USA)
Bacillus licheniformis SB3086	Numerous pathogens, especially *Sclerotinia homeocarpa*	Ornamentals, turf	EcoGuard, Green Releaf (Novozymes Biologicals Inc., USA)
Bacillus subtilis	Pythium damping-off	Tomato	Cillus, Green-all G (Greenbiotech Co, Korea)
	Rhizoctonia solani, Fusarium spp., *Alternaria* spp. and *Aspergillus* spp. pathogens	Root rots and seedling diseases generally	Kodiak, Epic, Concentrate, Kodiak HB, Quantum 4000, System 3 (Gustafson LLC, USA)
	Fusarium spp., *Verticillium* spp., *Rhizoctonia solani,* and *Pythium* spp. pathogens	Various vegetable and field crops	PHYTOVIT WG (Prophyta Biologischer Pflanzenschutz GmbH, Germany)
B. subtilis GB03	*Fusarium* spp., *Phytophthora* spp., *Pythium* spp., *Rhizoctonia solani*	Ornamentals, turf, dry and snap bean, cotton, peanut, soybean, wheat and barley	Companion, Kodiak (Growth Products Ltd., USA; Gustafson LLC, USA; Bayer CropScience LP, USA)
B. subtilis MBI600	*Alternaria* spp., *Aspergillus* spp., *Fusarium* spp., *Pythium* spp., *Rhizoctonia solani*	Alfalfa, dry/snap beans, peanut, soybean, field crops, turf, cotton	HiStick N/T, Pro-mix, Subtilex; Subtilex HB (Becker Underwood Inc., USA; Premier Horticulture Inc., Mexico)
B. subtilis subsp. amyloliquefaciens FZB24	*Fusarium* spp., *Rhizoctonia solani*	Various vegetables and ornamentals	Taegro, Tae-Technical (Earth Biosciences Inc., USA)

B. subtilis, Bacillus circulans, Bacillus amyloliquefaciens, Paenibacillus polymyxa (Mixture)	Damping-off (bacterial) diseases	All crops	Hydroguard (American Agritech, USA)
Burkholderia cepacia[a]	*Pythium* spp., *Fusarium* spp., *Rhizoctonia solani*, nematodes	Alfalfa, beans, clover, cotton, peas, wheat, vegetables and others	Deny, Blue Circle (Stine Microbial Products, USA)
	Pythium spp., *Fusarium* spp., *Rhizoctonia solani*,	Maize, vegetables, cotton	Intercept (Soil Technologies Corp., USA)
Erwinia carotovora, non-pathogenic	Bacterial soft rots	Vegetables, cruciferous plants, rice	Biokeeper (Central Glass Co. Ltd., Japan)
Paenibacillus polymyxa AC-1	Damping-off	Cucumber	NH, Topseed (Greenbiotech Co., Korea)
Pseudomonas sp.	Growth promotion, various seed- and soil-borne diseases	Vegetables, potato, cereals, etc.	Proradix (Sourcon-Padena, Germany)
Pseudomonas aureofaciens TX-1	Various turf grass pathogens	Turf grass diseases	Spotless (Turf Science Laboratories Inc., USA)
Pseudomonas chlororaphis MA 342	*Drechslera* spp., *Septoria* spp., *Fusarium* spp.	Cereal seed-borne diseases	Cedomon (BioAgri AB, Sweden)
P. chlororaphis 63–28	*Pythium* spp. *Rhizoctonia solani, Fusarium oxysporum*	Stem and root rots, and wilt disease in various crop plants	AtEze (Eco Soil Systems Inc., USA)
Pseudomonas fluorescens	Frost damage, *Erwinia amylovora*	Fruit, potato, tomato, berries	BlightBan A506 (NuFarm Inc., USA)
Pseudomonas solanacearum, non-pathogenic	*Pseudomonas solanacearum*	*P. solanacearum* rots in vegetables	PSSOL (Natural Plant Protection, France)
Pseudomonas syringae	*Botrytis* spp., *Penicillium* spp., *Mucor* spp.	Fruit, potato	Bio-save (EcoScience Corp., USA)
Streptomyces colombiensis WYE20	*Rhizoctonia solani*	Turf	Mycocide (KIBC Co., Korea)

(Continued)

Table 3.1 Continued.

Antagonist	Target pathogen(s)/Activity	Disease/Host	Product name and source
Streptomyces goshikiensis WYE324	*Rhizoctonia solani*	Rice, turf	Safegrow (KIBC Co., Korea)
Streptomyces griseoviridis K61	Various soil-borne pathogens	Vegetable and ornamental soil-borne diseases	Mycostop (Verdera Oy, Finland)
Streptomyces lydicus WYCD108	Various soil-borne pathogens	Root rots in many crops, turf and ornamentals	Acinovate (Natural Industries Inc., USA)
Fungi			
Aspergillus flavus AF36	*Aspergillus flavus* (aflatoxin +)	Cotton	AF36 (Arizona Cotton Research and Protection Council, USA)
A. flavus NRRL 21882	*Aspergillus flavus* (aflatoxin +)	Peanut	Afla-guard (Circle One Global Inc., USA)
Coniothyrium minitans CON/M/91–08	*Sclerotinia minor, S. sclerotiorum*	Protected vegetable and field crops	Contans WG (Prophyta Biologischer Pflanzenschutz GmbH, Germany; Sylvan Bio Products Inc., USA) and Intercept WG (Encore Technologies, MN, USA)
C. minitans KONI	*Sclerotinia minor, S. sclerotiorum*	Glasshouse crops and amenity areas	KONI (Bioved Ltd., Szigetszentmiklos, Hungary)
Fusarium oxysporum Fo47	*Fusarium oxysporum*	Asparagus, basil, carnation, cyclamen, gerbera, tomato	Fusaclean (Natural Plant Protection, France)
Gliocladium catenulatum J1446	*Pythium* spp., *Rhizoctonia solani*, and numerous other pathogens	Damping-off of vegetables, herbs and ornamentals and numerous other plants	Prestop Mix, Prestop WP, Primastop (Verdera Oy, Finland)
Gliocladium (Trichoderma) virens G-21	*Pythium ultimum, Rhizoctonia solani*	Damping-off of bedding plants, greenhouse crops and ornamentals	SoilGard 12 G formerly GlioGard (Certis Inc., Columbia, MD, USA)
Pythium oligandrum	Numerous diseases	Numerous crops	Polyversum (Biopreparaty Ltd., Czech Republic)

Trichoderma spp.	Soil-borne fungal pathogens	Turf, glasshouse crops and field crops	Trich-A-Soil (Becker Underwood Pty Ltd., Australia)
	Sclerotium cepivorum, Pyrenochaeta	Onion	Tenet (Agrimm Tecnologies, New Zealand)
	Soil-borne plant pathogens	Ornamentals, fruit, turf, olive, vine	Trichomic (AMC Chemical, Spain)
	Armillaria	Tree seedlings	Arborguard (Biodiscovery Ltd., New Zealand)
Trichoderma harzianum T-22 (KRL-AG2)	*Fusarium* spp., *Pythium* spp., *Cylindrocarpon* spp., *Thielaviopsis* spp. *Rhizoctonia solani, Sclerotinia homeocarpa*	Range of crops, ornamentals and turf	T-22 HC, T-22 Planter Box, T-22 Granules PlantShield HC, RootShield drench and granules, TurfShield, TRIANUM-P, TRIANUM-G (Bio-Works Inc, Fairport, NY, USA; Koppert, the Netherlands)
T. harzianum	Various fungi	Legumes and leaf vegetables	Supresivit (Borregaard and Reitzel, Denmark or Fytovita, Czech Republic)
	Pythium spp., *Rhizoctonia solani* seedling diseases	Numerous crops	Eco-T (Plant health Products, South Africa)
T. harzianum GBF-0208	Numerous pathogens	Vegetables, bulbs, turf	Green-all T WP (Green Biotech Co. Ltd., Korea)
T. harzianum + *Trichoderma viride*	*Fusarium* spp., *Phytophthora* spp., *Pythium* spp. and *Rhizoctonia solani*	Field crops, vegetables, ornamental and turf	Trichodry, Trichoflow, Trichogrow, Trichopel R Trichopel (Agrimm Technologies, New Zealand)
T. harzianum T 35 + *T. harzianum* TH315	*Fusarium* spp., *Pythium* spp., *Rhizoctonia solani, Sclerotium rolfsii*	Seedlings diseases on a range of crops and potato	Root Pro and Root-Protato (Mycontrol, Hamovil, Israel)
T. harzianum + *T. polysporum*	Various root-infecting fungi	Glasshouse crops	BINAB-T W P(Bio-Innovation Eftr AB, Bredholmen, Sweden; or Svenska Predator AB, Sweden; or Bayer, Sweden)
T. viride	*Fusarium* spp.; *Pythium* spp.; *Rhizoctonia solani*	Damping-off, foot rots and collar rots of several plants	Ecoderma (Margo Biocontrols Private Ltd., Bangalore, India)

[a] *Burkholderia cepacia* is no longer available in North America because of concerns over the potential for human pathogenicity of some strains of this species.

Table 3.2 Examples of viruses, bacteria and fungi registered or commercially marketed as biological control agents for control of foliar plant pathogens and timber decay fungi.

Antagonist	Target pathogen(s)/Activity	Disease/Host	Product name and source
Viruses			
Zucchini yellow mosaic virus (Mild strain)	*Zucchini yellow mosaic virus*	Curcubits	Agroguard Z (Bio-Oz Biotecnologies, Israel)
Bacteria			
Bacillus pumilus QST2808	Numerous foliar pathogens, especially soybean rust	Numerous field crops especially soybean	Ballad Plus, Sonata (Agraquest, USA)
Bacillus subtilis	*Botrytis cinerea*	Tomato	Cillus, Green-all G (Greenbiotech Co., Korea)
B. subtilis BD170	*Erwinia amylovora*	Apple	Biopro (Bio-Protect, Germany)
B. subtilis IK-1080	*Botrytis cinerea*, powdery mildew	Vegetables and fruit	Botokira Wettable Powder (Idemitsu Kosan Inc., Japan)
B. subtilis, QST 713	Numerous fungal and bacterial pathogens	Various crops, including fruit trees, curcubits, tomato, hops, grapes and turf	Rhapsody , Serenade, ASO, Serenade Garden, Serenade MAX (AgraQuest Inc., USA)
Paenibacillus polymyxa AC-1	Powdery mildew	Cucumber	NH, Topseed (Greenbiotech Co., Korea)
Pseudomonas fluorescens A506	*Erwinia amylovora* and russeting inducing bacteria	Stone fruit, tomato, potato, strawberry, blueberry	Blight Ban (NuFarm Inc., USA)
P. fluorescens A506 + *P. fluorescens* 1629RS + *Pseudomonas syringae* 742RS	Frost-forming bacteria	Fruits, almond, potato, tomato	Frostban A (Frost Technology Corp., USA)
P. fluorescens A506	Frost-forming bacteria	Fruits, almond, potato, tomato	Frostban B (Frost Technology Corp., USA)
P. syringae 742RS	Frost-forming bacteria	Fruits, almond, potato, tomato	Frostban C (Frost Technology Corp., USA)
P. fluorescens 1629RS	Frost-forming bacteria	Fruits, almond, potato, tomato	Frostban D (Frost Technology Corp., USA)

P. fluorescens A506	*Erwinia amylovora* and russeting inducing bacteria	Stone fruit, tomato, potato, strawberry, blueberry	Blight Ban (NuFarm Inc., USA)
Streptomyces colombiensis WYE20	*Botrytis*, powdery mildew	Pumpkin, strawberry	Mycocide (KIBC Co., Korea)
Fungi			
Aureobasidium pullulans	*Botrytis cinerea, Monilinia* sp.	Blossom infection of strawberry, cherry and damson	BoniProtect forte (Bio-protect, Germany)
	Erwinia amylovora	Pome fruits especially apples, pears and quinces	BlossomProtect (Bio-protect, Germany)
Ampelomyces quisqualis M10	Powdery mildew	Apples, curcubits, grapes, ornamentals, strawberries, tomatoes	AQ10 biofungicide, M10 (Ecogen Inc., USA)
A. quisqualis 94013	Powdery mildew	Cucumber, tomato, pepper, grape, strawberry	Green-all Q, Q-fect (Greenbiotech Co. Ltd., Korea)
Phlebiopsis gigantea	*Heterobasidion annosum*	Rot on pine and spruce trees	PG Suspension (Omex Environmental, UK) and RotStop (Verdera Oy, Finland)
Trichoderma atroviride LC52	*Botrytis cinerea*	Vine, tomato	Sentinel (Agrimm Technologies, New Zealand)
Trichoderma harzianum	*Botrytis cinerea, Eutypa lata*	Cucumber, tomato, grape	Eco-77 (Plant Health Products, South Africa)
T. harzianum + *Trichoderma polysporum*	*Chondrostereum purpureum*, timber decay fungi,	Silver leaf and pruning wounds, timber decay	Binab T Pellets (Binab Bio-innovation EFTR AB, Sweden)
T. harzianum + *Trichoderma viride*	*Armillaria, Eutypa lata, Botryosphaeria*	Vine, pome and stone fruit	Vinevax, Vinevax Bio-dowel, Vinevax Bio-injection, Vinevax pruning wound dressing (Agrimm Technologies, New Zealand)
Ulocladium oudemansii	*Botrytis cinerea*	Grapes, blackcurrants, ornamental flowers	BOTRY-Zen (Botry-Zen Ltd., New Zealand)
	Sclerotinia sclerotiorum	Kiwi fruit	

Botrytis cinerea and *Erwinia amylovora*, respectively (Peng *et al.*, 1992; Thomson *et al.*, 1992; Yu & Sutton, 1997). As part of a 3-year field study, beehives were equipped with dispensers containing the commercial *B. subtilis* product Serenade (Dedej *et al.*, 2004). The honey bees were effective in spreading the BCA to blueberry flowers to suppress mummy berry disease caused by *Monilinia vaccinii-corymbosi*. Further research on dissemination of BCAs is warranted.

3.4 Commercial products available and uses

The number of products for control of plant diseases continues to increase although there is an ongoing flux as some products appear and others are removed from the market. A list of bacterial, fungal and viral products currently either registered or marketed as BCAs for diseases of soil-borne, seed-borne, or foliar pathogens, timber decay fungi or post-harvest pathogens is given in Tables 3.1–3.3. There are more than 50 bacterial products, 50 fungal products and a single viral product. The majority of both bacteria and fungal products are sold for control of seed- or soil-borne pathogens (>70 in total), with fewer for foliar pathogens and timber decay fungi (>30 in total) and less than 10 for post-harvest pathogens. Most contain individual microorganisms although there are some products that contain microorganism mixtures, and some individual microorganism strains are marketed in several different products expanding the potential market of a single active ingredient. Bacterial products are dominated by *Bacillus* species reflecting their ease of growth and production of long-lived spores mentioned earlier. Fungal products are dominated by *Trichoderma* spp., which are also easy to produce, generally have a low toxicity and can sporulate well. Brief summaries of a few of these commercial products are given below, based largely on company website information with no independent corroboration of the data necessarily, to provide an indication of the variety of products on the market and their characteristics. These are discussed in terms of activity in soil and root, aerial and post-harvest microbiomes which may be considered reasonably well-defined habitats that have distinct physicochemical properties containing characteristic microbial communities (Whipps *et al.*, 1988).

3.4.1 Soil and root microbiomes

3.4.1.1 Bacterial products

Bacillus subtilis: Bayer CropScience LP (USA) market, America's first biological fungicide seed treatment, Kodiak, containing spores of *Bacillus subtilis* GB03 (http://www.bayercropscienceus.com) for control of *Alternaria, Aspergillus, Fusarium* and *Rhizoctonia* spp. that attack root systems of a number of plants, including seed and pod vegetables, cotton, peanut, soybean, wheat, barley and maize. The spore concentrate can be applied through standard slurry or mist seed treatment equipment, with fungicides if required, and the bacterium colonises the root system providing control over the whole growing season. A more vigorous root system may be established and with legumes, may increase nodulation by nitrogen-fixing bacteria.

Streptomyces griseoviridis: Verdera Oy (Finland) produce a long-established biofungicide, Mycostop, a dry powder that contains spores of *Streptomyces griseoviridis* K61.

It is sold for control of seed- and soil-borne diseases that cause stem and root rots. These include diseases caused by *Alternaria, Fusarium, Phytophthora* and *Pythium* on a number of glasshouse vegetables and ornamentals. It also has some suppressive effect on *Botrytis* and *Rhizoctonia* (http://www.verdera.com). Mycostop is applied as a suspension in the form of a drench, spray or via drip irrigation or alternatively as a dry powder seed treatment. Bulbs, corms and cuttings can be dipped in the spore suspension. The bacterium colonises the root, excluding pathogens, and also produces antibiotics. In addition, it produces metabolites that stimulate plant growth.

3.4.1.2 Fungal products

Coniothyrium minitans: This fungus is a mycoparasite of sclerotia of *Sclerotinia sclerotiorum* and *Sclerotinia minor* and two products containing conidia of different isolates of this BCA are available: Contans from Prophyta, Germany (http://www.prophyta.com) and KONI from Bioved, Hungary (http://www.bioved.eu). Contans is only available as a wettable granule formulated on glucose, but KONI is marketed as both a wettable granule formulated on perlite and glucose, and also as a wettable powder formulated on glucose alone. Both products have to be applied and incorporated into the soil in advance of the crop to enable the mycoparasite to degrade the sclerotia; this may take several months. All plants susceptible to Sclerotinia rots are suitable for treatment including peanut, oilseeds, vegetable crops and ornamentals. Much of the scientific background concerning *C. minitans* as a BCA has been reviewed recently (Whipps *et al.*, 2007)

Trichoderma harzianum: There are numerous products that contain different strains of *Trichoderma harzianum* (Table 3.1) but one isolate, KRL-AG2 (T-22), sold by BioWorks Inc., USA has been used and developed for several markets in US horticulture and agriculture in a number of different forms (http://www.bioworksinc.com). This BCA when applied to soil, planting mixes or turf, colonises plant roots and provides protection against root pathogens such as *Cylindrocarpon, Fusarium, Pythium, Rhizoctonia* and *Thielaviopsis.* It can also provide plant-growth promotion directly. RootShield granules are largely targeted at glasshouse and nursery use, and can be incorporated, top-dressed, broadcast, or applied in-furrow or to planting holes for use on flowers, bedding plants, ornamentals, vegetables, pome and stone fruit, trees and tree nuts. RootShield drench is a wettable powder formulation also targeted largely at glasshouse crops largely as a spray or drench. Another formulation is PlantShield HC, marketed for the same greenhouse and transplant crops as RootShield but applied as a soil drench, greenhouse foliar spray, as well as a dry powder dips or dusts for bulbs, and even to hydroponics. For agriculture, T-22 Planter Box is applied as a coating on seeds and seed pieces, an in-furrow spray, and as a transplant starter and T-22 HC as a broadcast or in-furrow treatment, both for use on field and row crops, hay and forage crops, bulbs and vegetables. In Europe, T-22 is sold by Koppert, the Netherlands as TRIANUM-G and TRIANUM-P, wettable granule and wettable powder formulations, respectively (http://koppert.nl). TRIANUM is marketed for use on vegetables, soft fruit, herbs, bulbs, ornamentals, perennials, turf and arboriculture but rather than being sold as a BCA *per se,* it is said to increase resistance of plants to stress caused by diseases, sub-optimal feeding and watering regimes of climatic conditions, as well as increasing nutrient uptake, presumably avoiding the need for registration as a BCA. This demonstrates the potential to maximise the range of target uses and commercial returns that may be possible with a BCA with a broad spectrum of activity. Much of the scientific background concerning the use of this isolate has been extensively reviewed (Harman, 2000, 2006).

3.4.2 Aerial microbiomes

3.4.2.1 Bacterial products

Bacillus subtilis: AgraQuest Inc. (USA) produce several commercial products containing strain QST713 of *Bacillus subtilis* sold for control of a wide range of bacterial and fungal diseases, including bacterial leaf spots, anthracnose, damping-off and root rots, botrytis and powdery mildews (http://www.agraquest.com). Examples of products include Serenade and Rhapsody. Serenade is available in two formulations, Serenade MAX, a concentrated wettable powder, and Serenade ASO, an aqueous suspension. Both formulations are used to control diseases affecting vegetables, fruit, nut and vine crops. In contrast, Rhapsody is an aqueous suspension and is primarily used to control diseases affecting bedding plants, pot plants, cut flowers, turf and container-grown ornamentals. The bacterium QST713 is presumed to work via multiple modes of action, including antibiosis, competition, parasitism and induction of SAR.

3.4.2.2 Fungal products

Ampelomyces quisqualis: Formulation AQ10, produced by Ecogen Inc. (USA) first obtained registration in 1995 for use in control of grape-vine powdery mildew (Kiss *et al.*, 2004). It is formulated as a water-dispersible granule and acts as a mycoparasite on powdery mildew mycelia. Although spraying with AQ10 was as effective as conventional fungicides providing that at least two sprays were applied when disease incidence was still low (<10% leaf infection), reduced efficacy often occurred at low relative humidity. This was overcome by spraying early in the morning or late afternoon, and by incorporation of a mineral oil-based surfactant as a wetting agent (Daoust & Hofstein, 1996). Subsequently, the product was registered for use in the control of powdery mildew on a range of other fruits, vegetables and ornamentals in conjunction with the wetting agent, Add-Q (Bélanger & Labbé, 2002). Currently, the use of AQ10 is recommended as part of an integrated disease control programme, using a reduced regime of conventional fungicides to minimise the development of fungicide resistance.

Phlebiopsis (Peniophora) gigantea: a common wood-rotting saprotroph, which has been used successfully in Europe as a biocontrol agent of conifer root and butt rott, caused by *Heterobasidion annosum*, for almost 50 years (Pratt *et al.*, 2000). The fungus is applied to stump surfaces and competes with *H. annosum* for the woody resource within conifer stumps. Three commercial products are available: PG Suspension (Omex Environmental Ltd.) in the United Kingdom, PG IBL (Biofood s.c.) in Poland, and Rotstop (Verdera Oy) in Finland and other Scandinavian countries. Formulations of the products consist of oidia maintained in a sucrose suspension, sawdust, or a wettable powder, respectively.

Ulocladium oudemansii: BOTRY-zen, a biocontrol product containing the fungus *Ulocladium oudemansii*, is registered for use in New Zealand (http://www.botryzen.co.nz). The product was first developed for early-season control of *Botrytis cinerea* in grapes (Elmer & Reglinski, 2006), and is formulated as a water-dispersible granule. *U. oudemansii* colonises the grape flower debris left on the developing grape bunches faster than *B. cinerea* and reduces infection. BOTRY-zen has been used as a foliar spray on New Zealand vineyards since 1997 to control botrytis over a wide range of disease pressure conditions, with control equivalent to standard fungicide programmes. The product is also registered for use on kiwifruit, blackcurrants and ornamental flowers. BOTRY-zen can also be used to control *Sclerotinia sclerotiorum* on kiwifruit, with two applications at flowering required to reduce sclerotinia infection and fruit loss.

3.4.3 Post-harvest microbiomes

3.4.3.1 Bacterial products

Pseudomonas syringae: Jet Harvest Solutions (USA) produce two commercial products, Bio-Save 10 LP and Bio-Save 11 LP, containing strains ESC-10 and ESC-11 of the bacterium *Pseudomonas syringae*, respectively. Both products are formulated as freeze-dried powders and are applied post-harvest. Bio-Save 10 LP is used on citrus fruits to reduce blue and green mould as well as sour rot, and on cherries to control blue and grey mould. Another use is on pome fruits in cold and controlled atmosphere storage for control of blue and grey mould, and mucor rot (*Mucor piriformis*). It can also be used on potatoes in cold storage to control dry rot and silver scurf, caused by *Fusarium sambucinum* and *Helminthosporium solani*, respectively. In contrast, Bio-Save 11 LP is mainly used on sweet potatoes to reduce soft rot in storage caused by *R. stolonifer*.

3.4.3.2 Fungal products

Aureobasidium pullulans: BoniProtect, a commercial product containing the yeast *Aureobasidium pullulans* on a carrier, is registered in Germany (http://www.bio-protect.de). The product is used on apples to control storage rots, including grey and blue mould caused by *Botrytis cinerea* and *Penicillium expansum*, respectively. BoniProtect works best when applied as a spray up to three times every 2 weeks starting 5 weeks before harvest. A similar product, BoniProtect forte, also containing *A. pullulans*, is used to control pre-harvest diseases of stone fruit and strawberries, caused by *Monilinia* spp. and *B. cinerea*, respectively.

Candida oleophila: The yeast, *Candida oleophila*, is primarily used to reduce green and blue mould of citrus and pome fruits, caused by *Penicillium digitatum* and *P. italicum*. It is produced by Ecogen Inc. (USA) and is marketed as Aspire in the United States and Israel. The product is most effective when used in combination with reduced application rates of thiabendazole, and often provides control equivalent to the full-rate fungicide treatment. It also reduces sour rot, caused by *Geotrichum candidum* (Droby *et al.*, 1998).

Metschnikowia fruticola: Shemer is a commercial product based on the yeast *Metschnikowia fruticola*. The product was registered in Israel in 2005, and is a water-dispersible granule formulation, which can be applied through spray or drench application systems in the field prior to harvest or in the packhouse (http://www.agrogreen.co.il). It is used to prevent the development of rots caused by a wide range of fungal pathogens, such as *P. digitatum* and *P. italicum* on citrus, *P. expansum* on pome and soft fruits, and *B. cinerea* and *Rhizopus stolonifer* on grapes and strawberries. Integration of Shemer with physical control measures such as hot-water wash, heat or modified atmosphere increases efficacy (Blachinsky *et al.*, 2007). The mode of action of the yeast is believed to be through competition.

3.5 Factors affecting variable efficacy and constraints on commercial developments

Inconsistency in efficacy of potential BCAs when evaluated in large-scale glasshouse or field trials is one of the major constraints in biological disease control. This can arise from various causes, especially extrinsic factors of the environment, reflecting the biological nature of the BCA. The BCA must first survive potential stresses of formulation and application procedures, and then remain active at the target site during the period when effective control is required. In addition, it must survive fluctuations in the natural environment, especially temperature, as well as the action of indigenous and competitive microbiota. Consequently, poor disease control at the scale-up stages of evaluation is always likely to be high (Whipps & Lumsden, 2001). In an attempt to resolve this problem and increase the number of BCAs reaching the market, it is recommended that all selection, screening and development processes adopt an ecological approach which takes into account the extrinsic factors of the environment of use (Whipps, 1997a, b, c; Whipps & Lumsden, 2001). It is unfortunate that most BCAs are only active under particular environmental conditions. Consequently, biological disease control in environmentally controlled structures, such as glasshouses and polytunnels, tends to be more successful and cost-effective compared to large-scale field application.

Economical, mass production of stable inoculum and appropriate formulation is imperative for the successful development of BCAs. Potential BCAs must also be easy to use and cost-effective, or they will never reach the market or be used by growers. Currently, many fungicides are relatively cheap and more effective than BCAs, and are unlikely to be substituted for by BCAs unless they are withdrawn from the market. Very few growers or extension workers know how to store and use BCAs, which often results in inadequate disease control and subsequent poor sales. Clearly, there is need to train growers on how to use BCAs effectively and integrate their use into crop protection programmes.

Another constraint to the development of bacteria and fungi as commercial BCAs has been poor long-term storage stability. Good long-term stability, preferably for 18–24 months at room temperature (*ca.* 21°C), is required to improve market competitiveness. Despite the hurdles in obtaining stability, considerable progress has been made, with stability of most current commercial products often being achieved by mixing propagules with various additives during formulation (Jones & Burges, 1998). Improved stability can also be achieved by treatment before formulation, for example, by appropriate growth conditions during production and by processing after production, such as drying. Furthermore, regulation of water availability in the formulation is important for stability.

Application technology can have a significant impact on the efficacy of BCAs. Unfortunately, this has often been neglected in the past, especially for the application of BCAs to aerial microbiomes, resulting in poor efficacy. Targeted delivery, deposition and coverage of the infection court are essential for good disease control. In laboratory experiments, Scherm *et al.* (2004) reported significant activity of the commercial product Serenade (*B. subtilis* QST713) against blueberry flower infection by *Monilinia vaccinii-corymbosis*. However, disease suppression was unsatisfactory when the *B. subtilis*-based product was applied in the field with a standard sprayer. This was likely due to low and variable coverage of the stigmatic infection court, which presents a difficult spray target. In a recent laboratory study, air-assisted electrostatic spraying significantly increased deposition of *B. subtilis* QST713 and coverage on the stigmatic surfaces of detached blueberry flower clusters compared to conventional hydraulic spraying (Scherm *et al.*, 2007). The increased deposition and coverage together with the excellent bacterial survival in the formulated product bodes well for electrostatic application of the product for disease control in the field.

One of the major economic hurdles in the commercialisation of BCAs is in risk assessment of toxicity and environmental impact of the organism, and its formulation (Brimmer & Boland, 2003; Winding *et al.*, 2004; Scherwinski *et al.*, 2007). Extensive trials are essential to generate data for registration purposes to show that potential commercial BCAs are safe both to humans and to other non-target organisms. Quality and efficacy data as well as additional technical protocols are also required by the registration authorities. All this can be extremely time-consuming and very expensive to generate as well as the cost for the assessment process itself. High registration costs have clearly been responsible for delaying or preventing the commercial development of BCAs in the past, especially by small–medium-size enterprises (SMEs) which are the main producers of BCAs. This has led to a large number of products appearing on the market which actually work by controlling plant pathogens but are claimed to be soil conditioners, plant-growth promoters or biofertilisers that do not require registration. However, without toxicological and efficacy data, safe use cannot be assured and consistent disease control and crop growth are not always observed (Cook *et al.*, 1996). Some regulatory authorities have recognised these problems and are encouraging legal registration and use of BCAs in a number of ways. For example, in the United Kingdom, the Pesticides Safety Directorate (PSD) offers a lower charging scheme for BCAs than chemicals, and in the United States the EPA offers a more flexible case-by-case and/or 'fast-track' approach. Despite the introduction of these schemes, these problems still remain. However, some BCAs have gone through the entire registration process and are available for sale and use in a number of countries (Tables 3.1–3.3).

Large-scale use of commercial products is still limited because of variability and inconsistency in terms of disease control. Coupled with a very competitive market with chemical pesticides, manufacturers of BCAs are finding it increasingly difficult to make sufficient profit from sale of commercial products to maintain the costs of registration. Unfortunately, this has resulted in the withdrawal of a number of products from the market. For example, Trichodex (*Trichoderma harzianum* T-39) introduced in 1993 for control of *B. cinerea* on grapes and greenhouse crops in Europe and Israel was withdrawn from the market in 2005 due to insufficient sales and increased registration costs.

Table 3.3 Examples of bacteria and fungi registered or commercially marketed as biological control agents for control of post-harvest plant pathogens.

Antagonist	Target pathogen(s)/Activity	Disease/Host	Product name and source
Bacteria			
Pseudomonas syringae ESC-10	*Botrytis cinerea, Helminthosporium* spp., *Fusarium* spp., *Penicillium* spp., *Mucor* spp., *Geotrichum candidum*	Storage rots of pome fruit, citrus, cherry, potato	Bio-Save 10LP, 110 (Jet Harvest solutions, USA)
P. syringae ESC-11	*Botrytis cinerea, Helminthosporium* spp., *Fusarium* spp., *Penicillium* spp., *Mucor* spp., *Geotrichum candidum*	Storage rots of pome fruit, potato	Bio-Save 11LP (Jet Harvest solutions, USA)
Fungi			
Aureobasidium pullulans	*Pezicula* sp., *Nectria galligena, Botrytis cinerea, Penicillium expansum, Monilinia fructigena*	Storage rots of apples	BoniProtect (BioProtect, Germany)
Candida oleophila	*Botrytis* spp., *Penicillium* spp.	Storage rots of pome fruit	Aspire (Ecogen Inc. USA)
Cryptococcus albidus	*Botrytis cinerea, Penicillium expansum*	Storage rots of apples and pears	Yield*Plus* (Anchor Yeast, South Africa)
Metschnikowia fruticola	*Aspergillus niger, Botrytis cinerea, Penicillium* spp., *Rhizopus stolonifera*	Storage rots of pome and stone fruit, grapes and strawberry	Shemer (AgroGreen, Israel)

3.6 Future research directions and conclusions

Great strides have been made over the last few years in understanding and utilising BCAs. Molecular biology has played a big part in mode of action studies and, as these techniques advance further, additional novel information that may help the practical application of BCAs is likely to be generated. For example, analysis of expressed sequence tags (ESTs) has provided novel information on genes expressed by *Trichoderma harzianum* (Liu & Yang, 2005; Vizcaino *et al.*, 2006), and genes differentially expressed during growth in the presence of cell walls of host fungi or during mycoparasitism itself have been identified in several systems (Carpenter *et al.*, 2005; Massart & Jijakli, 2006; Muthumeenakshi *et al.*, 2007). Differential gene expression has also been used to identify novel genetic markers associated with biocontrol activities in *Bacillus subtilis* (Joshi & Gardener, 2006). This can be scaled up further with gene chip technology or transcriptomics, already applied to monitor broad changes in gene expression due to ISR in *Arabidopsis* (Verhagen *et al.*, 2004). The increasing availability of whole genome sequences of BCAs, such as that for *Pseudomonas fluorescens* PF-5 (Loper *et al.*, 2007) and *Trichoderma harzianum* (underway) is also likely to shed new insights into biocontrol. Proteomic studies are also revealing new information on proteins expressed during growth of *Trichoderma atroviride* on cell walls of *Rhizoctonia solani* (Grinyer *et al.*, 2005) and during three-way interactions between *T. atroviride*, bean plants and *Botrytis cinerea* and *Rhizoctonia solani* (Marra *et al.*, 2006) and are also likely to provide new concepts to improve biocontrol activity. Understanding more about signalling between microorganisms during biocontrol may also allow a new way of controlling disease (Molina *et al.*, 2003; Uroz *et al.*, 2003; Dong *et al.*, 2004; Lutz *et al.*, 2004; Pierson & Pierson, 2007).

If genes active in biocontrol are identified and sequenced it is also possible to constitutively over express these genes and potentially enhance biocontrol activity. Recently, this concept has been tested by introducing two genes, a β-1,3-glucanase and a β-1,6-glucanase into *Trichoderma virens*, and despite a decrease in growth and sporulation, the double over-expression transformants still provided enhanced bioprotection of cotton seedlings against *Pythium ultimum, Rhizoctonia solani* and *Rhizopus oryzae* (Djonovic *et al.*, 2007). An endochitinase gene from *Trichoderma harzianum* has also been transformed into potato and tobacco and these transgenic plants exhibited increased disease resistance (Lorito *et al.*, 1998). Nevertheless, all genetic modification procedures are subject to strict regulation relating to use in the environment. Consequently, even though they may show great promise they are unlikely to become commercialised as they would face a double legislation hurdle associated with use both as a pesticide and genetically modified organism.

More practical areas for future research relate to the continuum of production, formulation, application and ecology of BCAs. Any advances that can improve quantity, quality and shelf life of inocula would be welcome, particularly bacteria such as *Pseudomonas* that do not form spores. Recently, for the first time, both *Pseudomonas* and *Trichoderma* isolates have been simultaneously applied to seed via drum priming and found to survive and proliferate on roots similarly to when applied individually (Bennett & Whipps, 2007). This procedure may be a way to apply and maintain multiple BCAs in a commercially relevant process. Combining different strains of BCAs has frequently been suggested as a way ahead for biocontrol as they may express different biocontrol traits, give higher levels of protection, reduce variability, and increase the range of pathogens suppressed (Dunne *et al.*, 1998; Guetsky *et al.*, 2002; Jetiyanon & Kloepper, 2002;

de Boer *et al.*, 2003). Other approaches include integration with other control strategies such as cultural procedures and chemical applications (Whipps, 2001; Someya *et al.*, 2007) and use of endophytic bacteria and fungi which may exhibit different patterns of behaviour to those found external to the plant (Wulff *et al.*, 2003; Sessitsch *et al.*, 2004). Thus, understanding more about the natural ecology of any potential BCA may enable a more rational approach to production and use (Whipps, 1997b).

In conclusion, despite legislation hurdles and perceived frequent inconsistency of effect, more BCAs are on the market than ever before. Much is known about the way they work and this knowledge continues to increase. Often, the market for BCAs is small as they generally have a restricted host range, but some are beginning to have global significance which bodes well for sustainable disease control in the future as chemical pesticides become scarcer.

3.7 References

Altomare C, Norvell WA, Bjorkman T, Harman GE, 1999. Solubilization of phosphates and micronutrients by the plant-growth-promoting and biocontrol fungus *Trichoderma harzianum* Rifai 1295–22. *Applied and Environmental Microbiology* **65,** 2926–2933.

Baek JM, Howell CR, Kenerley CM, 1999. The role of an extracellular chitinase from *Trichoderma virens* Gv29–8 in the biocontrol of *Rhizoctonia solani*. *Current Genetics* **35,** 41–50.

Bakker P, Pieterse CMJ, van Loon LC, 2007. Induced systemic resistance by fluorescent *Pseudomonas* spp. *Phytopathology* **97,** 239–243.

Bakker PAHM, Ran LX, Pieterse CMJ, van Loon LC, 2003. Understanding the involvement of rhizobacteria-mediated induction of systemic resistance in biocontrol of plant diseases. *Canadian Journal of Plant Pathology* **25,** 5–9.

Barak R, Elad Y, Mirelman D, Chet I, 1985. Lectins – a possible basis for specific recognition in the interaction of *Trichoderma* and *Sclerotium rolfsii*. *Phytopathology* **75,** 458–462.

Bargabus RL, Zidack NK, Sherwood JE, Jacobsen BJ, 2002. Characterisation of systemic resistance in sugar beet elicited by a non-pathogenic, phyllosphere-colonizing *Bacillus mycoides*, biological control agent. *Physiological and Molecular Plant Pathology* **61,** 289–298.

Bargabus RL, Zidack NK, Sherwood JE, Jacobsen BJ, 2003. Oxidative burst elicited by *Bacillus mycoides* isolate Bac J, a biological control agent, occurs independently of hypersensitive cell death in sugar beet. *Molecular Plant–Microbe Interactions* **16,** 1145–1153.

Basnayake WVS, Birch RG, 1995. A gene from *Alcaligenes denitrificans* that confers albicidin resistance by reversible antibiotic binding. *Microbiology-UK* **141,** 551–560.

Bélanger RR, Labbé C, 2002. Control of powdery mildews without chemicals: Prophylactic and biological alternatives for horticultural crops. In: Belanger RR, Dik AJ, Carver TLW, eds. *The Powdery Mildews: A Comprehensive Treatise*. St. Paul: APS Press, 256–267.

Benhamou N, Bélanger RR, Rey P, Tirilly Y, 2001. Oligandrin, the elicitin-like protein produced by the mycoparasite *Pythium oligandrum*, induces systemic resistance to Fusarium crown and root rot in tomato plants. *Plant Physiology and Biochemistry* **39,** 681–696.

Benhamou N, Chet I, 1996. Parasitism of sclerotia of *Sclerotium rolfsii* by *Trichoderma harzianum*: Ultrastructural and cytochemical aspects of the interaction. *Phytopathology* **86,** 405–416.

Benhamou N, Gagné S, Le Quéré D, Dehbi L, 2000. Bacterial-mediated induced resistance in cucumber: Beneficial effect of the endophytic bacterium *Serratia plymuthica* on the protection against infection by *Pythium ultimum*. *Phytopathology* **90,** 45–56.

Benhamou N, Kloepper JW, Quadt-Hallman A, Tuzun S, 1996. Induction of defense-related ultrastructural modifications in pea root tissues inoculated with endophytic bacteria. *Plant Physiology* **112,** 919–929.

Benhamou N, Kloepper JW, Tuzun S, 1998. Induction of resistance against Fusarium wilt of tomato by combination of chitosan with an endophytic bacterial strain: Ultrastructure and cytochemistry of the host response. *Planta* **204,** 153–168.

Benhamou N, Rey P, Chérif M, Hockenhull J, Tirilly Y, 1997. Treatment with the mycoparasite *Pythium oligandrum* triggers induction of defense-related reactions in tomato roots when challenged with *Fusarium oxysporum* f. sp. *radicis-lycopersici*. *Phytopathology* **87,** 108–122.

Bennett AJ, Whipps JM, 2007. Dual application of beneficial microorganisms to seed during drum priming. *Applied Soil Ecology* **38,** 83–89.

Blachinsky D, Lazare M, Shamai I, Antonov A, Husid A, Bercovitz A, Feldman K, El-ad B, Prengler M, Keren-Zur M, 2007. Commercial applications of "Shemer" for control of pre- and postharvest diseases. *Phytoparasitica* **35,** 202–203.

Bloemberg GV, Lugtenberg BJJ, 2001. Molecular basis of plant growth promotion and biocontrol by rhizobacteria. *Current Opinion in Plant Biology* **4,** 343–350.

Bolwerk A, Lagopodi AL, Lugtenberg BJJ, Bloemberg GV, 2005. Visualization of interactions between a pathogenic and a beneficial *Fusarium* strain during biocontrol of tomato foot and root rot. *Molecular Plant–Microbe Interactions* **18,** 710–721.

Bolwerk A, Lagopodi AL, Wijfjes AHM, Lamers GEM, Chin-A-Woeng TFC, Lugtenberg BJJ, Bloemberg GV, 2003. Interactions in the tomato rhizosphere of two *Pseudomonas* biocontrol strains with the phytopathogenic fungus *Fusarium oxysporum* f. sp *radicis-lycopersici*. *Molecular Plant–Microbe Interactions* **16,** 983–993.

Brimmer TA, Boland GJ, 2003. A review of the non-target effects of fungi used to biologically control plant diseases. *Agriculture Ecosystems & Environment* **100,** 3–16.

Brodhagen M, Henkels MD, Loper JE, 2004. Positive autoregulation and signaling properties of pyoluteorin, an antibiotic produced by the biological control organism *Pseudomonas fluorescens* Pf-5. *Applied and Environmental Microbiology* **70,** 1758–1766.

Brunner K, Peterbauer CK, Mach RL, Lorito M, Zeilinger S, Kubicek CP, 2003. The Nag1 *N*-acetylglucosaminidase of *Trichoderma atroviride* is essential for chitinase induction by chitin and of major relevance to biocontrol. *Current Genetics* **43,** 289–295.

Burges HD, 1998. *Formulation of Microbial Biopesticides, Beneficial Microorganisms, Nematodes and Seed Treatments*. Dordrecht: Kluwer Academic Publishers.

Carpenter MA, Stewart A, Ridgway HJ, 2005. Identification of novel *Trichoderma hamatum* genes expressed during mycoparasitism using subtractive hybridisation. *FEMS Microbiology Letters* **251,** 105–112.

Carsolio C, Benhamou N, Haran S, Cortés C, Gutiérrez A, Chet I, Herrera-Estrella A, 1999. Role of the *Trichoderma harzianum* endochitinase gene, ech42, in mycoparasitism. *Applied and Environmental Microbiology* **65,** 929–935.

Castoria R, Caputo L, De Curtis F, De Cicco CV, 2003. Resistance of postharvest biocontrol yeasts to oxidative stress: A possible new mechanism of action. *Phytopathology* **93,** 564–572.

Chancey ST, Wood DW, Pierson LS, 1999. Two-component transcriptional regulation of *N*-acyl-homoserine lactone production in *Pseudomonas aureofaciens*. *Applied and Environmental Microbiology* **65,** 2294–2299.

Chen CQ, Bélanger RR, Benhamou N, Paulitz TC, 2000. Defense enzymes induced in cucumber roots by treatment with plant growth-promoting rhizobacteria (PGPR) and *Pythium aphanidermatum*. *Physiological and Molecular Plant Pathology* **56,** 13–23.

Cheng YL, McNally DJ, Labbe C, Voyer N, Belzile F, Bélanger RR, 2003. Insertional mutagenesis of a fungal biocontrol agent led to discovery of a rare cellobiose lipid with antifungal activity. *Applied and Environmental Microbiology* **69,** 2595–2602.

Chernin L, Ismailov Z, Haran S, Chet I, 1995. Chitinolytic *Enterobacter agglomerans* antagonistic to fungal plant pathogens. *Applied and Environmental Microbiology* **61,** 1720–1726.

Chernin LS, de la Fuente L, Sobolev V, Haran S, Vorgias CE, Oppenheim AB, Chet I, 1997. Molecular cloning, structural analysis, and expression in *Escherichia coli* of a chitinase gene from *Enterobacter agglomerans*. *Applied and Environmental Microbiology* **63,** 834–839.

Chet I, Benhamou N, Haran S, 1998. Mycoparasitism and lytic enzymes. In: Harman GE, Kubicek CP, eds. *Trichoderma* and *Gliocladium*, Vol. 2. London: Taylor and Francis, 153–172

Chin-A-Woeng TFC, Thomas-Oates JE, Lugtenberg BJJ, Bloemberg GV, 2001. Introduction of the phzH gene of *Pseudomonas chlororaphis* PCL1391 extends the range of biocontrol ability of phenazine-1-carboxylic acid-producing *Pseudomonas* spp. strains. *Molecular Plant–Microbe Interactions* **14,** 1006–1015.

Chin-A-Woeng TFC, van den Broek D, Lugtenberg BJJ, Bloemberg GV, 2005. The *Pseudomonas chlororaphis* PCL1391 sigma regulator psrA represses the production of the antifungal metabolite phenazine-1-carboxamide. *Molecular Plant–Microbe Interactions* **18,** 244–253.

Collins DP, Jacobsen BJ, 2003. Optimizing a *Bacillus subtilis* isolate for biological control of sugar beet cercospora leaf spot. *Biological Control* **26,** 153–161.

Compant S, Duffy B, Nowak J, Clement C, Barka EA, 2005a. Use of plant growth-promoting bacteria for biocontrol of plant diseases: Principles, mechanisms of action, and future prospects. *Applied and Environmental Microbiology* **71,** 4951–4959.

Compant S, Reiter B, Sessitsch A, Nowak J, Clement C, Barka EA, 2005b. Endophytic colonization of *Vitis vinifera* L. by plant growth promoting bacterium *Burkholderia* sp strain PsJN. *Applied and Environmental Microbiology* **71,** 1685–1693.

Cook RJ, Bruckart WL, Coulson JR, Goettel MS, Humber RA, Lumsden RD, Maddox JV, McManus ML, Moore L, Meyer SF, Quimby PC, Stack JP, Vaughn JL, 1996. Safety of microorganisms intended for pest and plant disease control: A framework for scientific evaluation. *Biological Control* **7,** 333–351.

Corbell N, Loper JE, 1995. A global regulator of secondary metabolite production in *Pseudomonas fluorescens* Pf-5. *Journal of Bacteriology* **177,** 6230–6236.

Daoust RA, Hofstein R, 1996. *Ampelomyces quisqualis*, a new biofungicide to control powdery mildew in grapes. In: *Brighton Crop Protection Conference, Pests and Diseases*. Farnham, UK: British Crop Protection Council, 33–40.

Dawe AL, Nuss DL, 2001. Hypoviruses and chestnut blight: Exploiting viruses to understand and modulate fungal pathogenesis. *Annual Review of Genetics* **35,** 1–29.

de Boer M, Bom P, Kindt F, Keurentjes JJB, van der Sluis I, van Loon LC, Bakker PAHM, 2003. Control of Fusarium wilt of radish by combining *Pseudomonas putida* strains that have different disease-suppressive mechanisms. *Phytopathology* **93,** 626–632.

de Meyer G, Audenaert K, Höfte M, 1999. *Pseudomonas aeruginosa* 7NSK2-induced systemic resistance in tobacco depends on in planta salicylic acid accumulation but is not associated with PR1a expression. *European Journal of Plant Pathology* **105,** 513–517.

de Souza JT, de Boer M, de Waard P, van Beek TA, Raaijmakers JM, 2003. Biochemical, genetic, and zoosporicidal properties of cyclic lipopeptide surfactants produced by *Pseudomonas fluorescens*. *Applied and Environmental Microbiology* **69,** 7161–7172.

de Vrije T, Antoine N, Buitelaar RM, Bruckner S, Dissevelt M, Durand A, Gerlagh M, Jones EE, Luth P, Oostra J, Ravensberg WJ, Renaud R, Rinzema A, Weber FJ, Whipps JM, 2001. The fungal biocontrol agent *Coniothyrium minitans*: Production by solid-state fermentation, application and marketing. *Applied Microbiology and Biotechnology* **56,** 58–68.

Dedej S, Delaphane KS, Scherm H, 2004. Effectiveness of honey bees in delivering the biocontrol agent *Bacillus subtilis* to blueberry flowers to suppress mummy berry disease. *Biological Control* **31,** 422–427.

Défago G, 1993. 2,4-Diacetylphloroglucinol, a promising compound in biocontrol. *Plant Pathology* **42,** 311–312.

Dekkers LC, Mulders IHM, Phoelich CC, Chin-A-Woeng TFC, Wijfjes AHM, Lugtenberg BJJ, 2000. The sss colonization gene of the tomato-*Fusarium oxysporum* f. sp *radicis-lycopersici* biocontrol strain *Pseudomonas fluorescens* WCS365 can improve root colonization of other wild-type *Pseudomonas* spp. bacteria. *Molecular Plant–Microbe Interactions* **13,** 1177–1183.

Delany IR, Walsh UF, Ross I, Fenton AM, Corkery DM, O'Gara F, 2001. Enhancing the biocontrol efficacy of *Pseudomonas fluorescens* F113 by altering the regulation and production of 2,4-diacetylphloroglucinol – improved *Pseudomonas* biocontrol inoculants. *Plant and Soil* **232,** 195–205.

Djonovic S, Vittone G, Mendoza-Herrera A, Kenerley CM, 2007. Enhanced biocontrol activity of *Trichoderma virens* transformants constitutively coexprerssing β-1,3- and β-1,6-glucanase genes. *Molecular Plant Pathology* **8,** 469–480.

Dong YH, Zhang XF, Xu JL, Zhang LH, 2004. Insecticidal *Bacillus thuringiensis* silences *Erwinia carotovora* virulence by a new form of microbial antagonism, signal interference. *Applied and Environmental Microbiology* **70,** 954–960.

Droby S, Cohen L, Daus A, Weiss B, Horev B, Chalutz E, Katz H, Keren-Tzur M, Shachnai A, 1998. Commercial testing of aspire™: A yeast preparation for the biological control of postharvest decay of citrus. *Biological Control* **12,** 97–101.

Duffy B, Keel C, Défago G, 2004. Potential role of pathogen signaling in multitrophic plant–microbe interactions involved in disease protection. *Applied and Environmental Microbiology* **70,** 1836–1842.

Duijff BJ, Gianinazzi-Pearson V, Lemanceau P, 1997. Involvement of the outer membrane lipopolysaccharides in the endophytic colonization of tomato roots by biocontrol *Pseudomonas fluorescens* strain WCS417r. *New Phytologist* **135,** 325–334.

Duijff BJ, Meijer JW, Bakker PAHM, Schippers B, 1993. Siderophore-mediated competition for iron and induced resistance in the suppression of Fusarium wilt of carnation by fluorescent *Pseudomonas* spp. *Netherlands Journal of Plant Pathology* **99,** 277–289.

Duijff BJ, Pouhair D, Olivain C, Alabouvette C, Lemanceau P, 1998. Implication of systemic induced resistance in the suppression of fusarium wilt of tomato by *Pseudomonas fluorescens* WCS417r and by nonpathogenic *Fusarium oxysporum* Fo47. *European Journal of Plant Pathology* **104,** 903–910.

Dunne C, Crowley JJ, Moenne-Loccoz Y, Dowling DN, de Bruijn FJ, O'Gara F, 1997. Biological control of *Pythium ultimum* by *Stenotrophomonas maltophilia* W81 is mediated by an extracellular proteolytic activity. *Microbiology-UK* **143,** 3921–3931.

Dunne C, Moenne-Loccoz Y, McCarthy J, Higgins P, Powell J, Dowling DN, O'Gara F, 1998. Combining proteolytic and phloroglucinol-producing bacteria for improved biocontrol of Pythium-mediated damping-off of sugar beet. *Plant Pathology* **47,** 299–307.

Edwards SG, Seddon B, 2001. Mode of antagonism of *Brevibacillus brevis* against *Botrytis cinerea in vitro. Journal of Applied Microbiology* **91,** 652–659.

Elad Y, Kapat A, 1999. The role of *Trichoderma harzianum* protease in the biocontrol of *Botrytis cinerea. European Journal of Plant Pathology* **105,** 177–189.

Elmer PAG, Reglinski T, 2006. Biosuppression of *Botrytis cinerea* in grapes. *Plant Pathology* **55,** 155–177.

Filonow AB, 1998. Role of competition for sugars by yeasts in the biocontrol of gray mold of apple. *Biocontrol Science and Technology* **8,** 243–256.

Fravel D, Olivain C, Alabouvette C, 2003. *Fusarium oxysporum* and its biocontrol. *New Phytologist* **157,** 493–502.

Fravel DR, 2005. Commercialization and implementation of biocontrol. *Annual Review of Phytopathology* **43,** 337–359.

Fravel DR, Connick WJ Jr, Lewis JA, 1998. Formulation of microorgansisms to control plant diseases. In: Burges HD, ed. *Formulation of Microbial Biopesticides*. Dordrecht: Kluwer Academic Publishers, 187–202.

Fridlender M, Inbar J, Chet I, 1993. Biological control of soilborne plant pathogens by a β-1,3 glucanase producing *Pseudomonas cepacia. Soil Biology & Biochemistry* **25,** 1211–1221.

Friel D, Pessoa NMG, Vandenbol M, Jijakli MH, 2007. Separate and combined disruptions of two exo-β-1,3-glucanase genes decrease the efficiency of *Pichia anomala* (strain K) biocontrol against on apple. *Molecular Plant–Microbe Interactions* **20,** 371–379.

Fuchs JG, Moenne-Loccoz Y, Défago G, 1997. Nonpathogenic *Fusarium oxysporum* strain Fo47 induces resistance to Fusarium wilt in tomato. *Plant Disease* **81,** 492–496.

Gomez-Gomez L, Boller T, 2000. FLS2: An LRR receptor-like kinase involved in the perception of the bacterial elicitor flagellin in Arabidopsis. *Molecular Cell* **5,** 1003–1011.

Grevesse C, Lepoivre P, Jijakli MH, 2003. Characterization of the exoglucanase-encoding gene PaEXG2 and study of its role in the biocontrol activity of *Pichia anomala* Strain K. *Phytopathology* **93,** 1145–1152.

Grinyer J, Hunt S, McKay M, Herbert BR, Nevalainen H, 2005. Proteomic response of the biological control fungus *Trichoderma atroviride* to growth on the cell walls of *Rhizoctonia solani. Current Genetics* **47,** 381–388.

Guetsky R, Shtienberg D, Elad Y, Fischer E, Dinoor A, 2002. Improving biological control by combining biocontrol agents each with several mechanisms of disease suppression. *Phytopathology* **92,** 976–985.

Haas D, Défago G, 2005. Biological control of soil-borne pathogens by fluorescent pseudomonads. *Nature Reviews Microbiology* **3,** 307–319.

Haas D, Keel C, 2003. Regulation of antibiotic production in root-colonizing *Pseudomonas* spp. and relevance for biological control of plant disease. *Annual Review of Phytopathology* **41,** 117–153.

Hall FR, Menn JJ, 1999. *Biopesticides: Use and Delivery*. Totowa, NJ: Humana Press.

Hammer PE, Hill DS, Lam ST, Van Pee KH, Ligon JM, 1997. Four genes from *Pseudomonas fluorescens* that encode the biosynthesis of pyrrolnitrin. *Applied and Environmental Microbiology* **63,** 2147–2154.

Hanson LE, Howell CR, 2004. Elicitors of plant defense responses from biocontrol strains of *Trichoderma virens. Phytopathology* **94,** 171–176.

Harman GE, 1991. Seed treatments for biological control of plant disease. *Crop Protection* **10,** 166–171.

Harman GE, 2000. Myths and dogmas of biocontrol – changes in perceptions derived from research on *Trichoderma harzianum* T-22. *Plant Disease* **84,** 377–393.

Harman GE, 2006. Overview of mechanisms and uses of *Trichoderma* spp. *Phytopathology*, **96,** 190–194.

Harman GE, Howell CR, Viterbo A, Chet I, Lorito M, 2004. *Trichoderma* species – opportunistic, avirulent plant symbionts. *Nature Reviews Microbiology* **2,** 43–56.

Hashidoko Y, Nakayama T, Homma Y, Tahara S, 1999. Structure elucidation of xanthobaccin A, a new antibiotic produced from *Stenotrophomonas* sp. strain SB-K88. *Tetrahedron Letters* **40,** 2957–2960.

Heiniger U, Rigling D, 1994. Biological control of chestnut blight in Europe. *Annual Review of Phytopathology* **32,** 581–599.

Herr LJ, 1995. Biological control of *Rhizoctonia solani* by binucleate *Rhizoctonia* spp. and hypovirulent *R. solani* agents. *Crop Protection* **14,** 179–186.

Hervás A, Trapero-Casas JL, Jimenez-Diaz RM, 1995. Induced resistance against Fusarium wilt of chickpea by nonpathogenic races of *Fusarium oxysporum* f. sp. *ciceris* and nonpathogenic isolates of *Fusarium oxysporum*. *Plant Disease* **79,** 1110–1116.

Heungens K, Parke JL, 2000. Zoospore homing and infection events: Effects of the biocontrol bacterium *Burkholderia cepacia* AMMDR1 on two oomycete pathogens of pea (*Pisum sativum* L.). *Applied and Environmental Microbiology* **66,** 5192–5200.

Howell CR, 2002. Cotton seedling preemergence damping-off incited by *Rhizopus oryzae* and *Pythium* spp. and its biological control with *Trichoderma* spp. *Phytopathology* **92,** 177–180.

Howell CR, 2006. Understanding the mechanisms employed by *Trichoderma virens* to effect biological control of cotton diseases. *Phytopathology* **96,** 178–180.

Howell CR, Hanson LE, Stipanovic RD, Puckhaber LS, 2000. Induction of terpenoid synthesis in cotton roots and control of *Rhizoctonia solani* by seed treatment with *Trichoderma virens*. *Phytopathology* **90,** 248–252.

Howell CR, Puckhaber LS, 2005. Study of the characteristics of "P" and "Q" strains of *Trichoderma virens* to account for differences in biological control efficacy against cotton seedling diseases. *Biological Control* **33,** 217–222.

Howell CR, Stipanovic RD, 1995. Mechanisms in the biocontrol of *Rhizoctonia solani*-induced cotton seedling disease by *Gliocladium virens* – antibiosis. *Phytopathology* **85,** 469–472.

Huang ZY, Bonsall RF, Mavrodi DV, Weller DM, Thomashow LS, 2004. Transformation of *Pseudomonas fluorescens* with genes for biosynthesis of phenazine-1-carboxylic acid improves biocontrol of rhizoctonia root rot and *in situ* antibiotic production. *FEMS Microbiology Ecology* **49,** 243–251.

Iavicoli A, Boutet E, Buchala A, Metraux JP, 2003. Induced systemic resistance in *Arabidopsis thaliana* in response to root inoculation with *Pseudomonas fluorescens* CHA0. *Molecular Plant–Microbe Interactions* **16,** 851–858.

Jabaji-Hare S, Chamberland H, Charest PM, 1999. Cell wall alterations in hypocotyls of bean seedlings protected from Rhizoctonia stem canker by a binucleate *Rhizoctonia* isolate. *Mycological Research* **103,** 1035–1043.

Jabaji-Hare S, Neate SM, 2005. Nonpathogenic binucleate *Rhizoctonia* spp. and benzothiadiazole protect cotton seedlings against Rhizoctonia damping-off and Alternaria leaf spot in cotton. *Phytopathology* **95**, 1030–1036.

Jackson MA, 1997. Optimizing nutritional conditions for the liquid culture production of effective fungal biological control agents. *Journal of Industrial Microbiology & Biotechnology* **19,** 180–187.

Jacobsen BJ, 2006. Biological control of plant diseases by phyllosphere applied biological control agents. In: Bailey MJ, Lilley AK, Timms-Wilson TM, Spencer-Phillips, PTN eds. *Microbial Ecology of Aerial Plant Surfaces*. Wallingford: CABI, 133–147.

Janisiewicz WJ, Korsten L, 2002. Biological control of postharvest diseases of fruits. *Annual Review of Phytopathology* **40,** 411–441.

Jeffries P, Young TWK, 1994. *Interfungal Parasitic Relationships*. Wallingford: CABI Publishing.

Jetiyanon K, Kloepper JW, 2002. Mixtures of plant growth-promoting rhizobacteria for induction of systemic resistance against multiple plant diseases. *Biological Control* **24,** 285–291.

Jones KA, Burges HD, 1998. Technology of formulation and application. In: Burges HD, ed. *Formulation of Microbial Biopesticides*. Dordrecht: Kluwer Academic Publishers, 7–30.

Joshi R, Gardener BBM, 2006. Identification and characterization of novel genetic markers associated with biological control activities in *Bacillus subtilis*. *Phytopathology* **96,** 145–154.

Kageyama K, Nelson EB, 2003. Differential inactivation of seed exudate stimulation of *Pythium ultimum* sporangium germination by *Enterobacter cloacae* influences biological control efficacy on different plant species. *Applied and Environmental Microbiology* **69,** 1114–1120.

Kamensky M, Ovadis M, Chet I, Chernin L, 2003. Soil-borne strain IC14 of *Serratia plymuthica* with multiple mechanisms of antifungal activity provides biocontrol of *Botrytis cinerea* and *Sclerotinia sclerotiorum* diseases. *Soil Biology & Biochemistry* **35,** 323–331.

Kapat A, Zimand G, Elad Y, 1998. Effect of two isolates of *Trichoderma harzianum* on the activity of hydrolytic enzymes produced by *Botrytis cinerea*. *Physiological and Molecular Plant Pathology* **52,** 127–137.

Kessel GJT, Kohl J, Powell JA, Rabbinge R, van der Werf W, 2005. Modeling spatial characteristics in the biological control of fungi at leaf scale: Competitive substrate colonization by *Botrytis cinerea* and the saprophytic antagonist *Ulocladium atrum*. *Phytopathology* **95,** 439–448.

Kilic-Ekici O, Yuen GY, 2003. Induced resistance as a mechanism of biological control by *Lysobacter enzymogenes* strain C3. *Phytopathology* **93,** 1103–1110.

Kim B, Moon SS, Hwang BK, 1999. Isolation, identification, and antifungal activity of a macrolide antibiotic, oligomycin A, produced by *Streptomyces libani*. *Canadian Journal of Botany* **77,** 850–858.

Kiss L, Russell JC, Szentivanyi O, Xu X, Jeffries P, 2004. Biology and biocontrol potential of *Ampelomyces* mycoparasites, natural antagonists of powdery mildew fungi. *Biocontrol Science and Technology* **14,** 635–651.

Kloepper JW, Ryu CM, Zhang SA, 2004. Induced systemic resistance and promotion of plant growth by *Bacillus* spp. *Phytopathology* **94,** 1259–1266.

Kloepper JW, Tuzun S, Kuc JA, 1992. Proposed definitions related to induced disease resistance. *Biocontrol Science and Technology* **2,** 349–351.

Larkin RP, Fravel DR, 1999. Mechanisms of action and dose–response relationships governing biological control of fusarium wilt of tomato by nonpathogenic *Fusarium* spp. *Phytopathology* **89,** 1152–1161.

Leeman M, den Ouden EM, van Pelt JA, Dirkx FPM, Steijl H, Bakker PAHM, Schippers B, 1996. Iron availability affects induction of systemic resistance to Fusarium wilt of radish by *Pseudomonas fluorescens*. *Phytopathology* **86,** 149–155.

Leeman M, van Pelt JA, den Ouden FM, Heinsbroek M, Bakker PAHM, Schippers B, 1995. Induction of systemic resistance against Fusarium-wilt of radish by lipopolysaccharides of *Pseudomonas fluorescens*. *Phytopathology* **85,** 1021–1027.

Lindow SE, Leveau JH, 2002. Phyllosphere microbiology. *Current Opinion in Biotechnology* **13,** 238–243.

Liu PG, Yang Q, 2005. Identification of genes with a biocontrol function in *Trichoderma harzianum* mycelium using the expressed sequence tag approach. *Research in Microbiology* **156,** 416–423.

Loper JE, Buyer JS, 1991. Siderophores in microbial interactions on plant surfaces. *Molecular Plant–Microbe Interactions* **4,** 5–13.

Loper JE, Henkels MD, 1999. Utilization of heterologous siderophores enhances levels of iron available to *Pseudomonas putida* in the rhizosphere. *Applied and Environmental Microbiology* **65,** 5357–5363.

Loper JE, Kobayashi DY, Paulsen IT, 2007. The genomic sequence of *Pseudomonas fluorescens* Pf-5: Insights into biological control. *Phytopathology* **97,** 233–238.

Lorito M, Woo SL, Fernandez IG, Colucci G, Harman GE, Pintor-Toro JA, Filippone E, Muccifora S, Lawrence CB, Zoina A, Tuzun S, Scala F, 1998. Genes from mycoparasitic fungi as a source for improving plant resistance to fungal pathogens. *Proceedings of the National Academy of Sciences USA* **95,** 7860–7865.

Lu ZX, Tombolini R, Woo S, Zeilinger S, Lorito M, Jansson JK, 2004. *In vivo* study of *Trichoderma*–pathogen–plant interactions, using constitutive and inducible green fluorescent protein reporter systems. *Applied and Environmental Microbiology* **70,** 3073–3081.

Lugtenberg BJJ, Leveau JH, 2007. Biocontrol of plant pathogens: Principles, promises, and pitfalls. In: Pinton R, Varanini Z, Nannipieri P, eds. *The Rhizosphere – Biochemistry and Organic Substances at the Soil–Plant Interface*. Boca Raton, FL: CRC Press, 267–296.

Lutz MP, Wenger S, Maurhofer M, Défago G, Duffy B, 2004. Signaling between bacterial and fungal biocontrol agents in a strain mixture. *FEMS Microbiology Ecology* **48,** 447–455.

Markovich NA, Kononova GL, 2003. Lytic enzymes of *Trichoderma* and their role in plant defense from fungal diseases: A review. *Applied Biochemistry and Microbiology* **39,** 341–351.

Marra R, Ambrosino P, Carbone V, Vinale F, Woo SL, Ruocco M, Ciliento R, Lanzuise S, Ferraioli S, Soriente I, Gigante S, Turra D, Fogliano V, Scala F, Lorito M, 2006. Study of the three-way interaction between *Trichoderma atroviride*, plant and fungal pathogens by using a proteomic approach. *Current Genetics* **50,** 307–321.

Massart S, Jijakli MH, 2006. Identification of differentially expressed genes by cDNA-amplified fragment length polymorphism in the biocontrol agent *Pichia anomala* (strain Kh5). *Phytopathology* **96,** 80–86.

Maurhofer M, Baehler E, Notz R, Martinez V, Keel C, 2004. Cross talk between 2,4-diacetylphloroglucinol-producing biocontrol pseudomonads on wheat roots. *Applied and Environmental Microbiology* **70,** 1990–1998.

McQuilken MP, Halmer P, Rhodes DJ, 1998. Application of microorganisms to seeds. In: Burges HD, ed. *Formulation of Microbial Biopesticides*. Dordrecht: Kluwer Academic Publishers, 256–285.

Mendoza-Mendoza A, Pozo MJ, Grzegorski D, Martinez P, Garcia JM, Olmedo-Monfil V, Cortes C, Kenerley C, Herrera-Estrella A, 2003. Enhanced biocontrol activity of *Trichoderma* through inactivation of a mitogen-activated protein kinase. *Proceedings of the National Academy of Sciences USA* **100,** 15965–15970.

Metraux JP, Signer H, Ryals J, Ward E, Wyssbenz M, Gaudin J, Raschdorf K, Schmid E, Blum W, Inverardi B, 1990. Increase in salicylic acid at the onset of systemic acquired resistance in cucumber. *Science* **250,** 1004–1006.

Meziane H, van der Sluis I, van Loon LC, Höfte M, Bakker PAHM, 2005. Determinants of *Pseudomonas putida* WCS358 involved in inducing systemic resistance in plants. *Molecular Plant Pathology* **6,** 177–185.

Milner JL, Raffel SJ, Lethbridge BJ, Handelsman J, 1995. Culture conditions that influence accumulation of zwittermicin A by *Bacillus cereus* UW85. *Applied Microbiology and Biotechnology* **43,** 685–691.

Milner JL, Silo-Suh L, Lee JC, He HY, Clardy J, Handelsman J, 1996. Production of kanosamine by *Bacillus cereus* UW85. *Applied and Environmental Microbiology* **62,** 3061–3065.

Mitchell R, Hurwitz E, 1965. Suppression of *Pythium debaryanum* by lytic rhizosphere bacteria. *Phytopathology* **55,** 156–158.

Molina L, Constantinescu F, Michel L, Reimmann C, Duffy B, Défago G, 2003. Degradation of pathogen quorum-sensing molecules by soil bacteria: A preventive and curative biological control mechanism. *FEMS Microbiology Ecology* **45,** 71–81.

Montesinos E, 2003. Development, registration and commercialization of microbial pesticides for plant protection. *International Microbiology* **6,** 245–252.

Morello JE, Pierson EA, Pierson LS, 2004. Negative cross-communication among wheat rhizosphere bacteria: Effect on antibiotic production by the biological control bacterium *Pseudomonas aureofaciens* 30–84. *Applied and Environmental Microbiology* **70,** 3103–3109.

Moricca S, Ragazzi A, Mitchelson KR, Assante G, 2001. Antagonism of the two-needle pine stem rust fungi *Cronartium flaccidum* and *Peridermium pini* by *Cladosporium tenuissimum in vitro* and *in planta*. *Phytopathology* **91,** 457–468.

Morissette DC, Seguin P, Jabaji-Hare SH, 2006. Expression regulation of the endochitinase-encoding gene sechi44 from the mycoparasite *Stachybotrys elegans*. *Canadian Journal of Microbiology* **52,** 1103–1109.

M'Piga P, Bélanger RR, Paulitz TC, Benhamou N, 1997. Increased resistance to *Fusarium oxysporum* f. sp. *radicis-lycopersici* in tomato plants treated with the endophytic bacterium *Pseudomonas fluorescens* strain 63–28. *Physiological and Molecular Plant Pathology* **50,** 301–320.

Mukherjee PK, Latha J, Hadar R, Horwitz BA, 2003. TmkA, a mitogen-activated protein kinase of *Trichoderma virens*, is involved in biocontrol properties and repression of conidiation in the dark. *Eukaryotic Cell* **2,** 446–455.

Mukherjee PK, Latha J, Hadar R, Horwitz BA, 2004. Role of two G-protein alpha subunits, TgaA and TgaB, in the antagonism of plant pathogens by *Trichoderma virens*. *Applied and Environmental Microbiology* **70,** 542–549.

Muthumeenakshi S, Sreenivasaprasad S, Rogers CW, Challen MP, Whipps JM, 2007. Analysis of cDNA transcripts from *Coniothyrium minitans* reveals a diverse array of genes involved in key processes during sclerotial mycoparasitism. *Fungal Genetics and Biology* **44,** 1262–1284.

Nakayama T, Homma Y, Hashidoko Y, Mizutani J, Tahara S, 1999. Possible role of xanthobaccins produced by *Stenotrophomonas* sp. strain SB-K88 in suppression of sugar beet damping-off disease. *Applied and Environmental Microbiology* **65,** 4334–4339.

Nelson EB, Chao WL, Norton JM, Nash GT, Harman GE, 1986. Attachment of *Enterobacter cloacae* to hyphae of *Pythium ultimum* – possible role in the biological control of Pythium preemergence damping-off. *Phytopathology* **76,** 327–335.

Nielsen TH, Sørensen D, Tobiasen C, Andersen JB, Christophersen C, Givskov M, Sorensen J, 2002. Antibiotic and biosurfactant properties of cyclic lipopeptides produced by fluorescent *Pseudomonas* spp. from the sugar beet rhizosphere. *Applied and Environmental Microbiology* **68,** 3416–3423.

Nielsen TH, Sørensen J, 2003. Production of cyclic lipopeptides by *Pseudomonas fluorescens* strains in bulk soil and in the sugar beet rhizosphere. *Applied and Environmental Microbiology* **69,** 861–868.

Nowak-Thompson B, Chaney N, Wing JS, Gould SJ, Loper JE, 1999. Characterization of the pyoluteorin biosynthetic gene cluster of *Pseudomonas fluorescens* Pf-5. *Journal of Bacteriology* **181,** 2166–2174.

Olivain C, Alabouvette C, 1999. Process of tomato root colonization by a pathogenic strain of *Fusarium oxysporum* f. sp. *lycopersici* in comparison with a non-pathogenic strain. *New Phytologist* **141,** 497–510.

Olivain C, Humbert C, Nahalkova J, Fatehi J, L'Haridon F, Alabouvette C, 2006. Colonization of tomato root by pathogenic and nonpathogenic *Fusarium oxysporum* strains inoculated together and separately into the soil. *Applied and Environmental Microbiology* **72,** 1523–1531.

Ongena M, Daayf F, Jacques P, Thonart P, Benhamou N, Paulitz TC, Cornelis P, Koedam N, Bélanger RR, 1999. Protection of cucumber against Pythium root rot by fluorescent pseudomonads: Predominant role of induced resistance over siderophores and antibiosis. *Plant Pathology* **48,** 66–76.

Ongena M, Jourdan E, Schafer M, Kech C, Budzikiewicz H, Luxen A, Thonart P, 2005. Isolation of an *n*-alkylated benzylamine derivative from *Pseudomonas putida* BTP1 as elicitor of induced systemic resistance in bean. *Molecular Plant–Microbe Interactions* **18,** 562–569.

Ovadis M, Liu XG, Gavriel S, Ismailov Z, Chet I, Chernin L, 2004. The global regulator genes from biocontrol strain *Serratia plymuthica* IC1270: Cloning, sequencing, and functional studies. *Journal of Bacteriology* **186,** 4986–4993.

Palumbo JD, Yuen GY, Jochum CC, Tatum K, Kobayashi DY, 2005. Mutagenesis of β-1,3-glucanase genes in *Lysobacter enzymogenes* strain C3 results in reduced biological control activity toward Bipolaris leaf spot of tall fescue and Pythium damping-off of sugar beet. *Phytopathology* **95,** 701–707.

Paulsen IT, Press CM, Ravel J, Kobayashi DY, Myers GSA, Mavrodi DV, DeBoy RT, Seshadri R, Ren QH, Madupu R, Dodson RJ, Durkin AS, Brinkac LM, Daugherty SC, Sullivan SA, Rosovitz MJ, Gwinn ML, Zhou LW, Schneider DJ, Cartinhour SW, Nelson WC, Weidman J, Watkins K, Tran K, Khouri H, Pierson EA, Pierson LS, Thomashow LS, Loper JE, 2005. Complete genome sequence of the plant commensal *Pseudomonas fluorescens* Pf-5. *Nature Biotechnology* **23,** 873–878.

Pechy-Tarr M, Bottiglieri M, Mathys S, Lejbolle KB, Schnider-Keel U, Maurhofer M, Keel C, 2005. RpoN (sigma (54)) controls production of antifungal compounds and biocontrol activity in *Pseudomonas fluorescens* CHA0. *Molecular Plant–Microbe Interactions* **18,** 260–272.

Peng C, Sutton JC, Kevan PG, 1992. Effectiveness of honey bees for applying the biocontrolagent *Gliocladium roseum* to strawberry flowers to suppress *Botrytis cinerea*. *Canadian Journal of Plant Pathology* **14,** 117–129.

Picard K, Ponchet M, Blein JP, Rey P, Tirilly Y, Benhamou N, 2000. Oligandrin. A proteinaceous molecule produced by the mycoparasite *Pythium oligandrum* induces resistance to *Phytophthora parasitica* infection in tomato plants. *Plant Physiology* **124,** 379–395.

Pierson LS, Pierson EA, 2007. Roles of diffusible signals in communication among plant-associated bacteria. *Phytopathology* **97,** 227–232.

Pierson LS, Wood DW, Pierson EA, 1998. Homoserine lactone-mediated gene regulation in plant-associated bacteria. *Annual Review of Phytopathology* **36,** 207–225.

Pleban S, Chernin L, Chet I, 1997. Chitinolytic activity of an endophytic strain of *Bacillus cereus*. *Letters in Applied Microbiology* **25,** 284–288.

Pozo MJ, Baek JM, Garcia JM, Kenerley CM, 2004. Functional analysis of tvsp1, a serine protease-encoding gene in the biocontrol agent *Trichoderma virens*. *Fungal Genetics and Biology* **41,** 336–348.

Pratt JE, Niemi M, Sierota ZH, 2000. Comparison of three products based on *Phlebiopsis gigantea* for the control of *Heterobasidion annosum* in Europe. *Biocontrol Science and Technology* **10,** 467–477.

Raaijmakers JM, Leeman M, Vanoorschot MMP, Vandersluis I, Schippers B, Bakker PAHM, 1995. Dose–response relationships in biological control of Fusarium wilt of radish by *Pseudomonas* spp. *Phytopathology* **85,** 1075–1081.

Raaijmakers JM, Vlami M, de Souza JT, 2002. Antibiotic production by bacterial biocontrol agents. *Antonie Van Leeuwenhoek International Journal of General and Molecular Microbiology* **81,** 537–547.

Ran LX, Li ZN, Wu GJ, van Loon LC, Bakker PAHM, 2005a. Induction of systemic resistance against bacterial wilt in *Eucalyptus urophylla* by fluorescent *Pseudomonas* spp. *European Journal of Plant Pathology* **113**, 59–70.

Ran LX, van Loon LC, Bakker PAHM, 2005b. No role for bacterially produced salicylic acid in rhizobacterial induction of systemic resistance in Arabidopsis. *Phytopathology* **95,** 1349–1355.

Recorbet G, Bestel-Corre G, Dumas-Gaudot E, Gianinazzi S, Alabouvette C, 1998. Differential accumulation of β-1,3-glucanase and chitinase isoforms in tomato roots in response to colonization by either pathogenic or non-pathogenic strains of *Fusarium oxysporum*. *Microbiological Research* **153,** 257–263.

Reithner B, Brunner K, Schuhmacher R, Peissl I, Seidl V, Krska R, Zeilinger S, 2005. The G protein alpha subunit Tga1 of *Trichoderma atroviride* is involved in chitinase formation and differential production of antifungal metabolites. *Fungal Genetics and Biology* **42,** 749–760.

Rey P, Le Floch G, Benhamou N, Salerno MI, Thuillier E, Tirilly Y, 2005. Interactions between the mycoparasite *Pythium oligandrum* and two types of sclerotia of plant–pathogenic fungi. *Mycological Research* **109,** 779–788.

Rocha-Ramirez V, Omero C, Chet I, Horwitz BA, Herrera-Estrella A, 2002. *Trichoderma atroviride* G-protein alpha-subunit gene tga1 is involved in mycoparasitic coiling and conidiation. *Eukaryotic Cell* **1,** 594–605.

Romero D, de Vicente A, Rakotoaly RH, Dufour SE, Veening JW, Arrebola E, Cazorla FM, Kuipers OP, Paquot M, Perez-Garcia A, 2007. The iturin and fengycin families of lipopeptides are key factors in antagonism of *Bacillus subtilis* toward *Podosphaera fusca*. *Molecular Plant–Microbe Interactions* **20,** 430–440.

Rosenblueth M, Martinez-Romero E, 2006. Bacterial endophytes and their interactions with hosts. *Molecular Plant–Microbe Interactions* **19,** 827–837.

Rotem Y, Yarden O, Sztejnberg A, 1999. The mycoparasite *Ampelomyces quisqualis* expresses exgA encoding an exo-β-1,3-glucanase in culture and during mycoparasitism. *Phytopathology* **89,** 631–638.

Ryu CM, Farag MA, Hu CH, Reddy MS, Kloepper JW, Pare PW, 2004. Bacterial volatiles induce systemic resistance in Arabidopsis. *Plant Physiology* **134,** 1017–1026.

Ryu CM, Hu CH, Reddy MS, Kloepper JW, 2003. Different signaling pathways of induced resistance by rhizobacteria in *Arabidopsis thaliana* against two pathovars of *Pseudomonas syringae*. *New Phytologist* **160,** 413–420.

Sacherer P, Défago G, Haas D, 1994. Extracellular protease and phospholipase-C are controlled by the global regulatory gene GacA in the biocontrol strain *Pseudomonas fluorescens* CHA0. *FEMS Microbiology Letters* **116,** 155–160.

Scherm B, Ortu G, Muzzu A, Budroni M, Arras G, Migheli Q, 2003. Biocontrol activity of antagonistic yeasts against *Penicillium expansum* on apple. *Journal of Plant Pathology* **85,** 205–213.

Scherm H, Ngugi HK, Savelle AT, Edwards JR, 2004. Biological control of infection of blueberry flowers caused by *Monilinia vaccinii-corymbosi*. *Biological Control* **29,** 199–206.

Scherm H, Savelle AT, Law SE, 2007. Effect of electrostatic parameters on the viability of two bacterial biocontrol agents and their deposition on blueberry flower stigmas. *Biocontrol Science and Technology* **17,** 285–293.

Scherwinski K, Wolf A, Berg G, 2007. Assessing the risk of biological control agents on the indigenous microbial communities: *Serratia plymuthica* HRO-C48 and *Streptomyces* sp. HRO-71 as model bacteria. *BioControl* **52,** 87–112.

Schirmbock M, Lorito M, Wang YL, Hayes CK, Arisanatac I, Scala F, Harman GE, Kubicek CP, 1994. Parallel formation and synergism of hydrolytic enzymes and peptaibol antibiotics, molecular mechanisms involved in the antagonistic action of *Trichoderma harzianum* against phytopathogenic fungi. *Applied and Environmental Microbiology* **60,** 4364–4370.

Schisler DA, Slininger RJ, Behle RW, Jackson MA, 2004. Formulation of *Bacillus* spp. for biological control of plant diseases. *Phytopathology* **94,** 1267–1271.

Schnider-Keel U, Seematter A, Maurhofer M, Blumer C, Duffy B, Gigot-Bonnefoy C, Reimmann C, Notz R, Défago G, Haas D, Keel C, 2000. Autoinduction of 2,4-diacetylphloroglucinol biosynthesis in the biocontrol agent *Pseudomonas fluorescens* CHA0 and repression by the bacterial metabolites salicylate and pyoluteorin. *Journal of Bacteriology* **182,** 1215–1225.

Sessitsch A, Reiter B, Berg G, 2004. Endophytic bacterial communities of field-grown potato plants and their plant-growth-promoting and antagonistic abilities. *Canadian Journal of Microbiology* **50,** 239–249.

Shoresh M, Yedidia I, Chet I, 2005. Involvement of jasmonic acid/ethylene signaling pathway in the systemic resistance induced in cucumber by *Trichoderma asperellum* T203. *Phytopathology* **95,** 76–84.

Someya N, Tsuchiya K, Yoshida T, Tsujimoto-Noguchi M, Sawada H, 2007. Combined application of *Pseudomonas fluorescens* strain LRB3W1 with a low dosage of benomyl for control of cabbage yellows caused by *Fusarium oxysporum* f. sp. *conglutinans*. *Biocontrol Science and Technology* **17,** 21–31.

Steijl H, Niemann GJ, Boon JJ, 1999. Changes in chemical composition related to fungal infection and induced resistance in carnation and radish investigated by pyrolysis mass spectrometry. *Physiological and Molecular Plant Pathology* **55,** 297–311.

Steyaert JM, Ridgway HJ, Elad Y, Stewart A, 2003. Genetic basis of mycoparasitism: A mechanism of biological control by species of *Trichoderma*. *New Zealand Journal of Crop and Horticultural Science* **31,** 281–291.

Stinson AM, Zidack NK, Strobel GA, Jacobsen BJ, 2003. Mycofumigation with *Muscodor albus* and *Muscodor roseus* for control of seedling diseases of sugar beet and verticillium wilt of eggplant. *Plant Disease* **87,** 1349–1354.

Strobel GA, Dirkse E, Sears J, Markworth C, 2001. Volatile antimicrobials from *Muscodor albus*, a novel endophytic fungus. *Microbiology-UK* **147,** 2943–2950.

Takenaka S, Nakamura Y, Kono T, Sekiguchi H, Masunaka A, Takahashi H, 2006. Novel elicitin-like proteins isolated from the cell wall of the biocontrol agent *Pythium oligandrum* induce defence-related genes in sugar beet. *Molecular Plant Pathology* **7,** 325–339.

Tamietti G, Ferraris L, Matta A, Gentile IA, 1993. Physiological responses of tomato plants grown in Fusarium suppressive soil. *Journal of Phytopathology* **138,** 66–76.

Thomson SV, Hansen DR, Flint KM, 1992. Dissemination of bacteria antagonistic to *Erwinia amylovora* by honey bees. *Plant Disease* **76,** 1052–1056.

Timms-Wilson TM, Ellis RJ, Renwick A, Rhodes DJ, Mavrodi DV, Weller DM, Thomashow LS, Bailey MJ, 2000. Chromosomal insertion of phenazine-1-carboxylic acid biosynthetic pathway enhances efficacy of damping-off disease control by *Pseudomonas fluorescens*. *Molecular Plant–Microbe Interactions* **13,** 1293–1300.

Timmusk S, Wagner EGH, 1999. The plant-growth-promoting rhizobacterium *Paenibacillus polymyxa* induces changes in *Arabidopsis thaliana* gene expression: A possible connection between biotic and abiotic stress responses. *Molecular Plant–Microbe Interactions* **12,** 951–959.

Toyoda H, Hashimoto H, Utsumi R, Kobayash H, Ouchi S, 1988. Detoxification of fusaric acid by a fusaric acid resistant mutant of *Pseudomonas solanacearum* and its application to biological control of Fusarium wilt of tomato. *Phytopathology* **78,** 1307–1311.

Uroz S, D'Angelo-Picard C, Carlier A, Elasri M., Sicot C, Petit A, Oger P, Faure D, Dessaux Y, 2003. Novel bacteria degrading *N*-acylhomoserine lactones and their use as quenchers of quorum-sensing-regulated functions of plant pathogenic bacteria. *Microbiology-SGM* **149,** 1981–1989.

van den Boogert PHJF, Deacon JW, 1994. Biotrophic mycoparasitism by *Verticillium biguttatum* on *Rhizoctonia solani*. *European Journal of Plant Pathology* **100,** 137–156.

van Loon LC, Glick BR 2004. Increased plant fitness by rhizobacteria. In: Sandermann H, ed. *Ecological Studies: Molecular Ecotoxicology of Plants*, Vol. 170. Berlin: Springer-Verlag, 177–205.

van Peer R, Niemann GJ, Schippers B, 1991. Induced resistance and phytoalexin accumulation in biological control of Fusarium wilt of carnation by *Pseudomonas* sp. strain WCS417r. *Phytopathology* **81,** 728–734.

Verhagen BWM, Glazebrook J, Zhu T, Chang HS, van Loon LC, Pieterse CMJ, 2004. The transcriptome of rhizobacteria-induced systemic resistance in Arabidopsis. *Molecular Plant–Microbe Interactions* **17,** 895–908.

Vessey JK, 2003. Plant growth promoting rhizobacteria as biofertilizers. *Plant and Soil* **255,** 571–586.

Viterbo A, Ramot O, Chernin L, Chet I, 2002. Significance of lytic enzymes from *Trichoderma* spp. in the biocontrol of fungal plant pathogens. *Antonie Van Leeuwenhoek International Journal of General and Molecular Microbiology* **81,** 549–556.

Vizcaino JA, Gonzalez FJ, Suarez MB, Redondo J, Heinrich J, Delgado-Jarana J, Hermosa R, Gutierrez S, Monte E, Llobell A, Rey M, 2006. Generation, annotation and analysis of ESTs from *Trichoderma harzianum* CECT 2413. *BMC Genomics* **7,** 193–207.

Walker MJ, Birch RG, Pemberton JM, 1988. Cloning and characterization of an albicidin resistance Gene from *Klebsiella oxytoca*. *Molecular Microbiology* **2,** 443–454.

Waspi U, Blanc D, Winkler T, Ruedi P, Dudler R, 1998. Syringolin, a novel peptide elicitor from *Pseudomonas syringae* pv. *syringae* that induces resistance to *Pyricularia oryzae* in rice. *Molecular Plant–Microbe Interactions* **11,** 727–733.

Weller DM, 2007. *Pseudomonas* biocontrol agents of soilborne pathogens: Looking back over 30 years. *Phytopathology* **97,** 250–256.

Whipps JM, 1997a. Developments in the biological control of soil-borne plant pathogens. *Advances in Botanical Research* **26,** 1–134.

Whipps JM, 1997b. Ecological considerations involved in commercial development of biological control agents for soil-borne diseases. In: Van Elsas JD, Wellington EMH, Trevors JT, eds. *Modern Soil Microbiology*. New York: Marcel Dekker, 525–546.

Whipps JM, 1997c. Interactions between fungi and plant pathogens in soil and the rhizosphere. In: Gange AC, Brown VK, eds. *Multitrophic Interactions in Teresstrial Systems*. Oxford: Blackwell Science Publications, 47–63.

Whipps JM, 2001. Microbial interactions and biocontrol in the rhizosphere. *Journal of Experimental Botany* **52,** 487–511.

Whipps JM, 2004. Prospects and limitations for mycorrhizas in biocontrol of root pathogens. *Canadian Journal of Botany* **82,** 1198–1227.

Whipps JM, Bennett A, Challen M, Clarkson J, Coventry E, Muthumeenakshi S, Noble R, Rogers C, Sreenivasaprasad S, Jones EE, 2007. Control of sclerotial pathogens with the mycoparasite *Coniothyrium minitans*. In: Vurro M, Gressel J, eds. *Novel Biotechnologies for Biocontrol Agent Enhancement and Management*. Dordrecht: Springer, 223–241.

Whipps JM, Davies KG, 2000. Biocontrol of plant pathogens and nematodes by microorganisms. In: Gurr G, Wratten SD, eds. *Measures of Success in Biological Control*. Dordrecht: Kluwer Academic Publishers, 231–269.

Whipps JM, Gerhardson B, 2007. Biological pesticides for control of seed- and soil-borne plant pathogens. In: van Elsas JD, Jansson JK, Trevors JT, eds. *Modern Soil Microbiology*. Boca Raton, FL: CRC Press LLC, 479–501.

Whipps JM, Lewis K, Cooke RC, 1988. Mycoparasitism and plant disease control. In: Burge MN, ed. *Fungi in Biological Control Systems*. Manchester: Manchester University Press, 161–187.

Whipps JM, Lumsden RD, 2001. Commercial use of fungi as plant disease biological control agents: Status and prospects. In: Butt TM, Jackson CW, Magan N, eds. *Fungi as Biocontrol Agents: Progress, Problems and Potential*. Wallingford, Oxford: CABI Publishing, 9–22.

Wiest A, Grzegorski D, Xu BW, Goulard C, Rebuffat S, Ebbole DJ, Bodo B, Kenerley C, 2002. Identification of peptaibols from *Trichoderma virens* and cloning of a peptaibol synthetase. *Journal of Biological Chemistry* **277,** 20862–20868.

Wilhite SE, Lumsden RD, Straney DC, 1994. Mutational analysis of gliotoxin production by the biocontrol fungus *Gliocladium virens* in relation to suppression of Pythium damping-off. *Phytopathology* **84,** 816–821.

Wilson CL, Elghaouth A, Chalutz E, Droby S, Stevens C, Lu JY, Khan V, Arul J, 1994. Potential of induced resistance to control postharvest diseases of fruits and vegetables. *Plant Disease* **78,** 837–844.

Winding A, Binnercup SJ, Pritchard H, 2004. Non-target effects of bacterial biological control agents suppressing root pathogenic fungi. *FEMS Microbiology Ecology* **47,** 129–141.

Woo SL, Donzelli B, Scala F, Mach R, Harman GE, Kubicek CP, Del Sorbo G, Lorito M, 1999. Disruption of the ech42 (endochitinase-encoding) gene affects biocontrol activity in *Trichoderma harzianum* P1. *Molecular Plant–Microbe Interactions* **12,** 419–429.

Woo SL, Scala F, Ruocco M, Lorito M, 2006. The molecular biology of the interactions between *Trichoderma* spp., phytopathogenic fungi, and plants. *Phytopathology* **96,** 181–185.

Wright B, Rowse HR, Whipps JM, 2003. Application of beneficial microorganisms to seeds during drum priming. *Biocontrol Science and Technology* **13,** 599–614.

Wulff EG, van Vuurde JWL, Hockenhull J, 2003. The ability of the biological control agent *Bacillus subtilis*, strain BB, to colonise vegetable brassicas endophytically following seed inoculation. *Plant and Soil* **255,** 463–474.

Yedidia I, Benhamou N, Kapulnik Y, Chet I, 2000. Induction and accumulation of PR proteins activity during early stages of root colonization by the mycoparasite *Trichoderma harzianum* strain T-203. *Plant Physiology and Biochemistry* **38,** 863–873.

Yedidia I, Shoresh M, Kerem Z, Benhamou N, Kapulnik Y, Chet I, 2003. Concomitant induction of systemic resistance to *Pseudomonas spingae* pv. *lachrymans* in cucumber by *Trichoderma asperellum* (T-203) and accumulation of phytoalexins. *Applied and Environmental Microbiology* **69,** 7343–7353.

Yu H, Sutton JC, 1997. Effectiveness of bumblebees and honeybees for delivering inoculum of *Gliocladium roseum* to raspberry flowers to control *Botrytis cinerea*. *Biological Control* **10,** 113–122.

Zahir ZA, Arshad M, Frankenberger WT, 2004. Plant growth promoting rhizobacteria: Applications and perspectives in agriculture. *Advances in Agronomy* **81,** 97–168.

Zeilinger S, Schmoll M, Pail M, Mach RL, Kubicek CP, 2003. Nucleosome transactions on the *Hypocrea jecorina* (*Trichoderma reesei*) cellulase promoter cbh2 associated with cellulase induction. *Molecular Genetics and Genomics* **270,** 46–55.

Zhang L, Birch RG, 1996. Biocontrol of sugar cane leaf scald disease by an isolate of *Pantoea dispersa* which detoxifies albicidin phytotoxins. *Letters in Applied Microbiology* **22,** 132–136.

Zhang LH, Birch RG, 1997. The gene for albicidin detoxification from *Pantoea dispersa* encodes an esterase and attenuates pathogenicity of *Xanthomonas albilineans* to sugarcane. *Proceedings of the National Academy of Sciences USA* **94,** 9984–9989.

Zipfel C, Robatzek S, Navarro L, Oakeley EJ, Jones JDG, Felix G, Boller T, 2004. Bacterial disease resistance in *Arabidopsis* through flagellin perception. *Nature* **428,** 764–767.

Chapter 4
Induced resistance for plant disease control

Tony Reglinski[1] *and Dale Walters*[2]

[1]Plant and Food Research, Hamilton, New Zealand
[2]Scottish Agricultural College, Edinburgh, UK

4.1 Introduction

It is well known that treatment of plants with various agents, including virulent or avirulent pathogens, nonpathogens, cell wall fragments, plant extracts, and synthetic chemicals, can promote resistance to subsequent pathogen attack, both locally and systemically (Walters *et al.*, 2005). This phenomenon, known as induced resistance, results in reductions in lesion incidence and severity but rarely leads to complete control of pathogens (Kuc, 1982). Importantly, the expression of induced resistance does not require the presence of major pathogen-specific resistance genes, although the defences activated are associated with both R-gene-mediated resistance and non-host resistance (Heath, 1998). These include the oxidative burst, which can lead to cell death (Heath, 1998), thereby trapping the pathogen in dead cells, alterations to cell wall composition that can reduce or inhibit pathogen penetration, and synthesis and accumulation of antimicrobial compounds like phytoalexins (Garcion *et al.*, 2007). Induction of systemic resistance can result in the direct activation of defence genes, but can also lead to the phenomenon of priming, wherein plant defences are not directly activated by the inducing agent but instead are potentiated for enhanced expression upon subsequent pathogen attack (Benhamou *et al.*, 2000; Beckers & Conrath, 2007).

Induced resistance to microbial pathogens can be split broadly into systemic acquired resistance (SAR) and induced systemic resistance (ISR). SAR describes the phenomenon whereby plants develop a broad-spectrum systemic resistance to pathogenic attack following a localised infection or after topical treatment with the so-called 'activators' (e.g. acibenzolar-*S*-methyl, 2,6-dichloroisonicotinic acid) (Walters *et al.*, 2005; Hammerschmidt, 2007). The development of SAR is associated with elevated levels of salicylic acid (SA) locally and systemically and with the coordinate expression of a specific set of genes encoding pathogenesis-related (PR) proteins (Van Loon *et al.*, 2006). Activation of *PR* gene expression and SAR is dependent on transduction of the SA signal and this, in turn, requires the function of the regulatory protein NPR1 (Figure 4.1) (Shah *et al.*, 1997; Pieterse & Van Loon, 2007). It is important to note, however, that although expression of a set of *PR* genes, and *PR-1* in particular, is used as a marker for SAR

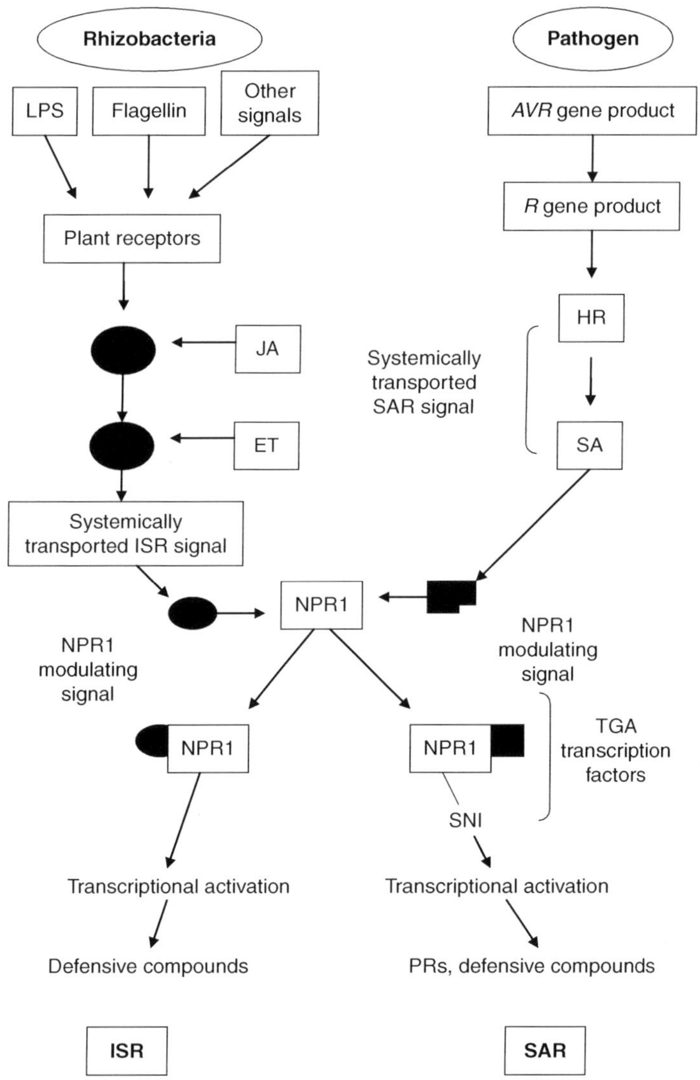

Figure 4.1 A model for the signal transduction network controlling induced systemic resistance (ISR) mediated by plant growth promoting rhizobacteria and pathogen-induced systemic acquired resistance (SAR) in *Arabidopsis thaliana*. LPS = lipopolysaccharide; PRs = pathogenesis-related proteins; *AVR* = avirulence gene product; *R* = resistance gene product; HR = hypersensitive response; SA = salicylic acid; JA = jasmonic acid; ET = ethylene; NPR1 = a regulatory protein involved in signalling in SAR and ISR in *A. thaliana;* SNI = transcriptional repressor of SAR genes; TGA transcription factors = family of transcription factors interacting with SA-induced NPR1. (From Walters & Heil (2007), with permission from Elsevier.)

induction, the role of these *PR* genes in the expression of resistance is unclear (Durrant & Dong, 2004) and induction of resistance is not always accompanied by expression of *PR-1* (Sparla *et al.*, 2004).

In contrast to SAR, ISR develops as a result of colonisation of plant roots by certain strains of plant growth-promoting rhizobacteria (PGPR) and is mediated by a jasmonic acid (JA) and ethylene (ET) sensitive pathway (Pieterse *et al.*, 1998; Pieterse & Van Loon, 2007). ISR is phenotypically similar to SAR in that it acts unspecifically against taxonomically different pathogens (Zehnder *et al.*, 2001; Pieterse & Van Loon, 2007). Further support for a general similarity of ISR and SAR came from the demonstration that a functioning NPR1 is required for the successful establishment of ISR (Figure 4.1) (Pieterse *et al.*, 1998) and from the finding that ISR can be associated with the accumulation of PR-proteins and/or phytoalexins and with changes in cell wall composition, all traits that also characterise SAR (Ramamoorthy *et al.*, 2001). Another feature of induced resistance that is common to both SAR and ISR is a phenomenon called priming, whereby plant defences are not directly activated by the inducing agent but instead are potentiated for enhanced expression upon subsequent pathogen attack (Benhamou *et al.*, 2000; Conrath *et al.*, 2006; Beckers & Conrath, 2007).

4.2 Induced resistance in practice

Numerous biotic and abiotic agents have been reported to activate plant defence responses thereby rendering treated plants more resistant to pathogenic infection (da Rocha & Hammerschmidt, 2005). The commercialisation of resistance 'activators' has created a new generation of crop protectants including, Bion®/Actigard® (acibenzolar-*S*-methyl (ASM), Syngenta), Milsana® (*Reynoutria sachalinensis* extract, KHH BioScience Inc. USA), Elexa® (chitosan, SafeScience, USA), and Messenger® (harpin protein, Eden Bioscience, USA). In this section, we describe the implementation of activators in different cropping systems and examine the contribution of induced resistance to disease management. Our discussion will focus primarily on the performance of products that were specifically developed as activators. Therefore, fungicides such as probenazole (Oryzemate®, Meiji Seika) and fosetyl-al (Aliette®, Bayer Crop Science), and 'beneficial microbes', including certain *Bacillus* spp., *Pseudomonas* spp., and *Trichoderma* spp., that have been shown to activate systemic resistance in plants will not be discussed in detail.

4.2.1 Fruit and vegetable production

4.2.1.1 Tomato

Tomato is a major global crop and production is estimated to have a farm gate value worth more than US$1 billion in the US alone (Florida Tomato Committee, 2005). Several pathogens can attack tomato leaves and fruit during production and disease management involves the integrated use of different bactericides and fungicides. A considerable number of studies have evaluated the use of inducing agents for disease control in tomato. Actigard® (ASM), in particular, has been the focus of intensive research as a management option to control bacterial spot (*Xanthomonas axonopodis* pv. *vesicatoria*) and bacterial speck (*Pseudomonas syringae* pv. *tomato*) (Louws *et al.*, 2001). In a series of glasshouse and field studies carried out over a four-year period in eastern North America, ASM was as effective or superior to copper-based bactericides against bacterial spot and bacterial speck (Louws *et al.*, 2001). The authors proposed that ASM should prove to be particularly useful in areas with copper-resistant pathogenic strains. Actigard is now established as a recommended component of spray schedules for management of bacterial spot and bacterial speck in North Carolina (Ivors & Louws, 2007) and in Florida (Olson & Simonne, 2005).

The inclusion of ASM in tomato spray programmes may have additional benefits since foliar applications of ASM have shown efficacy against bacterial wilt (*Ralstonia solanacearum*) (Anith *et al.*, 2004; Pradhanang *et al.*, 2005) and late blight (*Phytophthora infestans*) (Becktell *et al.*, 2005). The efficacy of ASM against bacterial wilt has been shown to be affected by the inherent resistance of different tomato cultivars to the disease and also by the bacterial inoculum concentration. For example, ASM was effective in 'Neptune', which exhibits moderate resistance to bacterial wilt, but not on the highly susceptible 'Solar Set', when the treated plants were challenged with high bacterial concentrations (10^7 CFU ml^{-1}). However, ASM controlled bacterial wilt on 'Solar Set' when lower bacterial inoculum concentrations (10^5 CFU ml^{-1}) were used (Anith *et al.*, 2004; Hacisalihoglu *et al.*, 2007).

At present, tomato spray programmes recommend the use of ASM in combination with other fungicides and bactericides (Ivors & Louws, 2007) on the premise that ASM will elevate host resistance whilst the other products will reduce pathogen inoculum levels. In an effort to identify 'more sustainable' methods to reduce pathogen populations, researchers have investigated ASM in combination with microbial antagonists (Anith *et al.*, 2004; Obradovic *et al.*, 2004, 2005) and antimicrobial natural products (Ji *et al.*, 2007). In studies on bacterial wilt, Anith *et al.* (2004) demonstrated that ASM, in combination with two *Bacillus* strains (BioYield™, Gustafson, LLC, Dallas), reduced disease incidence in 'Solar Set', whereas ASM alone did not. More recently, the integrated use of ASM in the field with thymol, applied as a preplant fumigant, reduced bacterial wilt incidence and increased tomato yield (Ji *et al.*, 2007). Furthermore, this combination treatment demonstrated potential against root-knot nematode (*Meloidogyne* spp.) by causing a reduction in root galling and juvenile nematode populations. The authors propose that ASM/thymol combinations may provide an alternative to methyl bromide for managing soilborne diseases.

Field and glasshouse studies on tomato bacterial spot demonstrated that applications of ASM in combination with a bacteriophage formulation (Agriphage™, OmniLytics, Inc., USA) can provide similar or better disease control than copper-mancozeb treatments (Obradovic *et al.*, 2004, 2005). However, the bacteriophage was inconsistent under glasshouse conditions and showed limited survival under natural conditions when unformulated (Balogh *et al.*, 2003). It has been proposed that formulated bacteriophage should be applied at dusk for optimum efficacy. In the same study, the integrated use of ASM with antagonistic bacteria (*Bacillus pumilus* B122 and *Pseudomonas fluorescens* B130) did not offer any benefit with respect to control of bacterial spot (Obradovic *et al.*, 2005). The integrated use of ASM in combination with reduced-risk insecticides and mulches has been developed for management of tomato spotted wilt virus (Momol *et al.*, 2004). The authors reported that these strategies are being widely adopted by growers in north Florida and South Georgia.

Induced resistance has also demonstrated efficacy against fungal and oomycete pathogens in tomatoes. Grey mould, caused by *Botrytis cinerea,* is a continual threat to tomato production and can infect leaves, stems, flowers and fruit. In growth chamber experiments, application of ASM retarded the development of *B. cinerea* lesions on 'Perkoz' leaves (Malolepsza, 2006). Postharvest application of ASM protected 'Ciliegino' tomatoes against wound inoculation with *B. cinerea* (Iriti *et al.*, 2007). Konstantinidou-Doltsinis *et al.* (2006) reported that Milsana induced resistance to powdery mildew (*Leveillula taurica* (Lev) Arn.) on 'Manthos' tomato and reduced foliar infection by 42–64% in glasshouse trials. The level of control was equivalent to that obtained using sulphur, but was not as effective as conventional fungicides. Foliar applications of Milsana and chitosan (Chitoplant®, ChiProGmbH-Germany) reduced powdery mildew severity on 'Bison' and 'Elpida' greenhouse tomatoes (Dafermos *et al.*, 2007). Disease severity was further reduced if plants were grown in soil amended with chitin. The authors suggested that the integrated use of these treatments, particularly on cultivars with low susceptibility, could be an alternative to sulphur for the management of powdery mildew in organic tomato production.

Bourbos & Barbopoulou (2006) reported that harpin (Messenger) provided over 98% control of late blight (*Phytophthora infestans*) in glasshouse 'Bella Dona' tomato and also increased fruit yield by over 47%. In contrast, twice-weekly foliar applications of harpin did not control late blight on 'Sunrise' glasshouse tomatoes as effectively as fungicide (Becktell *et al.*, 2005). Immersion of tomato seeds (Santa Cruz 'Kada') in a *Bacillus cereus* suspension before sowing significantly enhanced field resistance to late blight and also to foliar infections caused by *Alternaria solani* (early blight) and *Septoria lycopersici* (septoria leaf spot) (Silva *et al.*, 2004). Disease severity was further reduced by foliar application of chlorothalonil fungicide. The combination of these treatments enabled a reduction in the number of fungicide applications without loss of disease control or crop yield. More recently, chitosan and mustard seed extract (Tillekur®, Biofa, Germany) demonstrated potential to protect tomato seed against infection by the fungus *Didymella lycopersici* Kleb. (Kasselaki *et al.*, 2007). This is one of the most important seedborne diseases of tomato and causes serious reductions in germination and seedling survival.

4.2.1.2 Pepper

The efficacy of ASM as a potential management tool for Phytophthora root and crown rot (*Phytophthora capsici*) was demonstrated in glasshouse studies (Matheron & Porchas, 2002). In highly favourable disease conditions, ASM was less effective than the fungicide mefenoxam (Ridomil Gold® 44WP). However, in more moderate conditions, plants that received three ASM applications plus a soil treatment of mefenoxam survived significantly longer in naturally infested soil, and produced greater shoot growth, than those treated only with ASM or mefenoxam. Resistance to mefenoxam in *P. capsici* has been detected in North Carolina and it was suggested that the use of ASM may prolong the effectiveness of this fungicide for disease control. The efficacy of copper-based sprays to control bacterial spot in pepper, caused by *Xanthomonas axonopodis* [*campestris*] pv. *vesicatoria*, can be disappointing because of the occurrence of copper-resistant bacterial strains. Integrated use of ASM with copper hydroxide has shown potential as a management option to control bacterial spot in pepper (Romero *et al.*, 2001; Buonaurio *et al.*, 2002). In field studies, a tank mix containing ASM plus copper hydroxide had higher efficacy against leaf (69%) and fruit (67%) infection than either products alone (Buonaurio *et al.*, 2002). Seven or more applications of ASM significantly reduced bacterial spot in bell pepper but resulted in lower yields compared with copper plus mancozeb treated plots (Romero *et al.*, 2001). However, when ASM was applied in rotation with copper hydroxide, disease control was equal to that obtained with weekly applications of copper hydroxide plus maneb and there was no loss in yield. The effect of ASM on yield was cultivar specific and correlated with application frequency.

4.2.1.3 Apple

Field application of ASM has been shown to protect the apple cultivars 'Golden Delicious' (Brisset *et al.*, 2000), 'Jonathan' and 'Fuji' (Maxson-Stein *et al.*, 2002) from infection with *Erwinia amylovora,* the causal agent of fire blight. ASM application in each of these studies was initiated approximately one week pre-bloom with successive sprays at 7- or 14-day intervals. On 'Golden Delicious', blossom blight incidence was reduced by approximately 50% when ASM was applied twice before inoculation or twice before and twice after inoculation. On 'Jonathan', ASM significantly reduced the incidence of natural fire blight strikes associated with severe weather conditions at the end of petal fall. ASM was more effective when applied at weekly intervals than when applied fortnightly. Furthermore, there was a positive correlation between ASM concentration and the reduced spread of fire blight cankers on inoculated shoots of 'Fuji' apple. The combined use of ASM and streptomycin was more effective at limiting canker development than either treatment alone (Maxson-Stein *et al.*, 2002). It was suggested that the integrated use of ASM with streptomycin could reduce the need for streptomycin in the post bloom period and so reduce the risk of the development of streptomycin-resistant strains in the field.

Prohexadione-calcium (Pro-Ca, Regalis®; BASF, Limburgerhof, Germany), a growth regulator that is used to reduce shoot growth in pome fruit, has also been shown to activate host defences including PR-proteins (Bini *et al.*, 2008) and phenylpropanoid metabolism (Roemmelt *et al.*, 2003). Application of Pro-Ca to container-grown apple trees 'Idared' and 'Freedom', two weeks before shoot inoculation with *E. amylovora,* resulted in a significant reduction in the development of fire blight lesions (Buban *et al.*, 2004). Blossoms represent the primary site of infection for the fire blight pathogen. In field experiments, Pro-Ca and the growth regulatory compound, acylcyclohexanedione (TrixE, Moddus®, Syngenta), reduced the severity of blossom blight infections on 'Golden Delicious' and 'Pink Lady' when applied to open flowers 14 days before pathogen inoculation (Spinelli *et al.*, 2007). In the same study, experiments on 'Golden Delicious' and 'Abbe Fetel' pear plantlets demonstrated that the growth regulators treatments the migration of the pathogen in host tissues in a similar manner, and to a similar extent, to that observed following ASM application. Whilst the above results are encouraging, the use of different activators have not exhibited sufficient efficacy against fire blight to merit consideration as complete replacements for bactericides. However, they do provide viable alternatives to enable a reduction in the frequency and/or application concentrations of bactericides and fungicides that are currently used for pome fruit disease management.

Pro-Ca has also demonstrated efficacy against apple scab caused by *Venturia inaequalis.* Treatment of 'Golden Delicious' seedlings with Pro-Ca, 10 days before inoculation with *V. inaequalis,* afforded a significant reduction in apple scab symptoms on leaves when compared with untreated seedlings (Bini *et al.*, 2008). Furthermore, the expression of PR genes, in response to pathogen inoculation, was greater in seedlings treated with Pro-Ca than in non-treated seedlings. The authors offered a cautionary note regarding the elevation of these defence proteins in food products since Roux *et al.* (2006) found that transcripts for *PR-10* allergens significantly increased in apples in response to Pro-Ca treatment. The major allergen in apple, Mal Dd 1, is a member of the *PR-10* group of defence proteins (Pühringer *et al.*, 2000) and therefore it would be prudent to consider the impact of inducing agents on levels of potential allergens in fruit. Timing of activator applications will play an important role in determining their impact on PR accumulation in fruits.

The integrated use of activators and fungicides may offer a practical option for controlling apple scab. Flusilazole was more efficacious against *V. inaequalis* on apple seedlings that were treated with the activator dichloroisonicotinic acid (INA) than on untreated seedlings (Ortega *et al.*, 1998). The efficacy of INA against apple scab was not affected by the sensitivity of different pathogen strains to the triazole group of fungicides. Additive benefits from combining plant activators and fungicides in the field could enable a reduction in the number of fungicide applications, and possibly fungicide dose rate.

'Red Delicious' apples expressed enhanced resistance to postharvest blue mould (*Penicillium expansum*) after treatment with chitosan, harpin, or UV-C irradiation (de Capdeville *et al.*, 2002). Freshly harvested fruit, or fruit stored in controlled-atmosphere for 3 months, were treated either 24, 48 or 96 hours before wound inoculation with the pathogen. UV-C was the most effective treatment and reduced lesion development by over 60% compared with control fruit. Treatments were more effective on fresh fruit than on stored fruit and were most effective when applied 96 hours before inoculation. This latter finding is not surprising, since a time delay is required for the onset of resistance. UV-C has also been shown to directly inhibit microbial growth and a packhouse UV-C treatment may be an attractive commercial option for postharvest fruit treatment. Preharvest treatment of apple trees with harpin, before fruit inoculation, resulted in enhanced resistance to blue mould infection on 'McIntosh', 'Empire' and 'Red Delicious' apples (de Capdeville *et al.*, 2003). Higher concentrations of harpin resulted in greater control, but, there was no difference in disease control between fruit treated with harpin either four or eight days before inoculation. Salicylic acid (Yu & Zheng, 2006), gibberellic acid (Yu & Zheng, 2007) and chitosan (Yu *et al.*, 2007a) have been shown to augment the efficacy of the biocontrol yeast, *Cryptococcus laurentii* against postharvest blue mould in apples. The treatment combinations were applied to wounded fruit at least 2 hours before inoculation with *P. expansum.*

4.2.1.4 Pear

Pear fruits are susceptible to postharvest diseases including blue mould (*Penicillium expansum* Link) and black mould (*Alternaria alternata* (Fr:Fr.) Keissl.). Several non-fungicidal control options, including antagonistic biocontrol agents, heat treatment, antimicrobial natural products and resistance activators, have been investigated for their potential to control postharvest rots. Field applications of salicylic acid (SA) (Cao *et al.*, 2006) and ASM (Cao & Jiang, 2006) enhanced resistance to postharvest fungal infection in 'Yali' pears. In both trials, the pear trees were sprayed with the activator on three occasions, at 30, 60 and 90 days after flowering. Fruit were harvested at full commercial maturity (*ca* 120 days after flowering) and then surface sterilised before wound-inoculation with *P. expansum*, for SA-treated fruit (Cao *et al.*, 2006), and *P. expansum* and

A. alternata, for ASM-treated fruit (Cao & Jiang, 2006). After 17 days incubation at 25°C, disease incidence and severity on treated fruit were reduced by 23–42% compared with control fruit. Postharvest infiltration of Yali pears with ASM, 3–15 days before inoculation with *P. expansum*, resulted in a reduction in disease incidence and severity (Cao *et al.*, 2005). In field trials on 'Kosui' pear, four applications of ASM showed control efficacy against Japanese pear scab (*Venturia nashicola*) equal to that of the commercial fungicide (Ishii *et al.*, 1999).

The use of activators in combination with antagonistic yeast offers a promising method for controlling wound-invading pathogens in pears. Yu *et al.* (2007b) found that postharvest application of SA, together with the biocontrol yeast *Cryptococcus laurentii*, significantly reduced blue mould and grey mould (*B. cinerea*) in wound inoculated 'Shuijing' pears. The combination treatment was more effective than the individual components alone and provided greater disease control when applied 2 hours rather than 10 hours before pathogen inoculation. In related studies, combinations of *C. laurentii* plus gibberellic acid (Yu *et al.*, 2006) and *C. laurentii* plus cytokinin (Zheng *et al.*, 2007) have been shown to control blue mould in 'Shuijing' pears when applied to wound-inoculated fruit. Enzyme activity studies suggest that the 'enhanced resistance' associated with gibberellic acid and cytokinin may, in part, be linked to the delay of senescence in the pear fruit.

4.2.1.5 Citrus

A range of commercial inducing agents demonstrated efficacy against citrus scab (*Elsinoe fawcettii*), melanose (*Diaporthe citri*) and Alternaria brown spot (*Alternaria alternata*) in different citrus fruit under glass (Agostini *et al.*, 2003). The most consistent products were ASM, Re-Zist (Stoller Enterprises Inc., USA) and Aliette (fosetyl-Al, Aventis Crop Science, USA). In field studies, 16 applications of ASM failed to control Alternaria brown spot on 'Murcott' mandarins (Miles *et al.*, 2005). However, tank-mixing ASM with azoxystrobin improved the efficacy of the fungicide by over 50% and the level of disease control was comparable with an industry standard programme. In contrast, three or four ASM applications on 'Imperial' mandarin and 'Navel' orange reduced citrus black spot (*Guignardi citricarpa*) by *ca* 50% when compared with controls, but did not significantly improve fungicide efficacy when tank-mixed (Miles *et al.*, 2004).

In glasshouse studies, foliar application of ASM and harpin, three to seven days before inoculation, reduced the incidence of citrus bacterial spot (*Xanthomonas axonopodis* pv. *citrumelo*) and citrus canker (*X. axonopodis* pv. *citri*) on citrumelo leaves (Graham & Leite Jr., 2004). ASM reduced the incidence of both diseases by up to 80% and was found to be more effective and more consistent than harpin. Based on this activity under glass, ASM and harpin were then assessed in four orchard trials for their potential to complement the activity of copper-based fungicides on sweet oranges. In these trials, the inducing agents did not improve the efficacy of the copper programmes against citrus canker on fruit and leaves, and had no effect on fruit drop. For sweet oranges used in juice processing, prevention of fruit drop is more important than reduction of fruit blemishes. Postbloom fruit drop, caused by *Colletotrichum acutatum,* was reduced on sweet orange and grapefruit following application of SA, ReZist and ASM in multiyear screenhouse trials (Chen *et al.*, 2006; Liao *et al.*, 2006). Treatments were applied seven days after inoculation and fruit retention was measured up to 108 days later. The authors proposed that ReZist and ASM reduce fruit abscission by altering endogenous hormones levels.

Effective postharvest decay control is important for marketing fresh citrus. Field trials were conducted between 1999 and 2003 to evaluate effects of several chemicals, including the plant activators ASM, fosetyl-Al and phosphorous acid, on natural postharvest infections of citrus (Ritenour *et al.*, 2004). However, only phosphorous acid demonstrated significant efficacy and, in one experiment, reduced total decay by 36%.

4.2.1.6 Peach

Peach fruit usually have a very short postharvest shelf life and rely on the use of the modified atmosphere storage, refrigeration and fungicides to control postharvest disease. Liu *et al.* (2005a,b) investigated postharvest application of ASM as a possible method to improve resistance of 'Jiubao' peach fruit to *P. expansum,* the causal agent of blue mould. Harvested peaches were immersed for 5 minutes in a solution containing 200 mg$\,l^{-1}$ of ASM and then air dried. Sixty hours after treatment, the fruit were wound-inoculated with *P. expansum* and then incubated at 22°C at high humidity. After 7 days incubation, disease incidence on ASM-treated fruit was reduced by 50% and disease severity by 64% compared with control fruit. Effects of postharvest ASM treatment on fruit quality parameters such as taste and aroma are unknown and therefore further research is required before this treatment could be considered for commercial production.

4.2.1.7 Melon

Foliar application of ASM (50 mg$\,l^{-1}$) before flowering protected rock melons and Hami melons against natural postharvest infection by *Alternaria* spp., *Fusarium* spp. and *Rhizopus* spp. (Huang *et al.*, 2000). A further increase in disease resistance was achieved when ASM-treated fruit were dipped in guazatine at harvest. Benefits of combining treatments was also observed on fruit that were sprayed with ASM 2 weeks before harvest, and then dipped in guazatine after harvest (Bokshi *et al.*, 2006). Interestingly, postharvest disease severity was not further reduced on fruit that had received four ASM applications during the season. More recently, Bokshi *et al.* (2007) demonstrated that application of ASM 2 weeks before harvest, combined with a postharvest dip in hot (55°C) iodine, provided significantly greater control of postharvest rots than guazatine alone. They proposed that the integrated use of chemically induced resistance with hot iodine can offer an environmentally safe alternative to the conventional fungicide dip.

Harpin was investigated as a postharvest dipping treatment to protect Hami melons against pink rot, caused by *Tricothecium roseum* (Bi *et al.*, 2005). Fruit that were half immersed in harpin (90 mg$\,l^{-1}$) before inoculation expressed elevated resistance to decay in both the treated and the untreated halves. Disease suppression was evident for 8 and 6 days in '8601', and for 5 and 3 days in 'New Queen', in treated and untreated halves, respectively. Increasing harpin concentration to 200 mg$\,l^{-1}$ did not further enhance disease control. More recently, full immersion of '8601', and 'New Queen' fruit in harpin (90 mg$\,l^{-1}$) was shown to promote resistance to postharvest wound infections by *Alternaria alternata, Fusarium semitectum* and *T. roseum* (Yang *et al.,* 2007). Decay lesion diameter was reduced by *ca* 18–28% on fruit that were dipped for 10 minutes, 2 days before inoculation.

Foliar applications of harpin, in combination with chlorothalonil and azoxystrobin, were evaluated against downy mildew (*Pseudoperonospora cubensis*) and gummy stem blight (*Didymella bryoniae*) on 'Athena' melon (Keinath *et al.*, 2007a). Harpin did not control either disease but was associated with a 10–12% increase in fruit yield in one year out of two. The same authors conducted 12 field trials between 2002 and 2004 in which they evaluated harpin, ASM and *Reynoutria* extract (Milsana®), in combination with chlorothalonil, for their ability to control gummy stem blight and Alternaria leaf blight (*A. cucumerina*) in 'Athena' melon (Keinath *et al.*, 2007b). Results indicated that harpin and *Reynoutria* extract could be alternated with weekly chlorothalonil applications without loss of disease control. ASM was, however, less effective and had a negative effect on yield. Scheduling treatments according to the weather-based application programme (Melcast) generally reduced the number of spray applications, but did not always provide as effective disease control as the standard schedule, except when disease severity was low. It was concluded that, to reduce the amount of fungicide applied, growers could use reduced-risk fungicides (such as activators) in rotation with a reduced rate of fungicide on a weekly basis. This offers the potential to use compounds with shorter re-entry intervals compared with chlorothalonil (48 hours) and mancozeb (5 days) during the vital harvest period. This is particularly important for a crop such as melon, since harvest can be protracted and may last between two to four weeks.

4.2.1.8 Grape

Grey mould (*Botrytis cinerea*), powdery mildew (*Uncinula necator*), and downy mildew (*Plasmopara viticola*) can cause significant economic damage in vineyards worldwide. Increasingly stringent regulations governing chemical residues on grapes and in wines are restricting fungicide control options in conventionally managed vineyards. For example, in New Zealand and Australia, preharvest intervals have been extended for several highly effective fungicides. Similarly, changes are taking place in organic production in Europe where initiatives are being funded to find replacements for copper fungicides that are traditionally used to control downy mildew in grapevine. Resistance activators are amongst the options that are being considered to supplement or replace fungicides in both conventional and organic viticulture.

Ten field applications of 0.2% SA reduced downy mildew severity on leaves and on berries by *ca* 30% and *ca* 40%, respectively (Kast, 2000). However, SA can be phytotoxic and was reported to retard berry ripening in Shiraz grapes (Kraeva *et al.*, 1998). Regular applications of 5-chlorosalicylic acid (5CSA) provided control of botrytis bunch rot on Chardonnay grapes that was equivalent to that achieved by fungicide application (Reglinski *et al.*, 2005). However, as with SA above, 5CSA was phytotoxic and caused leaf chlorosis and a reduction in berry weight. Furthermore, three applications of 5CSA on 'Cabernet Sauvignon', between fruit-set and harvest, caused a decrease in wine quality (Duxbury *et al.*, 2004). Three applications of ASM after veraison induced resistance to *B. cinerea* in Merlot bunches (Iriti *et al.*, 2004). None of the bunches from ASM-treated vines exceeded 25% disease severity whereas over 90% of untreated bunches showed a disease severity of >50%. ASM has also demonstrated efficacy against downy mildew thus broadening its potential appeal for growers. Alternating the use of ASM with copper treatments during flowering allowed a reduction in copper use without any significant loss of downy mildew control on Cabernet Sauvignon vines (Dagostin *et al.*, 2006).

Formulations containing the non-protein amino acid β-aminobutyric acid (BABA) have been field evaluated for the control of downy mildew in grapes (Reuveni *et al.*, 2001). Two foliar sprays of BABA, or a mixture of BABA with reduced-rate fungicides, reduced downy mildew severity by over 90% in field-grown Chardonnay and Cabernet Sauvignon grapevines. BABA was as efficacious as metalaxyl-Cu (Ridomil-Copper) or dimethomorph + mancozeb (Acrobat Plus®). A BABA-Cu complex manufactured by Makhteshim-Agan, Israel, was highly effective and reduced downy mildew by 83–98% over two consecutive seasons indicating an additive interaction between BABA and fungicides.

Chitosan-based activators have shown some promise in viticulture. In the USA, eight applications of Elexa™ (a.i. chitosan, Safescience Inc., USA) over the season reduced the incidence of downy mildew by 50% and powdery mildew by 75% on 'Vidal' grapes compared with untreated controls (Schilder *et al.*, 2002). Powdery mildew incidence was reduced on vines treated with 0.01–0.1% chitosan, shortly after flowering and then again 14 days later (Gorbatenko *et al.*, 1996). Chitosan treatment has also demonstrated efficacy against infections caused by *B. cinerea*. A formulated chitosan solution, Chitogel® (Ecobulle, France) stimulated the growth of Chardonnay plantlets and induced resistance to challenge inoculation with *B. cinerea* (Ait Barka *et al.*, 2004). This treatment has potential to reduce the need for fungicides in grapevine propagation and at the same time enhance plant vigour.

Preharvest applications of 1% chitosan 21 days and/or 5 days before harvest, reduced botrytis bunch rot severity by 55% on 'Italia' table grapes compared with untreated controls (Romanazzi *et al.*, 2002). Combining preharvest chitosan application with postharvest UV-C irradiation provided greater control of botrytis on 'Autumn Black' and 'B36–55' table grapes than with either treatment alone (Romanazzi *et al.*, 2006). It was proposed that this integrated strategy could be an alternative to the use of sulphur dioxide for postharvest treatment of table grapes. A chitosan-based formulation ARMOUR-Zen® (Botry-Zen Ltd, New Zealand) has recently been introduced onto the New Zealand market for the control of botrytis bunch rot in wine grapes. However, the product label suggests that the active ingredient is primarily antifungal rather than an activator of host resistance.

Milsana® reduced both powdery mildew and botrytis bunch rot incidence, when applied to grape berries every 7–10 days (Schmitt *et al.*, 2002). Efficacy was equal to or better than sulphur and the copper-containing agent FW 450 (Dow Agro, USA). Four applications of Milsana®, applied between pre-bunch closure and harvest, reduced incidence of *B. cinerea* in grape clusters by 50% (Schilder *et al.*, 2002). Integrated use of Milsana® with Myco-Sin (diatomaceous earth, sulphuric basalt, silicic acid, and *Equisetum* extract) and a bacterial antagonist (*Bacillus brevis*), reduced incidence of *B. cinerea* on grape berries to 29.8%, compared with 89.7% on control plots (Schmitt *et al.*, 2002). The treatments were applied at 10-day intervals throughout the season. More recently, four field applications of harpin (Eden Bioscience Corp., USA), between early shoot growth and fruit set, reduced the incidence of Pierce's disease (*Xylella fastidiosa*) in 'Flame Seedless' grapes in California (Tubajika *et al.*, 2007). When averaged across two years, disease incidence was reduced by 26%, 63%, and 74% in vines treated with 160, 320, and 480 g of harpin ha^{-1}, respectively, compared with the untreated controls.

4.2.1.9 Onion

Several studies have shown that ASM has efficacy against bacterial and viral diseases in onions. Field trials by Gent *et al.* (2004) showed that four applications of ASM resulted in a 34% reduction in the incidence of Iris yellow spot virus (IYSV) and also increased the number of jumbo grade bulbs. Recent reports have suggested that the product will soon be registered in Colorado for use in onions to manage IYSV (Lang *et al.*, 2007). ASM has also demonstrated potential to replace copper-based sprays that are routinely used to manage xanthomonas leaf blight (*X. axonopodis* pv. *allii*) in onions (Gent & Schwartz, 2005). Four weekly applications of ASM controlled xanthomonas leaf blight as effectively as a copper oxychloride-mancozeb programme consisting of between nine and twelve applications (Gent & Schwartz, 2005). In the absence of disease, ten weekly applications of ASM reduced total bulb yield by up to 27%. However, this result should be treated with caution as no yield reduction was observed when four applications of ASM were applied at the recommended label rate. More recently, Lang *et al.* (2007) evaluated mixtures of bacteriophages and ASM to manage xanthomonas leaf blight under glasshouse and field conditions in Colorado. Two applications of ASM integrated with bi-weekly bacteriophage (AgriPhage®, OmniLytics, USA) applications reduced disease severity by up to 50% and were as effective as a multiple-application copper-based programme. While this result is encouraging, this approach would be more costly than the conventional programme and studies are required to investigate strategies that make ASM/bacteriophage combinations more economically attractive.

4.2.1.10 Potato

Foliar application of ASM, at 60 days of crop growth, reduced the severity of natural foliar infections, mainly early blight (*Alternaria solani*), in 'Sebago' potatoes (Bokshi *et al.*, 2003). However, the treatment did not induce resistance in tubers to postharvest inoculation with the dry rot fungus (*Fusarium semitectum*). In glasshouse studies on 'Coliban', ASM treatment significantly reduced the severity of early blight and powdery mildew (*Erysiphe cichoracearum*) on foliage and dry rot in tubers. Greater resistance to dry rot was observed when foliage was treated at 30 days of crop growth than at 60 days. In glasshouse studies, ASM treatment induced resistance to blackleg (*Pectobacterium carotovorum*) in 'Asterix' and 'Baronese' potatoes, but not 'Monalisa' (Benelli *et al.*, 2004). More recently, Collins *et al.* (2006) investigated whether activation of systemic resistance in potatoes would impact on rhizosphere populations in the field. Repeated foliar applications of ASM and harpin did not affect rhizosphere microbial populations, but did affect nematode densities. Non plant-parasitic nematodes increased in the harpin treatments and decreased with ASM treatment compared with the control. Both treatments resulted in a reduction of lesion nematodes (*Pratylenchus* spp.) and ASM reduced root knot nematodes (*Meloidogyne chitwoodi*) by harvest. The use of ASM plus harpin reduced the nematode infection index in comparison to the control. It is important to understand the ecological impact of induced resistance and further studies are necessary to determine potential effects of activators on soil microbial diversity.

4.2.1.11 Lettuce

Field applications of the commercial activator Oxycom™ (Redox Chemicals Inc., USA) reduced downy mildew (*Bremia lactucae*) and root rot (*Pythium* sp.) in lettuce (Kim *et al.*, 2001). In some cases, multiple Oxycom™ treatments compared favourably with the industry standard fungicide regime. Four applications of ASM significantly reduced the level of powdery mildew (*Erysiphe cichoracearum*) on nine different lettuce cultivars that differed in susceptibility to the pathogen (Matheron & Porchas, 2000a). The use of ASM-fungicide mixtures also proved to be highly effective thus demonstrating the potential to use ASM as part of a fungicide resistance management programme (Matheron & Porchas, 2000b). In field testing during 2001, ASM reduced bacterial spot incidence in lettuce, caused by *Xanthomonas campestris* pv. *vitians*, by between 30% and 50% and this was equivalent to that obtained using a mixture of copper sulphate (Cuprofix®) and manganese ethylene bisdithiocarbamate (Maneb) (Bull & Koike, 2005).

4.2.1.12 Cereals

ASM was originally registered as a plant tonic (Bion®) to control powdery mildew (*Blumeria graminis*) on wheat (Gorlach *et al.*, 1996). Unfortunately, the field performance of ASM in winter wheat was disappointing and the product exhibited variable field efficacy when compared with conventional fungicides. The integrated use of ASM with fungicides enabled a reduction in the frequency of fungicide applications in cereal trials in Denmark (Jorgensen *et al.*, 1997). Tank-mixing ASM with azoxystrobin provided better control of powdery mildew (*B. graminis* f. sp. *tritici*) and leaf blotch (*Septoria tritici*) on wheat than that achieved by either component alone (Stadnik & Buchenauer, 1999). However, the addition of the plant activator offered no benefit in relation to grain yield compared with fungicide only.

Iodus 40® (a.i. laminarin, Goëmar, France) is registered for use on wheat to control powdery mildew. Spray application with Iodus 40® (laminarin 1 $g l^{-1}$), 48 hours before inoculation reduced mildew infection by 55% on wheat plantlets (Renard-Merlier *et al*, 2007). The ethanolic extract of *Reynoutria sachalinensis*, Milsana®, has been shown to suppress powdery mildew on wheat by a combination of induced resistance and direct antifungal activity (Randoux *et al.*, 2006). A single spray to run-off, 48 hours before inoculation, reduced mildew infection on young glasshouse seedlings by 97%. A water-soluble extract from *R. sachalinensis* has also shown field efficacy against powdery mildew in wheat (Vechet *et al.*, 2005).

Sonnemann *et al.* (2005) reported that mycorrhizal fungi and pathogenic nematodes significantly modulated the effect of ASM-induced resistance against powdery mildew (*Blumeria graminis* f. sp. *hordei*) in glasshouse grown barley. ASM-treated plants showed increased susceptibility to infection at higher levels of root infection. Although the results can not be extrapolated to field conditions, they do highlight the need to consider the impact of soil biota on the efficacy of plant activators.

4.2.1.13 Cotton

Mondal *et al.* (2005) reported that ASM seed treatment could reduce the severity of black root rot disease (*Thielaviopsis basicola*) in cotton seedlings raised in soils that were naturally infested with the pathogen. Soaking cotton seed in ASM for 3 hours before sowing resulted in a 33% reduction in black root rot severity on tap roots. Disease severity was reduced by 20% and 24%, respectively, when ASM was applied as a seed dressing or as an in-furrow spray during sowing. Foliar application of ASM was, however, not effective and this may be because infection was established before treatment application. Nevertheless, the authors suggested that ASM seed treatment could provide an important component of an integrated management approach. Syngenta Crop Protection (Australia) have recently registered Bion® (a.i. ASM) as a seed treatment in Australia for the management of fusarium wilt (*F. oxysporum* f. sp. *vasinfectum*) and black root rot in cotton (http://www.syngenta.com.au/start.aspx). In trials, Bion® has reportedly decreased both the incidence of fusarium wilt by up to 42% and severity of black root rot by up to 50%.

4.2.2 Ornamentals

Cut flowers are a highly perishable commodity crop that relies on low temperature and high humidity to retard flower senescence during transportation to market. Unfortunately, these conditions may encourage the development of infections by fungal pathogens like *B. cinerea,* the most problematical postharvest pathogen to the cut-flower industry (Ngugi & Scherm, 2006). Infection commonly occurs during flower development but then remains symptomless until after harvest. Geraldton waxflower, an important export cut-flower crop in Australia, is susceptible to postharvest floral abscission caused by *B. cinerea* infection. Postharvest MeJA vapour treatment (1–100 μl MeJA/L) of 'Purple Pride' and 'Mullering Brook' waxflower sprigs reduce the severity of natural *Botrytis* infections on flowers but did not control infections on 'artificially' inoculated flowers (Eyre *et al.,* 2006). Arguably, of greater concern was the observation that MeJA treatment induced increased floral organ abscission in 'Purple Pride' and 'Mullering Brook' after 4–8 days. In field experiments, application of MeJA (500–750 μM), three days before harvest, slightly reduced postharvest *Botrytis* severity on 'My Sweet Sixteen' but not on 'Mullering Brook' (Dinh *et al.,* 2007). In the same study, applications of ASM (50–150 $mg\,l^{-1}$) and silicon (0.5–1.5 $g\,l^{-1}$) had no effect on disease severity or floral abscission. More recently, Dinh *et al.* (2008) investigated multiple MeJA applications as well as MeJA/ASM and MeJA/silicon combinations against *Botrytis* in Geraldton waxflower. The only treatment that consistently reduced *Botrytis* incidence and severity was application of MeJA (1 mM), 3 days before harvest, followed by a second MeJA (0.5–1 mM) application within 2–6 hours of harvest. However, despite the reduction in disease incidence, the reduction of pathogen-induced floral abscission was considered too small to be of any practical value.

Cut freesias are frequently rejected because of visible lesions on flower petals (petal specking) caused by *B. cinerea* infections. MeJA and ASM have been evaluated as preharvest and as postharvest treatments to protect freesias against petal specking (Darras *et al.*, 2005, 2006, 2007). Preharvest sprays of MeJA (0.2–0.6 mM) and ASM (1.43–5.71 mM), applied weekly from 28 days before harvest, afforded protection against postharvest inoculation with *B. cinerea* in 'Cote d'Azur' and 'Dukaat' flowers (Darras *et al.,* 2006). Efficacy of each compound was variable across different concentrations and incubation temperatures. For example, in the first year of the study, ASM (2.86 mM) was most effective on 'Cote d'Azur' flowers incubated at 5°C and reduced petal specking by 45%, whereas in the second year, ASM (1.43 mM) was most effective on 'Dukaat' flowers (51% reduction) at 20°C. In general, MeJA was more effective and more consistent than ASM and, when applied at 0.6 mM, reduced infection by up to 54% on 'Cote d'Azur' and 63% on 'Dukaat' flowers at 20°C.

When applied as postharvest treatments, ASM and MeJA have shown potential to offer some protection against natural and artificial infections on 'Cote d'Azur' flowers (Darras *et al.,* 2007). ASM spray application significantly reduced natural infections at 5°C and 12°C, but not at 20°C, but was not effective against artificial inoculation at any temperature. In contrast, MeJA was effective against both natural and artificial infections when applied as a spray or as a pulse treatment, and tended to be more effective at the higher incubation temperatures. This latter observation concurs with a previous findings that MeJA vapour (0.1 μl MeJA/L) treatment reduced petal specking on 'Cote d'Azur' flowers when incubated at 20°C and at 12°C and but not at 5°C (Darras *et al.,* 2005). Although ASM and MeJA show some promise for controlling petal specking in freesias, it is unlikely that efficacy is sufficient to warrant their practical implementation at present. The potentially synergistic effects of pre- and postharvest applications of ASM and MeJA merits investigation.

Grey mould on roses, caused by *B. cinerea,* can significantly reduce the ornamental value of the cut flowers. Postharvest treatment of six rose cultivars in a standard pulsing solution supplemented with 0.2 mM MeJA induced systemic protection against *B. cinerea* infection (Meir *et al.,* 1998). MeJA neither caused phytotoxicity on leaves and petals, nor impaired flower quality and longevity. More recently, this approach has been optimised by combining the pulse treatment (0.35 mM MeJA) with spray application of 0.5 mM MeJA (Meir *et al.,* 2005). This method suppressed grey mould development following both artificial and natural infections in eleven rose cultivars.

Fusarium wilt of cyclamen, caused by *Fusarium oxysporum* f. sp. *cyclaminis,* can be a devastating disease and economically limiting to the production of quality cyclamen. Treatment of 'Scarlet Red' seedlings with two applications of ASM (10–50 $mg\,l^{-1}$), at 7 days then again 1 day before inoculation, caused a dose-dependent suppression of fusarium wilt (Elmer, 2006a). ASM treatment delayed the onset of symptoms by up to 3 weeks, with some plants remaining symptomless for the duration of the study. In the absence of the pathogen, ASM caused a reduction in biomass, but not flower number or quality. ASM has also been field evaluated against fusarium corm rot (*Fusarium oxysporum* f. sp. *gladioli*) in gladiolus (Elmer, 2006b). Corms were soaked in ASM (50 $mg\,l^{-1}$) for 30 min before being sown in a field site with a recent history of fusarium corm rot. ASM reduced corm rot by 12%, compared with untreated controls, but this was not significant and was not as effective as chemical fungicides. Interestingly, despite relatively poor disease suppression, ASM-treated corms produced 48% more marketable flower spikes than untreated corms. Furthermore, there were interactions between ASM and fungicides. For example, ASM and azoxystrobin alone were ineffective in reducing symptoms but, when combined, they provided season-long suppression of corm rot. Twice-weekly foliar applications of ASM demonstrated efficacy against late blight on glasshouse petunias that was as effective as fungicide controls (Becktell *et al.,* 2005).

Integrated pest management is a cornerstone of floriculture (Daughtrey & Benson, 2005) and it is likely that activators will have a role to play in disease management by the ornamentals industry in the future. However, the problem of variable efficacy must be overcome before the industry will consider their practical implementation. For a more comprehensive review on the use of activators to manage postharvest disease in cut flowers, see Dinh & Joyce (2007).

4.2.3 Forestry

Induced resistance may have potential to contribute to disease management in forest nurseries (Reglinski & Dick, 2005). Foliar application of SA and 5-chlorosalicylic acid protected *Pinus radiata* seedlings against wound infection by *Sphaeropsis sapinea* for up to 32 days (Reglinski *et al.,* 1998). Interestingly, SA applications increased the biomass of *Pinus patula* seedlings, a trait that may aid establishment of young seedlings in plantation sites (San-Miguel *et al.*, 2003). *Banksia attenuata* seedlings expressed elevated resistance to stem inoculation with *Phytophthora cinnamomi* one week after treatment with 0.5 mM benzoic acid, applied either as a soil drench or foliar spray (Williams *et al.,* 2003). *Phytophthora cinnamomi* is also commonly shown to cause root rot in conifers. Phosphorous acid derivatives, also referred to as phosphonate or phosphite, have been shown to provide effective control of *Phytophthora*-related diseases in several plant species (Hardy *et al.,* 2001). These compounds exhibit direct antifungal activity and can activate host resistance against infection (Guest & Grant, 1991). Application of potassium phosphonate (Foli-R-Fos 200™, UIM Agrochemicals) caused a reduction in *P. cinnamomi* root rot in glasshouse-grown *P. radiata, Banksia integrifolia* and *Isopon cuneatus* (Ali *et al.*, 2000). Suppression of root rot was further enhanced when potassium phosphonate was sprayed in combination with ASM. All untreated seedlings died within 14 weeks of inoculation with *P. cinnamomi*, whereas all treated plants remained alive. Phosphite has also been shown to reduced infections caused by *P. cinnamomi* in mature *Banksia* sp. and *Eucalyptus marginata* trees when applied by trunk injection (Shearer & Fairman, 2007).

MeJA readily volatilises and is known to function as an airborne chemical signal among plants. Young Norway spruce seedlings were more resistant to infection by *P. ultimum* after exposure to gaseous MeJA (Kozlowski *et al.,* 1999). Mortality was reduced from 80% to 40% in seedlings that were exposed to 26 ppb MeJA for 3 days. Application to the bark of 30-year-old Norway spruce enhanced resin flow and induced resistance to inoculation with the blue stain fungus *Ceratocystis polonica* (Franceschi *et al.*, 2002). Induced resistance was expressed as a 50% reduction in lesion length in treated trees. More recently, radiata pine seedlings exhibited elevated resistance to wound infection by *S. sapinea* following one single foliar application of 4.5 mM MeJA (Gould *et al.*, 2008). The induced resistance response was accompanied by a concomitant reduction in seedling growth rate which the authors attributed to reallocation of resources towards active defence mechanisms.

Plant-growth-promoting rhizobacteria (PGPR), particularly *Bacillus* spp., and *Pseudomonas* spp., have been shown to enhance the rate and amount of seedling emergence, and also to stimulate seedling growth in conifers (Chanway, 1997; Enebak *et al.*, 1998). PGPR also have potential as biocontrol agents and have been shown to suppress plant disease by competing with pathogen populations for nutrients and space, and by inducing plant resistance to pathogen infection (Kloepper, 1993). Two bacterial strains (*Burkholderia cepacia* RAL3 & *Pseudomonas fluorescens* 64–3) have been identified that reduced disease of white spruce seedlings caused by *Fusarium* spp., and *Pythium* spp., in a commercial nursery and increased the survival of out-planted bare root white spruce in a re-forestation site (Reddy *et al.*, 1997). The treatments were applied by soaking either seeds or the seedling roots in bacterial suspension before planting.

Enebak & Carey (2004) reported on the effects of seed treatment with *Bacillus pumilus* strains on seedling growth and on fusiform rust infection (*Cronartium quercuum* f. sp. *fusiforme*) in bare root nurseries in Alabama and in Georgia, USA. In Georgia, seed treatment with *B. pumilus* strains T4 and SE34 promoted seedling growth whilst treatment with T4 significantly reduced infection by the rust fungus and resulted in the development of fewer galls. However, in Alabama there was no treatment effect. Such site specificity mirrors the findings reported for effects of PGPR on growth promotion and highlights the importance of understanding the impact of biotic and abiotic soil factors on PGPR-induced resistance.

4.3 Costs associated with induced resistance

Plants respond almost immediately to an inducing agent but, nevertheless, require a lag period of a few days to activate its full complement of inducible defences. Therefore, unlike constitutive resistance, where defences are always present, with induced resistance, the plant remains vulnerable to attack for a short period. The question therefore remains, why do plants have inducible defences? The continued and widespread existence of induced resistance suggests a selective advantage over constitutive resistance or that inducible defences can be more selectively targeted at an attacker than previously thought. One possible explanation for this selective advantage lies with fitness costs, where resistant plants would have decreased reproductive success (e.g. seed production) than non-resistant plants under conditions where there was no pathogen pressure. In other words, in addition to their positive effects, expression of defence genes also has negative effects (Heil & Baldwin, 2002). In fact, the various theories of plant defence all assume the existence of such costs. For example, the optimal defence theory focuses on the spatial and temporal distribution of the limited resources plants can invest in resistance (McKey, 1974, 1979; Rhoades, 1979), while the growth-differentiation balance hypothesis looks at the actual costs of resistance and highlights the compromise between growth and differentiation in growing plants attempting to defend themselves against attackers (Herms & Mattson, 1992). Costs include allocation costs arising from diversion of metabolites and energy from growth and other processes towards defence, as well as the negative effects (trade-offs) of the resistance on symbiotic associations and effects on resistance to insects (Gomez *et al.*, 2007; Walters & Heil, 2007; see also Section 4.4).

4.3.1 Allocation costs

Although there is much evidence that induced resistance to insects incurs costs (e.g. Zavala *et al.*, 2004), the situation with respect to pathogens is less clear (Walters & Heil, 2007). In some early work, Smedegaard-Petersen & Stolen (1981) demonstrated a

7% reduction in grain yield in barley plants inoculated with powdery mildew, compared with uninoculated control plants. These workers suggested that the reduction in grain yield was the result of the greatly increased rates of dark respiration, required to provide rapid resistance to mildew infection (Smedegaard-Petersen & Stolen, 1981). This contrasts with later work reporting yield increases associated with induced resistance to powdery mildew infection in barley (Steiner *et al.*, 1988; Oerke *et al.*, 1989) and the lack of any effect on yield in barley induced by application of yeast-derived elicitors (Reglinski *et al.*, 1994). However, because these studies were conducted in the presence of pathogen challenge, they cannot be used to quantify the costs associated with induced resistance. Subsequent work by Heil *et al.* (2000) showed that ASM, applied to wheat in the absence of pathogen pressure reduced plant growth and yield and provided a clear indication that use of ASM incurred allocation costs. In fact, similar results have been reported in other crop plants, including sunflower (Prats *et al.*, 2002), tobacco (Csinos *et al.*, 2001), cauliflower (Ziadi *et al.*, 2001), strawberry (Hukkanen *et al.*, 2007), melon (Buzi *et al.*, 2004) and cowpea (Latunde-Dada & Lucas, 2001). Further evidence that SAR is costly comes from studies on *Arabidopsis* mutants that overexpress SAR; such transformants usually have stunted growth and reduced seed yields (Bowling *et al.*, 1994; Greenberg *et al.*, 2000; Jirage *et al.*, 2001; Mauch *et al.*, 2001).

SAR might indeed be costly, but whether such costs are incurred will depend on environmental factors, both abiotic and biotic, and very likely will also be genotype-dependent, although this has received little attention to date. In wheat and *Arabidopsis,* whether or not costs were incurred following induction of resistance depended on nitrogen supply (Heil *et al.*, 2000; Dietrich *et al.*, 2005), while successful induction of resistance by ASM in barley depended on whether the plants were mycorrhizal or not (Sonnemann *et al.*, 2005). The importance of the rhizosphere community is becoming increasingly evident. For example, the root colonising basidiomycete (*Piriformospora indica*) has been shown to induce resistance in barley to powdery mildew and to root rot caused by *Fusarium culmorum* (Waller *et al.*, 2005). In addition, there was a significant increase in grain yield thus challenging the notion that an enhanced resistance status imposes a cost. Furthermore, a recent report by Nair *et al.* (2007) suggests that the presence of certain rhizobacteria may be able to compensate for some of the costs associated with induced resistance. In their study, growth of the leafy vegetable, amaranthus, was retarded when treated with ASM but not when plants were treated with a combination of the ASM and the rhizobacterial strain *Pseudomonas fluorescens* PN026R. It was proposed that growth retardation effect of ASM was compensated for by the growth promoting properties of the rhizobacteria.

As for the mechanisms by which SAR-associated costs are incurred, a number of workers have reported negative effects of resistance induction on plant carbon and nitrogen metabolism (e.g. Logemann *et al.*, 1995; Ramanujam *et al.*, 1998). Such reports have been confirmed by gene array studies, which have generally found that genes involved in photosynthesis and growth are downregulated during the expression of induced resistance (Scheideler *et al.*, 2002; Heidel *et al.*, 2004). It seems likely therefore that a switch from housekeeping to pathogen defence metabolism may be a prerequisite for the full commitment of a plant to transcriptional activation of resistance pathways (Logemann *et al.*, 1995).

4.4 Trade-offs associated with induced resistance

4.4.1 Trade-offs between pathogens with different lifestyles

In plant–pathogen interactions, distinct defence responses are activated, depending on the lifestyle of the attacking pathogen. As indicated earlier in this chapter, SA and JA play important signalling roles in these responses. It appears that SA induces defence against biotrophic pathogens, while JA mediates defences against necrotrophic pathogens (Glazebrook, 2005). Cross-talk between the two signalling pathways might help to fine-tune defence responses against a particular pathogen according to its mode of infection (Beckers & Spoel, 2006). Little is known however, about cross-talk between defence signalling in response to attempted infection by more than one pathogen. In some interesting recent work, Spoel *et al.* (2007) found that infection with the biotrophic pathogen *Pseudomonas syringae,* which induces SA-mediated defence, rendered *Arabidopsis thaliana* more susceptible to the necrotrophic pathogen *Alternaria brassicicola.* They found that this trade-off was restricted to plant tissue adjacent to the site of initial infection, since *A. brassicicola* infection in systemic tissue was not affected. Further, the trade-off only occurred with a virulent strain of *P. syringae*, and avirulent strains that induced programmed cell death (PCD), did not cause suppression of JA-dependent defence. The authors suggested that this might be advantageous to the plant by preventing growth of necrotrophic pathogens in tissue undergoing PCD.

4.4.2 Trade-offs with resistance to insects

Insect herbivory results in the induction of a series of events, including the generation and release of specific signals, the subsequent perception and transduction of those signals and, finally, activation of wound-related defence mechanisms (Figure 4.2) (Leon *et al.,* 2001). Thereafter, secondary signals are generated, leading to the further activation of local and systemic defences. These secondary signals include oxylipins (oxygenated fatty acids), including JA and its volatile methyl ester (MeJA), both of which, as indicated above, play an important role in regulating induced resistance to insect attack (Figure 4.2) (Bostock, 2005; Walters *et al.,* 2006). However, the situation is considerably more complex than the simple split between JA regulation of defence against insects and SA regulation of pathogen defence. For example, it is known that the nature of the activated signalling pathway depends on the particular plant/insect combination (Bostock, 2005), since both SA- and JA-responsive gene expression can be elicited by aphids and whiteflies, while methyl salicylate is generated by aphid attack in maize (Bernascone *et al.,* 1998; Walling, 2000). There are, nevertheless, several examples of negative cross-talk between the SA and JA signalling systems, especially for SA-mediated suppression of JA-inducible gene expression (Figure 4.2) (Van Wees *et al.*, 1999; Glazebrook *et al.*, 2003). Thus, activation of SA-dependent SAR has been shown to suppress JA signalling, thereby compromising the plants' ability to induce defences to insect attack (Stout *et al.*, 1999; Thaler *et al.*, 1999, 2002). For example, tobacco plants expressing TMV-induced SAR were more susceptible to grazing by the tobacco hornworm *Manduca sexta* than non-induced plants (Preston *et al.*, 1999), while application of the chemical activator acibenzolar-*S*-methyl (ASM) to field-grown tomato plants reduced resistance to the beet armyworm *Spodoptera exigua* (Thaler *et al.*, 1999). Interestingly, JA has also been shown

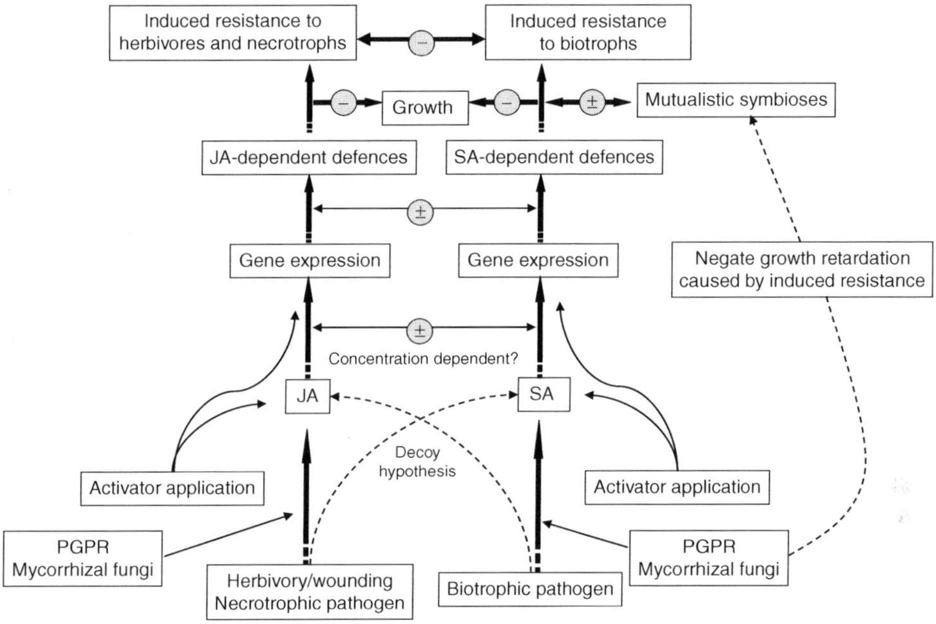

Figure 4.2 Plant resistance against pathogens and insects is regulated by interconnecting pathways in which salicylic acid (SA) and jasmonic acid (JA) play key signalling roles. The SA pathway is primarily activated in response to biotrophic pathogens whilst the JA-mediated responses are triggered by necrotrophic infection and insect attack. However, there are no absolutes and recent evidence indicates that plant defence is more complex than this simple dichotomy suggests. Cross-talk between the SA and JA pathways results in the activation of distinct sets of defence genes with corresponding trade-offs between pathogen and insect resistance. This cross-talk is believed to minimise expression of costly defences that divert resources from plant vegetative growth. Activation of defences in induced resistance to pathogens may also lead to trade-offs with mutualistic symbioses, for example nodulation or mycorrhizal establishment. Note that the use of exogenous elicitors by-passes recognition, although some have been reported to increase levels of JA or SA in plants. If exogenous elicitors activate induced resistance to both pathogens and insects, competition for limited resources can lead to phenotypically visible trade-offs.

to suppress SA-induced responses, although reports of this phenomenon are few (Nike *et al.*, 1998; Glazebrook *et al.*, 2003; Bostock, 2005).

So far, we have dealt with negative effects of interactions between pathogen and insect resistance. It is important to note therefore, that some workers could find no effect of induced resistance to pathogens on resistance to insects (Ajlan & Potter, 1992; Inbar *et al.*, 1998), while there are even reports of positive effects. For example, Stout *et al.* (1999) found that inoculation of tomato leaves with *Pseudomonas syringae* pv. *tomato* induced resistance against both *P. syringae* pv. *tomato* and the corn earworm *Helicoverpa zea*, while grazing of *Rumex obtusifolius* by the beetle *Gastrophysa viridula* reduced infection by a number of fungal pathogens (Hatcher & Paul, 2000).

These few examples of crosstalk between the JA and SA signalling pathways highlight the complex nature of signalling for disease and pest resistance. An interesting insight into how plants integrate insect- and pathogen-induced signals into specific defence responses was provided by De Vos *et al.* (2005). Using *Arabidopsis*, they tracked the dynamics of SA, JA and ET signalling following attack by pathogens and insect pests. When global gene expression profiles were compared, the workers found considerable overlap in the changes induced by pathogens and insects. Thus, all of the different pathogen and insect attackers stimulated JA biosynthesis, although most of the changes in JA-responsive gene expression were attacker specific (De Vos *et al.*, 2005). The authors suggest that although SA, JA and ET play a primary role in orchestrating plant defence, the final defence response is shaped by other regulatory mechanisms, for example cross-talk between different pathways. A recent study indicates that herbivorous nymphs of the silverleaf whitefly (*Bemisia tabaci*) may activate the SA signalling pathway as a decoy strategy to sabotage JA-dependent defences and so enhance insect performance (Zarate *et al.*, 2007).

4.4.3 Trade-offs with mutualistic symbioses

Since induced resistance is a broad-spectrum resistance against microorganisms, it is likely to exert an impact on plant interactions with mutualistic symbionts, including mycorrhizal fungi and nitrogen fixing *Rhizobia* and *Bradyrhizobia* bacteria (Figure 4.2). Perhaps surprisingly, this area of research has received little attention to date. However, some studies of the legume–*Rhizobium* symbiosis have shown that application of SA to the rooting substrate exerted a negative effect on nodule formation and/or functioning (Martínez-Abarca *et al.,* 1998; Ramanujam *et al.,* 1998; Lian *et al.,* 2000), while treatment of *Vicia faba* plants with ASM led to a reduction in the number and size of nodules compared with untreated controls (Heil, 2001). Although some studies have reported negative effects on colonisation of tobacco roots by the arbuscular mycorrhizal fungus *Glomus mosseae* in plants constitutively expressing β-1,3-glucanase (Vierheilig *et al.,* 1994; Glandorf *et al.,* 1997), other workers found no effect of ASM treatment on mycorrhizal infection of barley roots (Sonnemann *et al.,* 2002).

Interestingly, there are a number of reports indicating an effect of mycorrhizal infection on plant defence against pathogens. For example, colonisation of tomato roots by *G. mosseae* induced cell defence responses and localised and systemic resistance to *Phytophthora parasitica*, while infection of barley roots with *G. mosseae* induced systemic resistance to the take-all fungus *Gaeumannomyces graminis* f. sp. *tritici* (Cordier *et al.,* 1998; Khaosaad *et al.,* 2007). Mycorrhizal infection and colonisation has also been shown to modify the effectiveness of induced resistance. Thus, Sonnemann *et al.* (2005) found that at low and medium levels of colonisation of barley roots by *Glomus etunicatum*, ASM had either no effect or decreased foliar infection by powdery mildew, while high levels of mycorrhizal colonisation increased mildew infection.

4.5 Future prospects

Induced resistance offers the prospect of broad-spectrum disease control and yet, it remains on the fringes of mainstream crop protection. As indicated above, this is due in part to the variability in efficacy whenever activators are used. Such variability should not be surprising, since induced resistance is a host response and as such, will be influenced by genotype and environment. At present, our understanding of this area of induced resistance is woefully inadequate. For field crops, it is also likely that plants will already be induced, since they are continually interacting with microorganisms and insects. To what extent plants in the field are induced, and to what extent they can be further induced by application of elicitors, is not known. Nevertheless, as indicated above, there are many clear examples of induced resistance providing good levels of disease control in a range of different cropping systems, particularly in situations where disease levels are low to moderate. However, there must be a high probability that induced resistance will be of economic benefit to growers before it will find widespread acceptance. Inducing agents do provide a means to reduce fungicide inputs when used in rotation with fungicides or as tank mixes with reduced rates of fungicide and activators such as ASM are being integrated within existing disease management programmes. As our understanding of induced resistance increases, so will the opportunity to exploit the phenomenon for disease management within cropping systems.

4.6 Acknowledgements

Scottish Agricultural College receives financial support from the Scottish Government Rural and Environment Research and Analysis Directorate (RERAD).

4.7 References

Agostini JP, Bushong, PM, Timmer, LW, 2003. Greenhouse evaluation of products that induce host resistance for control of scab, melanose and Alternaria brown spot of citrus. *Plant Disease* **87,** 69–74.

Ait Barka E, Eullaffroy P, Clement C, Vernet G, 2004. Chitosan improves development, and protects *Vitis vinifera* L. against *Botrytis cinerea*. *Plant Cell Reports* **22,** 608–614.

Ajlan AM, Potter DA, 1992. Lack of effect of tobacco mosaic virus-induced systemic acquired resistance on arthropod herbivores in tobacco. *Phytopathology* **82,** 647–6451.

Ali Z, Smith I, Guest DI, 2000. Combinations of potassium phosphonate and Bion (acibenzolar-*S*-methyl) reduce root infection and dieback of *Pinus radiata*, *Banksia intergrifolia* and *Isopogon cuneatus* caused by *Phytophthora cinnamomi*. *Australasian Plant Pathology* **29,** 59–63.

Anith KN, Momol MT, Kloepper JW, Marois JJ, Olson SM, Jones JB, 2004. Efficacy of plant growth-promoting rhizobacteria, acibenzolar-*S*-methyl, and soil amendment for integrated management of bacterial wilt on tomato. *Plant Disease* **88**(6), 669–673.

Balogh B, Jones JB, Momol MT, Olson SM, Obradovic A, King P, Jackson LE, 2003. Improved efficacy of newly formulated bacteriophages for management of bacterial spot on tomato. *Plant Disease* **87,** 949–954.

Beckers GJ, Conrath U, 2007. Priming for stress resistance: From the lab to the field. *Current Opinion in Plant Biology* **10,** 425–431.

Beckers GJ, Spoel SH, 2006. Fine-tuning plant defence signalling: salicylate versus jasmonate. *Plant Biology* **8,** 1–10.

Becktell MC, Daughtrey ML, Fry WE, 2005. Epidemiology and management of petunia and tomato late blight in the greenhouse. *Plant Disease* **89**, 1000–1008.

Benelli AIH, Denardin ND, Forcelini CA, 2004. Action of acibenzolar-*S*-methyl applied on potato tubers and plants to prevent blackleg, caused by *Pectobacterium carotovorm* subsp. *atrosepticum atipica*. *Fitopatologia Brasileira* **29**, 263–267.

Benhamou N, Gagné S, Le Quéré D, Dehbi L, 2000. Bacterial-mediated induced resistance in cucumber: Beneficial effect of the endophytic bacterium *Serratia plymuthica* on the protection against infection by *Pythium ultimum*. *Phytopathology* **90,** 45–56.

Bernascone ML, Turlings TCJ, Ambrosetti L, Bassetti P, Dorn S, 1998. Herbivore-induced emissions of maize volatiles repel the corn leaf aphid. *Entomologia Experimentalis et Applicata* **87,** 133–142.

Bi Y, Tian SP, Zhao J, Ge YH, 2005. Harpin induces local and systemic resistance against *Trichothecium roseum* in harvested Hami melons. *Postharvest Biology and Technology* **38,** 183–187.

Bini F, Ragaini A, Bazzi C, 2008. Resistance responses induced by the plant growth retardant prohexadione-Ca in apple against scab infections. *Annals of Applied Biology* **152**, 19–27.

Bokshi AI, Morris SC, Deverall BJ, 2003. Effects of benzothiadiazole and acetylsalicylic acid on β-1,3-glucanase activity and disease resistance in potato. *Plant Pathology* **52**, 22–27.

Bokshi AI, Morris SC, McConchie RM, Deverall BJ, 2006. Pre-harvest application of 2, 6-dichloroisonicotinic acid, beta-aminobutyric acid or benzothiadiazole to control post-harvest storage diseases of melons by inducing systemic acquired resistance (SAR). *Journal of Horticultural Science and Biotechnology* **81**, 700–706.

Bokshi AI, Morris SC, McConchie R, 2007. Environmentally-safe control of postharvest diseases of melons (*Cucumis melo*) by integrating heat treatment, safe chemicals, and systemic acquired resistance. *New Zealand Journal of Crop and Horticultural Science* **35**, 179–186.

Bostock RM, 2005. Signal crosstalk and induced resistance: Straddling the line between cost and benefit. *Annual Review of Phytopathology* **43,** 545–580.

Bourbos VA, Barbopoulou EA, 2006. Effect of harpin Ea on the fruit production and control of *Phytophthora infestans* in greenhouse tomato. *Acta Horticulturae* **727,** 557–560.

Bowling SA, Guo A, Cao H, Gordon AS, Klessig DF, Dong X, 1994. A mutation in *Arabidopsis* that leads to constitutive expression of systemic acquired resistance. *Plant Cell* **6,** 1845–1857.

Brisset M-N, Cesbron S, Thomson SV, Paulin J-P, 2000. Acibenzolar-*S*-methyl induces the accumulation of defense-related enzymes in apple and protects from fire blight. *European Journal of Plant Pathology* **106,** 529–536.

Buban T, Foldes L, Fekete Z, Rademacher W, 2004. Effectiveness of the resistance inducer prohexadione-Ca against fireblight in shoots of apple trees inoculated with *Erwinia amylovora*. *Bulletin OEPP* **34,** 369–376.

Bull CT, Koike ST, 2005. Evaluating the efficacy of commercial products for management of bacterial leaf spot on lettuce. Online. *Plant Health Progress* doi:10.1094/PHP-2005–1121-01-RS.

Buonaurio R, Scarponi L, Ferrara M, Sidoti P, Bertona A, 2002. Induction of systemic acquired resistance in pepper plants by acibenzolar-*S*-methyl against bacterial spot disease. *European Journal of Plant Pathology* **108,** 41–49.

Buzi A, Chilosi G, De Sillo D, Magro P, 2004. Induction of resistance in melon to *Didymella bryoniae* and *Sclerotinia sclerotiorum* by seed treatments with acibenzolar-*S*-methyl and methyl jasmonate but not with salicylic acid. *Journal of Phytopathology* **152,** 34–42.

Cao J, Jiang W, 2006. Induction of resistance in Yali pear (*Pyrus bretschneideri* Rehd.) fruit against postharvest diseases by acibenzolar-*S*-methyl sprays on trees during fruit growth. *Scientia Horticulturae* **110**, 181–186.

Cao J, Jiang W, He H, 2005. Induced resistance in Yali pear (*Pyrus bretschneideri* Rehd.) fruit against infection by *Penicillium expansum* by postharvest infiltration of acibenzolar-*S*-methyl. *Journal of Phytopathology* **153**, 640–646.

Cao J, Zeng K, Jiang W, 2006. Enhancement of postharvest disease resistance in Ya Li pear (*Pyrus bretschneideri*) fruit by salicylic acid sprays on the trees during fruit growth. *European Journal of Plant Pathology* **114**, 363–370.

Chanway CP, 1997. Inoculation of tree roots with plant growth promoting soil bacteria: An emerging technology for reforestation. *Forest Science* **43,** 99–112.

Chen H-Q, Dekkers KL, Cao L, Burns JK, Timmer LW, Chung K-R, 2006. Evaluation of growth regulator inhibitors for controlling postbloom fruit drop (PFD) of citrus induced by the fungus *Colletotrichum acutatum*. *HortScience* **41,** 1317–1321.

Collins HP, Navare DA, Riga E, Pierce FJ, 2006. Effect of foliar applied plant elicitors on microbial and nematode populations in the root zone of potato. *Communications in Soil Science and Plant Analysis* **37**, 1747–1759.

Conrath U, Beckers GJM, Flors V, Garcia-Agustin P, Jakab G, Mauch F, Newman MA, Pieterse CMJ, Poinssot B, Pozo MJ, Pugin A, Schaffrath U, Ton J, Wendehenne D, Zimmerli L, Mauch-Mani B, 2006. Priming: Getting ready for battle. *Molecular Plant-Microbe Interactions* **19,** 1062–1071.

Cordier C, Pozo MJ, Barea JM, Gianinazzi S, Gianinazzi-Pearson V, 1998. Cell defense responses associated with localized and systemic resistance to *Phytophthora parasitica* induced in tomato by an arbuscular mycorrhizal fungus. *Molecular Plant-Microbe Interactions* **11**, 1017–1028.

Csinos AS, Pappu HR, McPherson RM, Stephenson MG, 2001. Management of tomato spotted wilt virus in flue-cured tobacco with acibenzolar-*S*-methyl and imidacloprid. *Plant Disease* **85,** 292–296.

da Rocha AB, Hammerschmidt R, 2005. History and perspectives on the use of disease resistance inducers in horticultural crops. *HortTechnology* **15**, 518–529.

Dafermos NG, Kasselaki AM, Malathrakis NE, Leifert C, 2007. Integration of fertility management, cultivar selection and alternative spray treatments to optimize control of foliar diseases of greenhouse grown tomatoes. In: *Improving Sustainability in Organic and Low Input Food Production Systems. Proceedings of the 3rd International Congress of the European Integrated Project Quality Low Input Food*, pp. 181–184.

Dagostin S, Ferrari A, Pertot I, 2006. Integration of different control measures to maximise disease control of *Plasmopara viticola* in Italian organic viticulture. *Proceedings from European Joint Organic Congress,* 30–31 May, Odense, Denmark, 338–339.

Darras AI, Terry LA, Joyce DC, 2005. Methyl jasmonate vapour treatment suppresses specking caused by *Botrytis cinerea* on cut Freesia hybrida L. flowers. *Postharvest Biology and Technology* **38**, 175–182.

Darras AI, Joyce DC, Terry LA, 2006. Acibenzolar-*S*-methyl and methyl jasmonate treatments of glasshouse-grown freesias suppress post-harvest petal specking caused by *Botrytis cinerea. Journal of Horticultural Science and Biotechnology* **81**, 1043–1051.

Darras AI, Joyce DC, Terry LA, Pompodakis NE, Dimitriadis CI, 2007. Efficacy of postharvest treatments with acibenzolar-*S*-methyl and methyl jasmonate against *Botrytis cinerea* infecting cut *Freesia hybrida* L. flowers. *Australasian Plant Pathology* **36**, 332–340.

Daughtrey ML, Benson DM, 2005. Principles of plant health management for ornamental plants. *Annual Review of Phytopathology* **43**, 141–169.

de Capdeville G, Wilson CL, Beer SV, Aist JR, 2002. Alternative disease control agents induce resistance to blue mold in harvested 'Red Delicious' apple fruit. *Phytopathology* **92**, 900–908.

de Capdeville G, Beer SV, Watkins CB, Wilson CL, Tedeschi LO, Aist JR, 2003. Pre- and post-harvest harpin treatments of apples induce resistance to blue mold. *Plant Disease* **87,** 39–44.

De Vos M, Van Oosten VR, Van Poecke RMP, Van Pelt JA, Pozo MJ, Mueller MJ, Buchala AJ, Metraux J-P, Van Loon LC, Dicke M, Pieterse CMJ, 2005. Signal signature and transcriptome changes of *Arabidopsis* during pathogen and insect attack. *Molecular Plant-Microbe Interactions* **18,** 923–937.

Dietrich R, Ploss K, Heil M, 2005. Growth responses and fitness costs after induction of pathogen resistance depend on environmental conditions. *Plant, Cell and Environment* **28,** 211–222.

Dinh S, Joyce DC, 2007. Prospects for cut-flower postharvest disease management with host defence elicitors. *Stewart Postharvest Review* **3**, article 4.

Dinh SQ, Joyce DC, Irving DE, Wearing AH, 2007. Field applications of three different classes of known host plant defence elicitors did not suppress infection of Geraldton waxflower by *Botrytis cinerea. Australasian Plant Pathology* **36**, 142–148.

Dinh S, Joyce DC, Irving DE, Wearing AH, 2008. Effects of multiple applications of chemical elicitors on *Botrytis cinerea* infecting Geraldton waxflower. *Australasian Plant Pathology* **37**, 87–94.

Durrant WE, Dong X, 2004. Systemic acquired resistance. *Annual Review of Phytopathology* **42**, 185–209.

Duxbury M, Hotter GS, Reglinski T, Sharpe N, 2004. Effect of chitosan and 5-chlorosalicylic acid on total phenolic content of grapes and wine. *American Journal of Enology and Viticulture* **55**, 191–194.

Elmer WH, 2006a. Effects of acibenzolar-*S*-methyl on the suppression of Fusarium wilt of cyclamen. *Crop Protection* **25**, 671–676.

Elmer WH, 2006b. Efficacy of preplant treatments of gladiolus corms with combinations of acibenzolar-*S*-methyl and biological or chemical fungicides for suppression of fusarium corm rot [*Fusarium oxysporum f. sp. gladioli*]. *Canadian Journal of Plant Pathology* **28**, 609–614.

Enebak SA, Carey WA, 2004. Plant growth-promoting rhizobacteria may reduce Fusiform rust infection in nursery-grown Lobolly pine seedlings. *Southern Journal of Applied Forestry* **28,** 185–188.

Enebak SA, Wei G, Kloepper JW, 1998. Effects of plant growth-promoting rhizobacteria on loblolly and slash pine seedlings. *Forest Science* **44,** 139–144.

Eyre JX, Faragher J, Joyce DC, Franz PR, 2006. Effects of postharvest methyl jasmonate treatments against *Botrytis cinerea* on Geraldton waxflower (*Chamelaucium uncinatum*). *Australian Journal of Experimental Agriculture* **46**, 717–723.

Florida Tomato Committee. 2005. *Annual Report of Florida Tomato Committee*. Maitland, FL.

Franceschi VR, Krekling T, Christiansen E, 2002. Application of methyl jasmonate on *Picea abies* (Pinaceae) stems induces defense-related responses in phloem and xylem. *American Journal of Botany* **89,** 578–586.

Garcion C, Lamotte O, Metraux J-P, 2007. Mechanisms of defence to pathogens: biochemistry and physiology. In: Walters D, Newton A, Lyon G, eds. *Induced Resistance for Plant Disease Control: A Sustainable Approach to Crop Protection.* Oxford: Blackwell Publishing, 109–132.

Gent DH, Schwartz HF, 2005. Management of Xanthomonas leaf blight of onions with a plant activator, biological control agents, and copper bactericides. *Plant Disease* **89,** 631–639.

Gent DH, Schwartz HF, Khosla R, 2004. Managing *Iris Yellow spot virus* of onion with cultural practices, host genotype, and novel chemical treatments. (Abstract) *Phytopathology* **94,** S34.

Glandorf DCM, Bakker PAH, Van Loon LC, 1997. Influence of the production of antibacterial and antifungal proteins by transgenic plants on the saprophytic soil microflora. *Acta Botanica Neerlandica* **46,** 85–104.

Glazebrook J, 2005. Contrasting mechanisms of defence against biotrophic and necrotrophic pathogens. *Annual Review of Phytopathology* **43,** 205–227.

Glazebrook J, Chen W, Estes B, Chang H-S, Nawrath C, Metraux J-P, Zhu T, Katagiri F, 2003. Topology of the network integrating salicylate and jasmonate signal transduction derived from global expression phenotyping. *Plant Journal* **34,** 217–228.

Gomez S, Latzel V, Verhulst YM, Stuefer JF, 2007. Costs and benefits of induced resistance in a clonal plant network. *Oecologia* **153**, 921–930.

Gorbatenko IY, Onishchuk IA, Krivtsov GG, Vanyushin BF, 1996. Eliciting and growth-regulating effects of chitosan on plants. *Biology Bulletin of the Russian Academy of Sciences* **23**, 327–330.

Gorlach J, Volrath S, Knauf-Belter G, Hengy G, Beckhove U, Kogel K-H, Oostendorp M, Staub T, Ward E, Kessmann H, Ryals J, 1996. Benzothiadiazole, a novel class of inducers of systemic acquired resistance, activates gene expression and disease resistance in wheat. *Plant Cell* **8**, 629–643.

Gould N, Reglinski T, Spiers M & Taylor JT, 2008. Physiological trade-offs associated with methyl jasmonate-induced resistance in *Pinus radiata*. *Canadian Journal of Forestry*. **38,** 677–684.

Graham JH, Leite-Junior RP, 2004. Lack of control of citrus canker by induced systemic resistance compounds. *Plant Disease* **88,** 745–750.

Greenberg JT, Silverman FP, Liang H, 2000. Uncoupling salicylic acid-dependent cell death and defense-related responses from disease resistance in the *Arabidopsis* mutant acd5. *Genetics* **156**, 341–350.

Guest D, Grant B, 1991. The complex action of phosphonates as antifungal agents. *Biological Reviews* **66**, 159–187.

Hacisalihoglu G, Ji P, Longo LM, Olson S, Momol TM, 2007. Bacterial wilt induced changes in nutrient distribution and biomass and the effect of acibenzolar-*S*-methyl on bacterial wilt in tomato. *Crop Protection* **26**, 978–982.

Hammerschmidt R, 2007. Introduction: definitions and some history. In: Walters D, Newton A, Lyon G, eds. *Induced Resistance for Plant Disease Control: A Sustainable Approach to Crop Protection.* Oxford: Blackwell Publishing, 1–8.

Hardy GES, Barrett S, Shearer BL, 2001. The future of phosphate as a fungicide to control the soil-borne plant pathogen *Phytophthora cinnamomi* in natural ecosystems. *Australasian Plant Pathology* **30**, 133–139.

Hatcher PE, Paul ND, 2000. Beetle grazing reduces natural infection of *Rumex obtusifolius* by fungal pathogens. *New Phytologist* **146**, 325–333.

Heath MC, 1998. Apoptosis, programmed cell death and the hypersensitive response. *European Journal of Plant Pathology* **104,** 117–124.

Heidel AJ, Clarke JD, Antonovics J, Dong XN, 2004. Fitness costs of mutations affecting the systemic acquired resistance pathway in *Arabidopsis thaliana*. *Genetics* **168,** 2197–2206.

Heil M, 2001. Induced systemic resistance (ISR) against pathogens – a promising field for ecological research. *Perspectives in Plant Ecology, Evolution and Systematics* **4,** 65–79.

Heil M, Baldwin IT, 2002. Fitness costs of induced resistance: Emerging experimental support for a slippery concept. *Trends in Plant Science* **7,** 61–67.

Heil M, Hilpert A, Kaiser W, Linsenmair KE, 2000. Reduced growth and seed set following chemical induction of pathogen defence: Does systemic acquired resistance (SAR) incur allocation costs? *Journal of Ecology* **88,** 645–654.

Herms DA, Mattson WJ, 1992. The dilemma of plants: To grow or to defend. *Quarterly Review of Biology* **67,** 283–335.

Huang Y, Deverall BJ, Tang WH, Wang W, Wu FW, 2000. Foliar application of acibenzolar-S-methyl and protection of postharvest rock melons and Hami melons from disease. *European Journal of Plant Pathology* **106**, 651–656.

Hukkanen AT, Kokko HI, Buchala AJ, McDougall GJ, Stewart D, Karenlampi SO, Karjalainen RO, 2007. Benzothiadiazole induces the accumulation of phenolics and improves resistance to powdery mildew in strawberries. *Journal of Agricultural and Food Chemistry* **55**, 1862–1870.

Inbar M, Doostdar H, Sonoda RM, Leibee GL, Mayer RT, 1998. Elicitors of plant defensive systems reduce insect densities and disease incidence. *Journal of Chemical Ecology* **24,** 135–149.

Ishii H, Tomita Y, Horio T, Narusaka Y, Nakazawa Y, Nishimura K, Iwamoto S, 1999. Induced resistance of acibenzolar-*S*-methyl (CGA 245704) to cucumber and Japanese pear diseases. *European Journal of Plant Pathology* **105,** 77–85.

Iriti M, Rossoni M, Borgo M, Faoro F, 2004. Benzothiadiazole enhances resveratrol and anthocyanin biosynthesis in grapevine, meanwhile improving resistance to *Botrytis cinerea*. *Journal of Agricultural and Food Chemistry* **52**, 4406–4413.

Iriti M, Mapelli S, Faoro F, 2007. Chemical-induced resistance against post-harvest infection enhances tomato nutritional traits. *Food Chemistry* **105**, 1040–1046.

Ivors K, Louws F, 2007. Foliar fungicide spray guide for tomatoes in NC. www.ces.ncsu.edu/fletcher/programs/plantpath/tomato-spray-guide

Ji PS, Momol MT, Rich JR, Olson SM, Jones JB, 2007. Development of an integrated approach for managing bacterial wilt and root-knot on tomato under field conditions. *Plant Disease* **91**, 1321–1326.

Jirage D, Zhou N, Cooper B, Clarke JD, Dong X, Glazebrook J, 2001. Constitutive salicylic acid-dependent signaling in cpr1 and cpr6 mutants requires PAD4. *Plant Journal* **26,** 395–407.

Jorgensen LN, Jensen B, Smedegaard-Petersen V, 1997. Bion – a compound for disease control in cereal based on induced resistance. In: *Proceedings of 14*th *Danish Plant Protection Conference* **8,** 35–48.

Kasselaki AM, Malathrakis NE, Goumas DE, Leifert C, 2007. Effect of alternative seed treatments on seed-borne fungal diseases in tomato. In: *Improving Sustainability in Organic and Low Input Food Production Systems. Proceedings of the 3rd International Congress of the European Integrated Project Quality Low Input Food*, 289–291.

Kast WK, 2000. Salicylic acid and phosphorous acid – Possible alternatives of copper? In: *Proceedings of the 6th International Congress on Organic Viticulture*, 25–26 August. Basel, Switzerland, 177–179.

Keinath AP, Holmes GJ, Everts KL, Egel DS, Langston JDB, 2007a. Evaluation of combinations of chlorothalonil with azoxystrobin, harpin, and disease forecasting for control of downy mildew and gummy stem blight on melon. *Crop Protection* **26**, 83–88.

Keinath AP, Everts KL, Langston Jr DB, Egel DS, Holmes GJ, 2007b. Multi-state evaluation of reduced-risk fungicides and Melcast against Alternaria leaf blight and gummy stem blight on muskmelon. *Crop Protection* **26**, 1251–1258.

Khaosaad T, Garcia-Garrido JM, Steinkellner S, Vierheilig H, 2007. Take-all disease is systemically reduced in roots of mycorrhizal barley plants. *Soil Biology and Biochemistry* **39**, 727–734.

Kim YC, Blee KA, Robins J, Anderson AJ, 2001. OxycomTM under field and laboratory conditions increases resistance responses in plants. *European Journal of Plant Pathology* **107,** 129–136.

Kloepper JW, 1993. Plant growth-promoting rhizobacteria as biological control agents. In: Metting F.B. Jr, ed. *Soil Microbial Ecology – Applications in Agricultural and Environmental Management.* New York: Marcel Dekker, pp. 255–274.

Kozlowski G, Buchala A, Métraux JP, 1999. Methyl jasmonate protects Norway spruce (*Picea abies* (L.) Karst.) seedlings against *Pythium ultimum* Trow. *Physiological and Molecular Plant Pathology* **55**, 53–58.

Konstantinidou-Doltsinis S, Markellou E, Kasselaki AM, Fanouraki MN, Koumaki CM, Schmitt A, Liopa-Tsakalidis A, Malathrakis NE, 2006. Efficacy of Milsana®, a formulated plant extract from *Reynoutria sachalinensis*, against powdery mildew of tomato (*Leveillula taurica*). *Biocontrol* **51**, 375–392.

Kraeva E, Andary C, Carbonneau A, Deloire A, 1998. Salicylic acid treatment of grape berries retards ripening. *Vitis* **37**, 143–144.

Kuc J, 1982. Induced immunity to plant disease. *Bioscience* **32,** 854–860.

Lang JM, Gent DH, Schwartz HF, 2007. Management of Xanthomonas leaf blight of onion with bacteriophages and a plant activator. *Plant Disease* **91**, 871–878.

Latunde-Dada AO, Lucas JA, 2001. The plant defence activator acibenzolar-*S*-methyl primes cowpea [*Vigna unguiculata* (L.) Walp.] seedlings for rapid induction of resistance. *Physiological and Molecular Plant Pathology* **58**, 199–208.

Leon J, Rojo E, Sanchez-Serrano JJ, 2001. Wound signalling in plants. *Journal of Experimental Botany* **52**, 1–9.

Liao H-L, Chen H., Chung K-R, 2006. Plant hormone inhibitors for reducing postbloom fruit drop (PFD) of citrus. *Proceedings of Florida State Horticultural Society* **119**, 78–81.

Lian B, Zhou X, Miransari M, Smith DL, 2000. Effects of salicylic acid on the development and root nodulation of soybean seedlings. *Journal of Agronomy and Crop Science* **185**, 187–192.

Liu H, Jiang W, Bi Y, Luo Y, 2005a. Postharvest BTH treatment induces resistance of peach (*Prunus persica* L. cv. Jiubao) fruit to infection by *Penicillium expansum* and enhances activity of fruit defense mechanisms. *Postharvest Biology and Technology* **35,** 263–269.

Liu HX, Wang BG, Bi Y, Luo YB, Jiang WB, 2005b. Improving disease resistance in peach fruit during storage using benzo-(1,2,3)-thiodiazole-7-carbothioic acid *S*-methyl ester (BTH). *Journal of Horticultural Science and Biotechnology* **80**, 736–740.

Logemann E, Wu S-C, Schröder J, Schmelzer E, Somssich IE, Hahlbrock K, 1995. Gene activation by UV light, fungal elicitor or fungal infection in *Petroselium crispum* is correlated with repression of cell-cycle-related genes. *Plant Journal* **8**, 865–876.

Louws FJ, Wilson M, Campbell HL, Cuppels DA, Jones JB, Shoemaker PB, Sahin F, Miller SA, 2001. Field control of bacterial spot and bacterial speck of tomato using a plant activator. *Plant Disease* **85,** 481–488.

Malolepsza U, 2006. Induction of disease resistance by acibenzolar-*S*-methyl and *o*-hydroxyethylorutin against *Botrytis cinerea* in tomato plants. *Crop Protection* **25**, 956–962.

Martínez-Abarca F, Herrera-Cervera JA, Bueno P, Sanjuan J, Bisseling T, Olivares J, 1998. Involvement of salicylic acid in the establishment of the *Rhizobium meliloti*-alfalfa symbiosis. *Molecular Plant-Microbe Interactions* **11**, 153–155.

Matheron ME, Porchas M, 2000a. Effect of cultivar and Actigard on development of powdery mildew on lettuce. University of Arizona College of Agriculture 2000 vegetable report.

Matheron ME, Porchas M, 2000b. Evaluation of fungicide performance for control of powdery mildew on lettuce in 2000. University of Arizona College of Agriculture 2000 vegetable report.

Matheron ME, Porchas M, 2002. Suppression of Phytophthora root and crown rot on pepper plants treated with acibenzolar-*S*-methyl. *Plant Disease* **86**, 292–297.

Mauch F, Mauch-Mani B, Gaille C, Kull B, Haas D, Reimmann C, 2001. Manipulation of salicylate content in *Arabidopsis thaliana* by the expression of an engineered bacterial salicylate synthase. *Plant Journal* **25,** 67–77.

Maxson-Stein K, He S-Y, Hammerschmidt R, Jones AL, 2002. Effect of treating apple trees with acibenzolar-*S*-methyl on fire blight and expression of pathogenesis-related protein genes. *Plant Disease* **86,** 785–790.

Meir S, Droby S, Davidson H, Alsevia S, Cohen L, Horev B, Philosoph-Hadas S, 1998. Suppression of botrytis rot in cut rose flowers by postharvest application of methyl jasmonate. *Postharvest Biology and Technology* **13,** 235–243.

Meir S, Droby S, Kochanek B, Salim S, Philosoph-Hadas S, 2005. Use of methyl jasmonate for suppression of botrytis rot in various cultivars of cut rose flowers. *Acta Horticulturae*, **669,** 91–98.

McKey D, 1974. Adaptive patterns in alkaloid physiology. *American Naturalist* **108,** 305–320.

McKey D, 1979. The distribution of secondary compounds within plants. In: Rosenthal GA, Janzen DH, eds. *Herbivores: Their Interactions with Secondary Plant Metabolites.* New York: Academic Press, 55–133.

Miles AK, Willingham SL, Cooke AW, 2004. Field evaluation of strobilurins and a plant activator for the control of citrus black spot. *Australasian Plant Pathology* **33,** 371–378.

Miles AK, Willingham SL, Cooke AW, 2005. Field evaluation of a plant activator, captan, chlorothalonil, copper hydroxide, iprodione, mancozeb and strobilurins for the control of citrus brown spot of mandarin. *Australasian Plant Pathology* **34,** 63–71.

Momol MT, Olson SM, Funderburk JE, Stavisky J, Marois JJ, 2004. Integrated management of tomato spotted wilt on field-grown tomatoes. *Plant Disease* **88**, 882–890.

Mondal AH, Nehl DB, Allen SJ, 2005. Acibenzolar-*S*-methyl induces systemic resistance in cotton against black root rot caused by *Thielaviopsis basicola. Australasian Plant Pathology* **34**, 499–507.

Nair CB, Anith KN, Sreekumar J, 2007. Mitigation of growth retardation effect of plant defense activator, acibenzolar-*S*-methyl, in amaranthus plants by plant growth-promoting rhizobacteria. *World Journal of Microbiology & Biotechnology* **23**, 1183–1187.

Nike T, Mitsuhara I, Seo S, Ohtsubo N, Ohashi Y, 1998. Antagonistic effect of salicylic acid and jasmonic acid on the expression of pathogenesis-related (PR) protein genes in wounded mature tobacco leaves. *Plant and Cell Physiology* **39**, 500–507.

Ngugi HK, Scherm H, 2006. Biology of flower-infecting fungi. *Annual Review of Phytopathology* **44**, 261–282.

Obradovic A, Jones JB, Momol MT, Balogh B, Olson SM, 2004. Management of tomato bacterial spot in the field by foliar applications of bacteriophages and SAR inducers. *Plant Disease* **88,** 736–740.

Obradovic A, Jones JB, Momol MT, Olson SM, Jackson LE, Balogh B, Guven K, Iriarte FB, 2005. Integration of biological control agents and systemic acquired resistance inducers against bacterial spot on tomato. *Plant Disease* **89,** 712–716.

Oerke EC, Steiner U, Schonbeck F, 1989. Zur Wirksamkeit der induzierten Resistenz unter praktischen Anbaubedingungen. V. Mehltaubefall und Ertag von Winter- und Sommergerste in Abhangigkeit von der Stickstoffdungung. *Zeitschrift fur Pflanzenkrankheiten und Pflanzenschutz* **96,** 140–153.

Olson SM, Simonne E, 2005. *Vegetable Production Handbook for Florida 2005–2006*. Vance Publishing, Florida.

Ortega F, Steiner U, Dehne HW, 1998. Induced resistance: a tool for fungicide resistance management. *Pesticide Science* **53,** 193–196.

Pieterse CMJ, van Wees SCM, van Pelt JA, Knoester M, Laan R, Gerrits N, Weisbeek PJ, Van Loon LC, 1998. A novel signalling pathway controlling induced systemic resistance in Arabidopsis. *Plant Cell* **10,** 1571–1580.

Pieterse CMJ, Van Loon LC, 2007. Signalling cascades involved in induced resistance. In: Walters D, Newton A, Lyon G, eds. *Induced Resistance for Plant Disease Control: A Sustainable Approach to Crop Protection*. Oxford: Blackwell Publishing, 65–88.

Pradhanang PM, Ji P, Momol MT, Olson SM, Mayfield JL, Jones JB 2005. Application of acibenzolar-*S*-methyl enhances host resistance in tomato against *Ralstonia solanacearum*. *Plant Disease* **89,** 989–993.

Prats E, Rubiales D, Jorrin J, 2002. Acibenzolar-*S*-methyl-induced resistance to sunflower rust (*Puccinia helianthi)* is associated with an enhancement of coumarins on foliar surface. *Physiological and Molecular Plant Pathology* **60**, 155–162.

Preston CA, Lewandowski C, Enyedi AJ, Baldwin IT, 1999. Tobacco mosaic virus inoculation inhibits wound-induced jasmonic acid-mediated responses within but not between plants. *Planta* **209**, 87–95.

Pühringer H, Moll D, Hoffmann-Sommergruber K, Watillon B, Katinger H, da Câmara Machado ML, 2000. The promoter of an apple Ypr10 gene, encoding the major allergen Mal d 1, is stress- and pathogen-inducible. *Plant Science* **152**, 35–50.

Ramamoorthy V, Viswanathan R, Raguchander T, Prakasam V, Samiyappan R, 2001. Induction of systemic resistance by plant growth promoting rhizobacteria in crop plants against pests and diseases. *Crop Protection* **20,** 1–11.

Ramanujam MP, Abdul Jaleel V, Kumara Velu G, 1998. Effect of salicylic acid on nodulation, nitrogenous compounds and related enzymes of *Vigna mungo*. *Biologia Plantarum* **41,** 307–311.

Randoux B, Renard D, Nowak E, Sanssene J, Courtois J, Durand R, Reignault P, 2006. Inhibition of *Blumeria graminis* f. sp. *tritici* germination and partial enhancement of wheat defenses by Milsana. *Phytopathology* **96,** 1278–1286.

Reddy MS, Funk LM, Covert DC, He DN, Pedersen EA, 1997. Microbial inoculants for sustainable forests. *Journal of Sustainable Forestry* **5,** 293–306.

Reglinski T, Dick M, 2005. Biocontrol of forest nursery pathogens. *New Zealand Journal of Forestry*, **50,** 14–21.

Reglinski T, Newton AC, Lyon GD, 1994. Assessment of the ability of yeast-derived elicitors to control powdery mildew in the field. *Journal of Plant Diseases and Protection* **101,** 1–10.

Reglinski T, Stavely FJL, Taylor JT, 1998. Induction of phenylalanine ammonia lyase activity and control of *Sphaeropsis sapinea* infection in *Pinus radiata* by 5-chlorosalicylic acid. *European Journal of Forest Pathology* **28,** 153–158.

Reglinski T, Elmer PAG, Taylor JT, Parry FJ, Marsden R, Wood PN, 2005. Suppression of Botrytis bunch rot in Chardonnay grapevines by induction of host resistance and fungal antagonism. *Australasian Plant Pathology* **34,** 481–488.

Renard-Merlier D, Randoux B, Nowak E, Farcy F, Durand R, Reignault P, 2007. Iodus 40, salicylic acid, heptanoyl salicylic acid and trehalose exhibit different efficacies and defence targets during a wheat/powdery mildew interaction. *Phytochemistry* **68,** 1156–1164.

Reuveni M, Zahavi T, Cohen Y, 2001. Controlling downy mildew (*Plasmopara viticola*) in field-grown grapevine with beta-aminobutyric acid (BABA). *Phytoparasitica* **29,** 125–133.

Rhoades DF, 1979. Evolution of plant chemical defense against herbivores. In: Rosenthal GA, Janzen DH, eds. *Herbivores: Their Interaction with Secondary Plant Metabolites*. New York: Academic Press, 4–53.

Ritenour MA, Pelosi RR, Burton MS, Stover EW, Dou H, McCollum TG, 2004. Assessing the efficacy of preharvest fungicide applications to control postharvest diseases of Florida citrus. *HortTechnology* **14**, 58–62.

Roemmelt S, Zimmermann N, Rademacher W, Treutter D, 2003. Formation of novel flavonoids in apple (Malus × domestica) treated with the 2-oxoglutarate-dependent dioxygenase inhibitor prohexadione-Ca. *Phytochemistry* **64**, 709–716.

Romanazzi G, Nigro E, Ippolito A, Di Venere D, Salerno M, 2002. Effects of pre- and postharvest chitosan treatments to control storage grey mold of table grapes. *Journal of Food Science* **67**, 1862–1867.

Romanazzi G, Gabler FM, Smilanick JL, 2006. Preharvest chitosan and postharvest UV irradiation treatments suppress gray mold of table grapes. *Plant Disease* **90**, 445–450.

Romero AM, Kousik CS, Ritchie DF, 2001. Resistance to bacterial spot in bell pepper induced by acibenzolar-*S*-methyl. *Plant Disease* **85,** 188–194.

Roux C, Lemarquand A, Orain G, Campion C, Simoneau P, Poupard P, 2006. Effects of the plant growth regulator prohexadione-calcium and the SAR-inducer acibenzolar-S-methyl on the quality of apples at harvest. *Journal of Horticultural Science and Biotechnology* **81**, 139–145.

San-Miguel R, Gutierrez M, Larque-Saavedra A, 2003. Salicylic acid increases the biomass accumulation of *Pinus patula. Southern Journal of Applied Forestry* **27,** 52–54.

Scheideler M, Schlaich NL, Fellenberg K, Beissbarth T, Hauser NC, Vingron M, Slusarenko AJ, Hoheisel JD, 2002. Monitoring the switch from housekeeping to pathogen defense metabolism in *Arabidopsis thaliana* using cDNA arrays. *Journal of Biological Chemistry* **277**, 10555–10561.

Schilder A, Gillet J, Sysak R, Wise J, 2002. Evaluation of environmentally friendly products for control of fungal diseases of grapes. In: *Proceedings of the 10th International Conference on Cultivation Technique and Phytopathological Problems in Organic Fruit-Growing and Viticulture*, Weinsberg, Germany, 163–167.

Schmitt A, Kunz S, Nandi S, Seddon B, Ernst A, 2002. Use of *Reyonoutria sachalinensis* plant extracts, clay preparations and *Brevibacillus brevis* against fungal diseases of grape berries. In: *Proceedings of the 10th International Conference on Cultivation Technique and Phytopathological Problems in Organic Fruit-Growing and Viticulture*, 4–7 February. Weinsberg, Germany, 146–151.

Shah J, Tsui F, Klessig DF, 1997. Characterization of a salicylic acid-insensitive mutant (*sai 1*) of *Arabidopsis thaliana,* identified in a selective screen utilizing the SA-inducible expression of the *tms2* gene. *Molecular Plant-Microbe Interactions* **10,** 69–78.

Shearer BL, Fairman RG, 2007. A stem injection of phosphite protects *Banksia* species and *Eucalyptus marginata* from *Phytophthora cinnamomi* for at least four years. *Australasian Plant Pathology* **36**, 78–86.

Silva HSA, Romeiro RS, Carrer-Filho R, Pereira JLA, Mizubuti ESG, Mounteer A, 2004. Induction of systemic resistance by *Bacillus cereus* against tomato foliar diseases under field conditions. *Journal of Phytopathology* **152,** 371–375.

Smedegaard-Petersen V, Stolen O, 1981. Effect of energy requiring defense reactions on yield and grain quality in powdery mildew *Erysiphe graminis* sp. *hordei* resistant *Hordeum vulgare* cultivar Sultan. *Phytopathology* **71,** 396–399.

Sonnemann I, Finkhaeuser K, Wolters V, 2002. Does induced resistance in plants affect the belowground community? *Applied Soil Ecology* **21**, 179–185.

Sonnemann I, Streicher NM, Wolters V, 2005. Root associated organisms modify the effectiveness of chemically induced resistance in barley. *Soil Biology and Biochemistry* **37**, 1837–1842.

Sparla F, Rotino L, Valgimigli MC, Pupillo P, Trost P, 2004. Systemic resistance induced by benzothiadiazole in pear inoculated with the agent of fire blight (*Erwinia amylovora*). *Scientia Horticulturae* **101,** 269–279.

Spinelli F, Vanneste JL, Ciampolini F, Cresti M, Rademacher W, Geider K, Costa G, 2007. Potential and limits of acylcyclohexanediones for the control of blossom blight in apple and pear caused by *Erwinia amylovora*. *Plant Pathology* **56**, 702–710.

Spoel SH, Johnson JS, Dong X, 2007. Regulation of trade-offs between plant defences against pathogens with different lifestyles. *Proceedings of the National Academy of Sciences USA* **104,** 18842–18847.

Stadnik MJ, Buchenauer H, 1999. Control of wheat diseases by a benzothiadiazole-derivative and modern fungicides. *Journal of Plant Disease and Protection* **106,** 466–475.

Steiner U, Oerke EC, Schonbeck F, 1988. Zur Wirksamkeit der induzierten Resistenz unter praktischen Anbaubedingungen. IV. Befall und Ertag von Wintergertsensorten mit induzierter Resistenz und nach Fungizibehandlung. *Zeitschrift fur Pflanzenkrankheiten und Pflanzenschutz* **95**, 506–517.

Stout MJ, Fidantsef AL, Duffey SS, Bostock RM, 1999. Signal interactions in pathogen and insect attack: systemic plant-mediated interactions between pathogens and herbivores of the tomato, *Lycopersicon esculentum. Physiological and Molecular Plant Pathology* **54,** 115–130.

Thaler JS, Fidantsef AL, Duffey SS, Bostock RM, 1999. Trade-offs in plant defense against pathogens and herbivores: A field demonstration of chemical elicitors of induced resistance. *Journal of Chemical Ecology* **25,** 1597–1609.

Thaler JS, Fidantsef AL, Bostock RM, 2002. Antagonism between jasmonate- and salicylate-mediated induced plant resistance: Effects of concentration and timing of elicitation on defense-related proteins, herbivore, and pathogen performance in tomato. *Journal of Chemical Ecology* **28,** 113–159.

Tubajika KM, Civerolo EL, Puterka GJ, Hashim JM, Luvisi DA, 2007. The effects of kaolin, harpin, and imidacloprid on development of Pierce's disease in grape. *Crop Protection* **26,** 92–99.

Van Loon LC, Rep M, Pieterse CMJ, 2006. Significance of inducible defense-related proteins in infected plants. *Annual Review of Phytopathology* **44,** 135–162.

Van Wees SCM, Luijendijk M, Smoorenburg I, Van Loon LC, Pieterse CMJ, 1999. Rhizobacteria-mediated induced systemic resistance (ISR) in Arabidopsis is not associated with a direct effect on expression of known defense-related genes but stimulates the expression of the jasmonate-inducible gene *Atvsp* upon challenge. *Plant Molecular Biology* **41**, 537–549.

Vechet L, Martinkova J, Sindelarova M, Burketova L, 2005. Compounds of natural origin inducing winter wheat resistance to powdery mildew (*Blumeria graminis* f. sp. *tritici*). *Plant, Soil and Environment* **51**, 469–475.

Vierheilig H, Alt M, Mohr U, Boller T, Wiemken A, 1994. Ethylene biosynthesis and activities of chitinase and β-1,3-glucanase in the roots of host and non-host plants of vesicular–arbuscular mycorrhizal fungi after inoculation with *Glomus mosseae*. *Journal of Plant Physiology* **143**, 337–343.

Waller F, Achatz B, Baltruschat H, Fodor J, Becker K, Fischer M, Heier T, Huckelhoven R, Neumann C, von Wettstein D, Franken P, Kogel K-H, 2005. The endophytic fungus *Piriformospora indica* reprograms barley to salt-stress tolerance, disease resistance, and higher yield. *Proceedings of the National Academy of Sciences* **102**, 13386–13391.

Walling LL, 2000. The myriad plant responses to herbivores. *Journal of Plant Growth Regulation* **19**, 195–216.

Walters D, Heil M, 2007. Costs and trade-offs associated with induced resistance. *Physiological and Molecular Plant Pathology* **71,** 3–17.

Walters D, Walsh D, Newton A, Lyon G, 2005. Induced resistance for plant disease control: Maximizing the efficacy of resistance elicitors. *Phytopathology* **95,** 1368–1373.

Walters DR, Cowley T, Weber H, 2006. Rapid accumulation of trihydroxyoxylipins and resistance to the bean rust pathogen *Uromyces fabae* following wounding in *Vicia faba*. *Annals of Botany* **97**, 779–784.

Williams M, Senaratna T, Dixon K, Sivasithamparam K, 2003. Benzoic acid induces tolarence to biotic stress caused by *Phytophthora cinnamomi* in *Banksia attenuata*. *Plant Growth Regulation* **41,** 89–91.

Yang B, Yonghong G, Yurong G, Jie Z, 2007. Postharvest harpin treatment suppresses decay and induces the accumulation of defense-related enzymes in Hami melons. *Acta Horticulturae*, **731**, 439–450.

Yu T, Zheng X, 2006. Salicylic acid enhances biocontrol efficacy of the antagonist *Cryptococcus laurentii* in apple fruit. *Journal of Plant Growth Regulation* **25,** 166–174.

Yu T, Zheng X, 2007. An integrated strategy to control postharvest blue and grey mould rots of apple fruit by combining biocontrol yeast with gibberellic acid. *International Journal of Food Science & Technology* **42,** 977–984.

Yu T, Wu P, Qi J, Zheng X, Jiang F, Zhan X, 2006. Improved control of postharvest blue mold rot in pear fruit by a combination of *Cryptococcus laurentii* and gibberellic acid. *Biological Control* **39,** 128–134.

Yu T, Li HY, Zheng XD, 2007a. Synergistic effect of chitosan and *Cryptococcus laurentii* on inhibition of *Penicillium expansum* infections. *International Journal of Food Microbiology* **114,** 261–266.

Yu T, Chen J, Chen R, Huang B, Liu D, Zheng X, 2007b. Biocontrol of blue and gray mold diseases of pear fruit by integration of antagonistic yeast with salicylic acid. *International Journal of Food Microbiology* **116,** 339–345.

Zarate SI, Kempema LA, Walling LL, 2007. Silverleaf whitefly induces salicylic acid defenses and suppresses effectual jasmonic acid defenses. *Plant Physiology* **143,** 866–875.

Zavala JA, Patankar AG, Gase K, Hui DQ, Baldwin IT, 2004. Manipulation of endogenous trypsin proteinase inhibitor production in *Nicotiana attenuata* demonstrates their function as antiherbivore defenses. *Plant Physiology* **134,** 1181–1190.

Zehnder GW, Murphy JF, Sikora EJ, Kloepper JW, 2001. Application of rhizobacteria for induced resistance. *European Journal of Plant Pathology* **107,** 39–50.

Zheng X, Yu T, Chen R, Huang B, Wu C, 2007. Inhibiting *Penicillium expansum* infection on pear fruit by *Cryptococcus laurentii* and cytokinin. *Postharvest Biology and Technology* **45,** 221–227.

Ziadi S, Barbedette S, Godard JF, Monot C, Le Corre D, Silue D, 2001. Production of pathogenesis-related proteins in the cauliflower (*Brassica oleracea var. botrytis*)-downy mildew (*Peronospora parasitica*) pathosystem treated with acibenzolar-*S*-methyl. *Plant Pathology* **50,** 579–586.

Chapter 5
The use of composts and compost extracts in plant disease control

Audrey Litterick[1] *and Martin Wood*[2]

[1]SAC Environmental, Craibstone Estate, Aberdeen, UK
[2]Earthcare Environmental Ltd., Reading, UK

5.1 Introduction

Most developed countries throughout the world are now attempting to reduce the amount of biodegradable waste that is sent to landfill. Composting is a viable option for treating such wastes, not least because the finished product is a bulky fertilizer with useful applications in agriculture, horticulture and landscaping. There is a considerable body of evidence to show that composts and liquid preparations made from them can suppress or control plant diseases. However, the effect of these composts and compost extracts (or compost teas) on the soil and leaf microflora and on plant diseases varies greatly, depending on the nature of the compost (or liquid preparation), its preparation method and on the context in which the material is used.

This chapter aims to define composts, compost extracts and teas, to outline preparation methods and to summarize the findings of recent work which demonstrates their effects on plant disease. The mechanisms which are responsible for disease suppression and control will then be discussed, the problems associated with using composts and liquid preparations will be reviewed and the future potential for using composts in crop production systems will be considered.

5.2 Definitions of composts, composting, compost extracts and compost teas

5.2.1 Composts and the composting process

Compost can be defined as solid particulate organic material that is the result of composting, that has been sanitized and stabilized and that confers beneficial effects when added to soil and/or used in conjunction with plants. Composting can be defined as a process of controlled biological decomposition of biodegradable materials under managed conditions that are predominantly aerobic and that allow the development of thermophilic temperatures as a result of biologically produced heat, in order to achieve compost that is sanitary and stable. Compost production is briefly described later in this paper, but the principles of composting and the factors affecting the composting process are widely described in the literature (e.g. Epstein, 1997; Zibilske, 1998).

5.2.2 Manures

Manures can be defined as animal excrement, which may contain large amounts of bedding. It is important to distinguish between true composts (which contain material which has been composted) and fresh, stacked or stored manures which have not undergone aerobic composting. Limited research has shown variable effects of fresh or stacked manures on the incidence and severity of plant disease. However, such studies are outside the scope of this chapter.

5.2.3 Compost extracts and teas

The terms 'compost extract' and 'compost tea' have in the past been used interchangeably. However, for the purposes of this review, they are treated as different products.

The term 'compost extract' has been used in the literature to define water extracts prepared using a wide range of different methods (Scheuerell & Mahaffee, 2002). The terms 'compost extract', 'watery fermented compost extract', 'compost steepage', 'amended extract' and 'compost slurry' have all been used to refer to non-aerated fermentations. The terms 'compost extract' (Weltzein, 1989), 'watery fermented compost extract' (Weltzein, 1991) and 'steepages' (Hoitink *et al.*, 1997) are approximate synonyms defined as a 1:5 to 1:10 (v:v) ratio of compost to water that is fermented without stirring at room temperature for a defined length of time. 'Amended extracts' can be defined as compost extracts that have been fermented with the addition of nutrients or microorganisms prior to application (Weltzein, 1991). The term 'compost slurry' has been used to describe non-aerated compost teas before the filtration process (Cronin *et al.*, 1996). Nowadays, it is common practice to refer to compost extracts as the 'filtered products of compost mixed with any solvent (usually water), but not fermented' (Scheuerell & Mahaffee, 2002). This convention will be adhered to in this chapter.

Compost tea is the term given by an increasing number of commercial growers (in the USA and Europe) to the filtered product of compost fermented in water (Brinton *et al.*, 1996; Scheuerell & Mahaffee, 2002; Litterick *et al.*, 2004). Compost teas are produced by re-circulating water through loose compost or a porous bag or box of compost suspended over or within a tank with the intention of maintaining aerobic conditions (Litterick *et al.*, 2004). The product of this method has also been termed 'aerated compost tea' and 'organic tea' (Riggle, 1996). Several companies have developed machinery for the preparation of compost teas in this way under highly aerated conditions (www.attra.ncat.org; www.soilfoodweb.com).

The term 'compost tea' has not however, always been associated with an aerated fermentation process (Brinton *et al.*, 1996). It is important to distinguish between compost teas prepared using aerated and non-aerated processes, therefore, the terms 'aerated compost tea' (ACT) and 'non-aerated compost tea' (NCT) are used in this chapter to refer to the two dominant compost fermentation methods. ACT will refer to any method in which the water extract is actively aerated during the fermentation process. NCT will refer to methods where the water extract is not aerated or receives minimal aeration during fermentation apart from during the initial mixing.

5.3 Production of composts and compost extracts/teas

5.3.1 Composts

Composting systems differ greatly in terms of their sophistication and therefore cost, depending on the nature of the material being composted, and on the end use of the compost being prepared. Relatively low-cost, open turned windrow systems are generally used to compost low-hazard materials such as green or garden waste (termed yard waste in the United States), whereas enclosed, or in-vessel systems are more suitable for food wastes and wastes containing animal by-products. Several countries (including those in the European Union) have legislation that ensures that the design and operation of the composting process will guarantee safe, effective composting of all animal by-products treated in this way.

In order for the composting process to proceed optimally, the input (or feedstock) materials must be sufficiently moist (40–60% moisture content), have a sufficiently open structure to allow air to penetrate the mass, and must have a suitable carbon to nitrogen (C:N) ratio (around 25:1 to 40:1). Some materials, such as green waste or garden waste have a suitable C:N ratio, whereas others such as straw (C:N ratio around 80:1) or vegetable wastes (C:N around 12:1) must be mixed with other materials in order to obtain a feedstock with a suitable C:N ratio. For example, the composted bark products that have been used in so many documented studies generally have mineral nitrogen added prior to composting in order to allow the process to proceed optimally.

Most of the composts used in studies reported here were made using some form of simple, outdoor, turned windrow process. Feedstocks used vary widely, and for this reason, are always defined where the effects of composts are discussed.

Vermicomposts are the digestion products of worms. There is little or no self-generated heating during their production. Due to the differences in their manufacture and properties and the fact that few studies have examined their effect on plant disease, they are not covered in this chapter.

5.3.2 Compost extracts/teas

The production of aerated and non-aerated compost teas involves compost being fermented in water for a specific time period. Both methods require a fermentation vessel, compost, water, incubation and filtration prior to application. Nutrients may be added before or after fermentation and additives or adjuvants may be added prior to application.

There is continuing debate regarding the benefits of aeration during compost tea production (Brinton *et al.*, 1996; Ingham, 1999; Ingham & Alms, 1999). Aerated production requires less time. Non-aerated methods are associated with low cost, low-energy input and numerous reports of successful plant disease control (Weltzein, 1991). Several consultants and scientists have suggested that NCTs can cause phytotoxicity and that the production of NCTs provides an ideal environment for the growth and reproduction of human pathogens, but there is little evidence to substantiate these claims at present.

The design of machinery which facilitates preparation of aerated compost tea varies. Designs include:

- showers of recirculated water which filter through bags of compost suspended over open tanks (Riggle, 1996)

- recirculated water which is directed through a vortex nozzle held above a tank (Ingham & Alms, 1999)
- injection of air through various distribution channels including a hollow propeller shaft (www.soilsoup.com), venturi nozzles (www.composttea.com) or fine bubble diffusion mats (www.growingsolutions.com)
- suspension of compost in stirred, recirculated or aerated liquid (usually water) in a fermentation vessel (Diver, 2001; www.symbio.co.uk/library/files/Xtractor%20brochure%202%20x%20A3. pdf).

NCT has generally been made by mixing one volume of compost with between four to ten volumes of water in an open container. The mixture is stirred as it is made up, then it is left for at least 3 days at 15–20°C with minimal or no stirring (Brinton *et al.*, 1996; Weltzein, 1991). Compost teas can be made in quantities ranging from a few litres to several thousand litres in a single batch depending on the size of the fermentation vessel. Nutrients are often added to compost tea at the start of the production process. Recent work has suggested that the addition of such fermentation nutrients can result in proliferation of human pathogens including *Escherichia coli* and *Salmonella enteritidis*, therefore further work is urgently required to determine how to optimize the production process without putting the manufacturer or user at risk (Ingram & Millner, 2007).

5.4 History of the use of composts and compost extracts in crop production

Organic amendments, including composts and liquid preparations made from them, have been applied for centuries for the purpose of crop nutrition and to improve soil quality, but their use has declined since the advent of agrochemicals (van Bruggen, 1995). Interest has been maintained through the organic sector, and in recent years the more general pressure to reduce reliance on pesticides has created a renewed interest in organic amendments relating to their potential for improving soil health and preventing or controlling pests and diseases.

Some of the earliest research on the benefits of compost was carried out by Sir Albert Howard in India during the early decades of the twentieth century. Howard, a botanist and mycologist, pioneered an integrated approach to agricultural research and by observing local farmers and carrying out basic research he became convinced that the benefits of improved plant varieties could only be fully realized if the soil was also improved. The use of 'manufactured humus' (compost) was seen to be essential in preventing disease in plants (and animals) not just in the tropics, but around the world. In 1943 Howard published his hugely influential book 'An Agricultural Testament' (Howard, 1943) and his ideas have formed the principles of the international Organic Movement which forbid the use of artificial fertilizers (and other synthetic chemical products) and encourage the use of composting.

Sprays based on compost extracts have been used for hundreds of years; there is evidence that the Romans used compost teas and the ancient Egyptians used preparations based on compost or manure extracts as long as 4000 years ago (Koepf, 1992). Interest in compost extracts and teas waned when pesticides became available in the mid-twentieth century, since pesticides tend to give more reliable control of most foliar diseases. However, the recent increase in sustainable and organic farming and problems relating to pesticide use has led to an increase in scientific papers and non-refereed publications relating to compost extracts and teas (Scheuerell & Mahaffee, 2002). A considerable amount of work has been carried out to develop improved methods for preparation and use of compost extracts and teas. Most of this work has been done in the United States and much of it by commercial companies.

5.5 Current use of composts and compost extracts/teas in crop production

Composts and manures are often used for the purposes of crop nutrition. Their application specifically for the purpose of preventing or controlling crop pests or diseases in field crops is comparatively rare at present, although growers in the United States commonly use composted bark in container growing media in order to reduce the incidence and severity of root and stem-base diseases. Compost teas are currently being applied by relatively small numbers of commercial growers, mainly to ornamental crops in the Netherlands and the United Kingdom, and to edible and ornamental crops and private gardens in the United States. Some growers are applying compost teas with the aim, or partial aim, of preventing disease, but most currently apply them as a general crop tonic. The science to support the reports of efficacy of compost teas in suppressing plant disease is limited at present. However, there are abundant anecdotal reports of the efficacy of compost tea in suppressing plant disease (Scheuerell & Mahaffee, 2002; www.attra.ncat.org), and a compost tea industry has developed, particularly in the United States, which makes largely unproven claims based on the ability of their compost teas and brewers (used to make the teas) to suppress plant diseases.

5.6 Crop and soil health

5.6.1 *Diseases in cropping systems*

Conventionally grown crops, such as cereals are frequently grown in long-term monocultures. This practice often results in heavy pest and/or disease pressure on the crop, and the farmer usually relies on pesticides to maintain crop health and yield in successive crops. Crop rotations can be used to reduce incidence and severity of pest and disease attack (Abawi & Widmer, 2000). However, modern crop rotations are often short, and contain species from a limited number of families, therefore there is a reliance on pesticides to minimize crop losses due to pests and diseases. The less mobile soil-borne diseases such as rhizoctonia root rot/stem canker of potatoes (*Rhizoctonia solani*) and clubroot of brassicas (*Plasmodiophora brassicae*) can be partially controlled through the use of balanced rotations, appropriate break crops and good soil husbandry. However, it is often difficult to control root-inhabiting pathogens that survive saprophytically in soil organic matter and exist for long periods in the absence of a host plant without using fungicides. These pathogens include *Pythium* spp., some *Fusarium* and *Phytophthora* spp. and *Sclerotium rolfsii*.

The selection of varieties that have a high degree of resistance to locally significant pests and diseases is an important element in any crop protection strategy. This is especially important for diseases such as late blight in potatoes (*Phytophthora infestans*). Intercropping with two or more crop types (e.g. brassicas with an understorey of clover) has also been shown to reduce foliar disease and pest attack, but this practice is rare in conventional cropping systems.

Fertilizer strategy can also be altered to optimize the crop yield/crop health balance. As a general rule, excessive nitrogen (N) fertilization tends to result in higher levels of foliar disease (van Bruggen, 1995; Walters & Bingham, 2007). Organic crops tend to have a reduced N supply compared to conventional crops, and consequently a higher dry matter content and lower N content. This may make them less susceptible to air-borne or foliar disease. Cooke (1993) has shown that foliar and stem-base cereal disease levels are generally lower in long-term organic fields than in recently converted fields. Soil-borne pathogens and root diseases are also generally lower in organic than in conventional systems (van Bruggen, 1995; van Bruggen & Termorshuizen, 2003).

Pesticide legislation varies between different countries and between different US states, but conventional farmers in most temperate countries have access to several hundred approved synthetic and natural pesticides for controlling pests, diseases and weeds. A number of naturally occurring fungicides, including sulphur, copper and some plant extract-based fungicides are permitted for use in organic farming under specific circumstances. Biological control agents of fungal diseases are used to a very limited extent, mainly in protected and high value crops.

5.6.2 Soil health

Soil health is central to any sustainable farming system where reliance on synthetic fertilizers and pesticides is minimized, but its potential has not yet been fully explored. Soil health has physical, chemical and biological components and is concerned with the idea that soil is a living dynamic organism that functions in a holistic way depending upon its condition or state. The biological component of soil health depends on the numbers, diversity and health of the macro, meso and microfauna and microflora present. It has been formally defined as 'the capacity of a soil to function within ecosystem boundaries, to sustain biological productivity, maintain environmental quality, and promote plant and animal health'. (Doran *et al.*, 1996). However it is now generally accepted that this definition should be reserved for soil quality, which is a broader concept. Soil health can be considered a part of ecosystem health and is associated with biological diversity and stability (van Bruggen & Semenov, 2000; van Bruggen & Termorshuizen, 2003). It is therefore thought likely that there are links between soil health, the ability of the biological community to suppress plant pathogens, populations of soil-borne plant pathogens and also disease incidence and severity (Abawi & Widmer, 2000).

Soil health monitoring is rarely practiced in Europe, but in some parts of the United States, farmers are using test kits to determine chemical, physical and biological components of soil health (www.attra.ncat.org; www.solvita.com). These have proved useful in demonstrating effects of management on soil health (Ditzler & Tugel, 2002). The biological indicators used are earthworm count and soil respiration. It is recognized that soil health depends on more than just these two parameters and that the presence and numbers of specific species are important. However evidence suggests that these basic soil health test kits provide useful guidance to help farmers with soil management decisions.

There is increasing evidence of the impact which composts and compost extracts and teas can have on soil health and the presence of beneficial microorganisms (Abawi & Widmer, 2000; Albiach *et al.*, 2000; van Bruggen & Semenov, 2000; van Bruggen & Termorshuizen, 2003). This is considered further in the following sections.

5.7 Effects of composts on plant disease

Composts have been applied to agricultural land world-wide for centuries, but their deliberate application for the purpose of preventing and controlling plant diseases is a more recent phenomenon. The majority of published work relating to the use of composts and compost extracts or teas for prevention and control of diseases involves the production of plants in containers in controlled environments (growth rooms or glasshouses) using composts mixed with soil, sand or peat. However, there is increasing interest in the potential for composts and similar materials to help prevent and control diseases in field crops and information on the use of composts in field trials is slowly increasing. Work to determine the effects of composts on plant diseases in container, glasshouse and field production systems has recently been comprehensively reviewed (Litterick *et al.*, 2004; Noble & Coventry, 2005).

5.7.1 Container media

Numerous studies have examined the potential for composts to control or suppress diseases in plants grown in media based partly on compost (along with sand, soil or peat), or more rarely, made solely from compost. Most work has concentrated on one or more of the root and stem-base pathogens: *Pythium ultimum, Phytophthora* spp., *Rhizoctonia solani* or *Fusarium oxysporum*. Reports concern the effects of composts on ornamentals (including hardy nursery stock and pot plants) and container-produced edible crop plants (such as radish, cucumber, tomato, pea, pepper and brassicas). Input materials (or feedstocks) in composts found to suppress plant pathogens include hardwood and softwood bark, animal manures (pig, poultry, horse and cattle), vegetable wastes, green (or yard) wastes, sewage sludge (or municipal solid waste) and fish wastes. Workers using these composts have demonstrated suppression of root and soil-borne pathogens (and occasionally foliar diseases) on a range of crops. However, levels of disease suppression were not consistent for individual feedstocks and they were found to differ both between and within studies.

It is difficult to compare different pieces of work, since experimental conditions varied widely, but it is possible to assess which feedstocks may be considered for some crop/disease applications. For example, *Rhizoctonia solani* and *Sclerotium rolfsii* were suppressed in container media containing composted separated cattle manure and composted grape marc (Gorodecki & Hadar, 1990). *Pythium aphanidermatum* damping-off was suppressed in container media containing composted liquorice roots (Hadar & Mandelbaum, 1986) and disease caused by *Pythium ultimum* and *Rhizoctonia solani* was controlled or considerably reduced in container media amended with composted organic household wastes (Schueler *et al.*, 1989), green waste (or yard trimmings, Scheuerell *et al.*, 2005) or hardwood bark (e.g. Nelson *et al.*, 1983; Stephens & Stebbins, 1985).

The most notable commercial use of composts to suppress plant disease concerns the use of composted bark in the United States container-produced ornamentals sector. The concept of using composted hardwood bark as a peat substitute to control soil-borne pathogens on ornamental crops was first suggested by Hoitink (1980). Growing media based on composted bark was shown to suppress several soil-borne plant pathogens including *Phytophthora, Pythium* and *Rhizoctonia* spp. which cause damage to plants grown in container media (Spring *et al.*, 1980; Nelson *et al.*, 1983).

A considerable amount of work has since been done in the United States to develop reliable suppressive container media containing composted bark for different crops and production systems (Hoitink, 1980; Kuter *et al.*, 1983; Nelson *et al.*, 1983; Hoitink & Fahy, 1986; Mandelbaum *et al.*, 1988; Hoitink *et al.*, 1997). It is now recognized that the control of root rots caused by some plant pathogens (e.g. *Phytophthora* and *Pythium*) on plants grown in compost-amended substrates can be as effective as if they were treated with modern synthetic fungicides (Ownley & Benson, 1991). These media are now widely used either exclusively or as part of an integrated strategy to prevent and control root and soil-borne pathogens of container produced ornamental plants in the United States. However, such practices are little understood and rarely used by European growers.

Most reports of disease suppression following application of composts to field soils have been carried out in containers under controlled laboratory or glasshouse conditions. For example, Serra-Wittling *et al.* (1996) studied the effect of municipal solid waste (MSW) compost on the half life time of flax seedlings grown in soil inoculated with *Fusarium oxysporum* f. sp. *lini*. It was found that the addition of 10, 20 and 30% (by volume) of MSW to soil increased the half-life time (HLT) of flax seedlings to 67, 87 and 103 days respectively in comparison to that of flax seedlings grown in inoculated soil with no added compost (HLT 41 days). Lumsden *et al.* (1983) added composted sewage sludge to pots of field soils naturally or artificially infested with one or more crown or root-rot pathogens. They found that the addition of composted sewage sludge to soil (10% by volume) significantly reduced incidence and severity of diseases including: *Aphanomyces euticles* on pea, *Rhizoctonia solani* on bean and cotton, *Phytophthora capsici* on pepper, *Sclerotinia minor* on lettuce and *Fusarium oxysporum* f. sp. *melonis* on melons.

It is recognized that the suppressiveness of individual composts used as components of container media may not be replicated in commercial crops grown in field soils. There are many examples where workers have found no disease suppression in field soils when using composts that had previously been shown to suppress disease under laboratory or glasshouse bioassays. This may be partly because the environment in which host, pathogens and beneficial organisms live is more variable and more difficult to control in the field. However, the large body of information that has built up from container media trials is already providing useful indications of which composts may help confer suppressiveness to field soils.

5.7.2 Field crops

Pathogen suppression in field-grown crops following application of composts is considerably less predictable than that in container media when plants have been produced under controlled conditions. There are examples of disease suppression following application of composts to field soils, but our understanding of the factors affecting suppression in field soils is less well developed than that in container production systems. The inconsistency in the level of disease suppression reported in field soils is probably due to differences in experimental conditions and in compost types used. Most research in this area focuses on the three root and soil-borne pathogens *Rhizoctonia, Pythium* and *Phytophthora* spp., but work also covers other pathogens, including *Colletotrichum, Fusarium, Macrophomina, Mycosphaerella, Sclerotinia, Thielaviopsis* and *Verticillium* spp.

A limited number of studies have been carried out in field plots. For example, Bulluck & Ristaino (2002) examined the effects of composted cotton-gin trash on the incidence of southern blight of tomatoes (*Sclerotium rolfsii*). Disease incidence was between 3 and 23% in infested plots treated with compost and was between 61 and 67% in infested plots with no compost. Coventry *et al.* (2006) found that application of onion waste compost in field studies, reduced the viability of the onion white rot pathogen, *Sclerotium cepivorum* and was as effective in reducing onion white rot as a standard fungicide treatment (tebuconazole). In a study of the effect of cattle manure composts on the incidence and severity of black scurf (*Rhizoctonia solani*) on organically grown potatoes, Tsror [Lakhim] *et al.* (2001) found that application of compost at 60 $m^3 ha^{-1}$ reduced disease incidence by between 18.6 and 62.3% in comparison to controls with no compost added. Tilston *et al.* (2005) reported that greenwaste compost applied at 100 or 150 $t\,ha^{-1}$ ameliorated the effects of take-all (caused by *Gaeumannomyces graminis*) during a cropping period with high disease pressure. Finally, Lewis *et al.* (1992) found that amendment of field plots with composted sewage sludge (7–10 $t\,ha^{-1}$) significantly reduced the incidence of damping-off of smooth skinned and wrinkled pea cultivars caused mainly by *Rhizoctonia solani* and *Pythium ultimum*.

It has been shown that composts applied to field soils can also reduce the incidence and severity of foliar diseases. For example, Weltzein (1990) found that the addition of composts based on horse manure and straw bedding to soils in pots reduced the incidence of *Erysiphe graminis* on wheat and barley and *Sphaerotheca fuliginea* on cucumber.

The studies discussed here report differences between disease incidence and/or severity on untreated soil-grown crops and crops grown in soil which has been treated with compost. Few comparisons have been made between the level of disease suppression achieved through the use of composts and that achieved following standard fungicide treatments. These studies are valid due to the decreasing number of pesticide active ingredients and to the increasing number of farmers and growers who choose their crops without pesticides. It is difficult to compare different studies, since there is rarely more than one report of disease suppression using composts for any given crop. The studies also differ in terms of the feedstocks used, climate, soil type and experimental protocol. Work to date has, however, covered a range of economically important pathogens and cropping systems and the results of this work will form a useful platform on which to base future studies. The variable and often low to moderate levels of disease suppression and control typically obtained through the use of composts in field soils suggest that the most important applications for composts would lie in sustainable and organic agriculture and horticulture, where pesticide use is minimal or absent.

There are considerably more reports of disease suppression following application of true composts as opposed to stacked or fresh manures (stacked manures are sometimes wrongly labelled as composts). A limited number of studies have shown that fresh or stacked manures have enhanced suppression of soil-borne diseases. For example, damping-off of radish and lesion development caused by *Rhizoctonia solani* was reduced in plot experiments in Iowa where manure was added in comparison with non-amended controls (Voland & Epstein, 1994). Similarly, fresh chicken manure has been shown to reduce survival of the pathogen *Phytophthora cinnamomi* and the incidence of symptoms on *Lupinus albus* seedlings (Aryantha *et al.*, 2000). The impact of manures on disease incidence and severity is however, much less predictable than that of composts and several workers have shown that the application of manures to soils can increase the incidence and severity of pests and diseases (Roy & Newhook, 1970; Aryantha *et al.*, 2000). Available evidence suggests that composts are likely to give more predictable, consistent control of plant diseases and that research efforts would be better directed towards developing composts as aids to crop protection in organic systems, rather than manures.

5.7.3 Turf

Considerable work has been done, mainly in the United States, to determine the effects of composts used on turf grass diseases. Most of this work relates to golf courses, with the result that compost is now widely used on US golf courses to control a range of fungal diseases.

When used as top dressings, composts have been consistently been shown to suppress disease in comparison with untreated turf or turf dressed with either sand or topsoil (Nelson & Boehm, 2002). Diseases suppressed or controlled include those caused by foliar pathogens [red thread (*Laetisaria fuciformis*), brown patch patch (*Rhizoctonia solani*), dollar spot (*Sclerotinia homeocarpa*), typhula blight (*Typhula incarnata*) and pythium blight (*Pythium aphanidermatum*)] and root pathogens [summer patch (*Magnaporthe poae*), pythium root rot (*Pythium graminicola*) and necrotic ringspot (*Leptosphaeria korrae*)]. Levels of disease suppression varied between 0 and 94%, depending on the disease, the amount of compost used, the type of compost, the experimental year and the researchers concerned. Feedstocks of composts which have successfully suppressed turf grass diseases include animal manures, industrial and municipal sludges, greenwaste (or yard trimmings), grass clippings and food residuals. Monthly applications of topdressing composed of as little as 20% compost by volume and applied at rates of 10 lbs compost/ 1000 ft^2 (488 kg compost ha^{-1}) have been sufficient to suppress disease. In the short term, disease control has often been shown to be less effective with composts than with fungicides. However, work has shown that in the longer term, turf quality and the level of disease control obtained using composts can be better than where fungicides are used.

Composts have also been tested on United States golf courses as 'winter covers' (applied at approximately 1.25 cm depth during the winter months only to protect the grass) and root zone amendments (compost is brushed in to the hollow cores in the turf root zone made using hollow tine aerators). Composts used in both applications have been shown to suppress disease. For example, seedling establishment was enhanced and Pythium root rot was suppressed where the turf root zone was treated with compost based on municipal biosolids or brewery sludge (Nelson & Boehm, 2002). Preliminary studies have shown that the application of compost as a winter cover can protect turf grasses from snow mould damage in winter and early spring. There is little evidence that composts are being used for turf grass applications on European golf courses or other sports turf at present, although preliminary research in the UK has indicated positive benefits (Lawson & Brundage, 2006).

5.7.4 Factors affecting disease suppression

Several direct and indirect factors affect the extent to which composts confer disease suppressiveness in field soils and growing media. It is acknowledged that the addition of organic amendments (including composts) to field soils often leads to improved soil structure, water penetration and drainage, enhanced soil health and greater complexity of microorganisms and soil food webs (van Bruggen, 1995). Most organic amendments commonly applied on farms and holdings are acknowledged to benefit soil health, particularly if applied regularly over an extended period of time (Doran *et al.*, 1996; Albiach *et al.*, 2000; van Bruggen & Termorshuizen, 2003). The health and quality of crops grown in soils of high health status is likely to be better than that of those grown in poor soil conditions, hence the addition of organic amendments may indirectly improve crop health. The application of organic amendments often results in increased populations of both soil microflora and soil fauna. However, it is thought that the complexity of the soil ecosystem at the lower trophic levels is most important in relation to the suppression of plant disease, since fungal and bacterial plant pathogens are affected most by microorganisms at these lower trophic levels (van Bruggen & Termorshuizen, 2003).

The following factors have been identified as determinants of success in biological control of plant pathogens with composts:

- feedstock type(s), and the composting system;
- level of compost maturity and the degree of decomposition of the organic matter;
- chemical and physical attributes of the compost, which affect the activity of beneficial and pathogenic organisms within it and the susceptibility of the host plant to disease (these are largely determined by the feedstock type(s), composting system, level of compost maturity and degree of decomposition of the organic matter);
- biological attributes of the compost, which play an important role in determining its suppressiveness (in particular, the presence of plant pathogens and biological control agents). Again, these are largely determined by the feedstock type(s), composting system, level of compost maturity and degree of decomposition of the organic matter);
- the presence of biological control agents (which can affect disease suppressiveness) added to the compost after composting.

The above factors are discussed in more detail in the following pages.

5.7.4.1 Feedstock type and composting system

Composts made from a wide variety of feedstocks have been used with success to control pathogens. In most cases, the feedstocks have been chosen due to their low cost and/or ease of procurement in the geographical area in question. However several workers have shown that composts made using similar methods but from different feedstocks may perform differently. For example, de Brito Alvarez *et al.* (1995) found that three out of four composts prepared from different feedstocks were associated with improved growth of tomato, whereas the fourth compost significantly depressed tomato plant growth. Aryantha *et al.* (2000) showed that fresh chicken manure or chicken manure composted for 5 weeks, suppressed root rot caused by *Phytophthora cinnamomi*. They also found that chicken manure composted for 2 weeks was less suppressive and that composts made from cow, sheep or horse manure did not suppress root rot caused by *P. cinnamomi*. The nature of microbial populations has been shown to depend on the nature of the feedstock. For example, composts prepared from lignocellulosic substances such as tree barks tend to become colonized primarily by *Trichoderma* spp. (Kuter *et al.*, 1983), which have been shown to control *Rhizoctonia solani* (Grebus *et al.*, 1994). In contrast, grape pomace, which contains high concentrations of sugars and relatively low levels of cellulosic substances tends to become colonized by *Aspergillus* and *Penicillium* spp., which have been shown to suppress *Sclerotium rolfsii* (Hadar & Gorodecki, 1991).

Several consultants and researchers (mainly in the United States) are now suggesting that it is possible to produce specific composts for particular crops, soils and host/pathogen combinations (e.g. www.soilfoodweb.com). The work to support these recommendations is not yet widely published in the scientific literature and further work is required to fully develop these theories into practice for different cropping systems.

Most composts that have been used successfully to prevent and control diseases have been produced aerobically. Even composts produced under aerobic conditions contain small pockets of anaerobic material. However, it is recognized that composts produced under predominantly anaerobic conditions contain a range of toxic end products including low molecular weight organic acids. Some of these composts can remain toxic to plants for months or years (Hoitink, 1980). The presence of these toxic products is less critical for composts used in field soil if incorporation occurs well ahead of planting.

5.7.4.2 Organic matter decomposition level (and compost maturity)

The quality of the organic matter with respect to available substrate will determine whether beneficial organisms and/or facultative saprotrophic pathogens can multiply in it. The effects of partially decomposed organic materials such as composts on soil microbial populations (and plant disease) are different from those of fresh plant materials. The addition of fresh or barely decomposed organic residues to soil has often been shown to cause temporary increases in populations of facultative saprotrophic pathogens such as *Rhizoctonia* and *Pythium* spp., since they can reproduce easily in such material. For example, damping-off caused by *Pythium* spp. was most severe 7 or 10 days after incorporation of cover crop residues (van Bruggen & Termorshuizen, 2003).

Reports of disease suppression by fresh or stacked manures are relatively rare in the literature. This may be partly due to the fact that some manures tend to provide unsuitable conditions for growth and proliferation of antagonistic microorganisms due to high salt content, high ammonium concentrations and in some cases lack of oxygen within manure stacks (Aryantha *et al.*, 2000). They can also cause phytotoxicity, particularly to sensitive horticultural crops (Roy & Newhook, 1970; Aryantha *et al.*, 2000).

The degree of maturity of composts is crucial in determining their disease suppressiveness. Fresh organic matter does not usually support biological disease control, even if it is inoculated with microbial species/strains of proven efficacy (De Ceuster *et al.*, 1999).

High concentrations of free nutrients in fresh crop residues inhibit the production of enzymes required for parasitism by biocontrol agents such as *Trichoderma* spp. (Hoitink *et al.*, 1993).

Composts must be sufficiently stable and colonized to a degree that allows microbiostasis to prevail. Immature composts frequently contain toxic compounds which affect the growth of crop plants and pre-dispose them to attack by pests and pathogens (Hoitink *et al.*, 1993; Hoitink & Boehm, 1999). The point during development of maturity at which most of the readily soluble sugars and other easily metabolized nutrients have been used up often coincides with the development of disease suppression.

On the other hand, excessively humidified organic matter, such as dark sphagnum peat, cannot support the activity of biocontrol agents. For example, the addition of older, more humidified peats to composted bark (a common practice in container production) reduced or eliminated its suppressiveness (Boehm & Hoitink, 1992). The maturity of bark compost has also been shown to affect the degree of suppressiveness (Nelson *et al.*, 1983). Composts that contain organic matter with properties in between the two extremes of decomposition are likely to best support biocontrol. There are several commercially available diagnostic tests available to determine compost maturity based on respiration rate (O_2 uptake or CO_2 evolution), C:N ratio and ammonium-N:nitrate-N ratio or a mixture of these. Others have determined the decomposition level of organic matter present in composts using nuclear resonance spectroscopy (NMR) or Fourier Transformed Infrared (FT-IR) procedures with the aim of relating it to the level of disease suppression recorded (Inbar *et al.*, 1989; Boehm *et al.*, 1997). However, researchers do not yet agree on practical guidelines that define the critical stage of decomposition. The temperature zone within the compost pile from where the composted hardwood bark is taken has also been shown to affect suppressiveness of the container medium (Chung & Hoitink, 1990).

5.7.4.3 Chemical and physical attributes of the compost

Several chemical and physical compost properties (the values of which are largely derived from the feedstock type (s), the composting system, the level of compost maturity and the degree of decomposition of the organic matter) are known to affect crop growth, crop health and the degree of compost suppressiveness (Hoitink *et al.*, 1993). These include cellulose and lignin content, C:N ratio, nutrient (especially N) content, electrical conductivity (content of soluble salts), pH, the presence of toxic compounds, particle size and air-filled porosity (especially in container media), moisture content and possibly also the chloride ion concentration.

Sufficient information has now become available on the disease-suppressive properties of composts to allow predictable biological control of diseases in some crop production systems. In particular, reliable control of *Pythium* and *Phytophthora* spp. can be achieved in container production systems, where the optimum chemical and physical properties of growing media have been documented and tested in detail (Nelson *et al.*, 1983; Hoitink & Fahy, 1986; Ownley & Benson, 1991). It has also been shown that the use of composts prepared from heterogeneous wastes that vary in salinity, N availability and degree of decomposition can lead to marked increases in disease incidence and severity, therefore quality control of composts is of prime importance where they are to be used as part of a disease control strategy.

The amount of N present in composts has been shown to affect disease suppressiveness. Phytophthora dieback and fireblight (caused by *Erwinia amylovora*) are two examples of diseases which are increased as a result of excessive N fertility (Hoitink *et al.*, 1986). Fusarium diseases tend to be increased by compost amendments that are high in ammonium N in particular. This means that they can be enhanced in field soils or container media amended with sewage sludge composts. Sewage sludge composts have a low C:N ratio and release predominantly ammonium N (Kato *et al.*, 1981). In contrast, composts with high C:N ratios immobilize N and suppress Fusarium diseases if colonized by appropriate microorganisms (Trillas-Gay *et al.*, 1986).

The total salt content in composts has been shown to affect the biological control of root rot caused by *Phytophthora* spp. on soybean (Hoitink *et al.*, 1993). Composted municipal solid waste applied 4 months ahead of planting (to allow for leaching of salts) increased soybean yield and controlled the root rot. Application of the same compost just prior to planting decreased soybean yield in comparison to the control.

Composts based on tree bark release toxic compounds (natural fungicides) that lyse zoospores and sporangia of *Phytophthora* spp. (Hoitink & Fahy, 1986). As the decomposition level of composts increases, the role of these natural fungicides in the overall suppressive effect on the pathogen decreases and the contribution of the biocontrol agents gradually increase.

A lack of oxygen around plant roots, for example due to an abundance of small particles/pores in the soil/growing medium/compost mix or to the presence of rapidly decomposing compost can predispose plants to attack by root pathogens such as *Phytophthora* spp. (Hoitink & Fahy, 1986).

5.7.4.4 Biological attributes of the compost

Very little information exists on the biology of compost-amended soils. A significant amount of literature now exists on the antagonists involved in suppression of plant pathogens in compost-amended media however, and this can be studied with a view to extending the use of disease-suppressive composts in field soils. The biology of composts depends primarily on the feedstocks used, the nature of the composting process (in particular the temperatures generated and their duration, the moisture and oxygen content) and on colonization of the compost after peak heating.

Fate of plant pathogens and biocontrol agents during composting

Disease-suppressive composts should by definition contain no plant pathogens. Eradication of plant pathogens present in the original feedstock occurs during the composting process as a result of exposure to high temperatures, release of toxic products during or after the self-heating process and microbial antagonism in the sub-lethal outer temperature zones of piles/windrows or later during curing. Most plant pathogens are killed by 30 minutes exposure to 55°C (Hoitink & Fahy, 1986). A few plant pathogens such as tobacco mosaic virus, the clubroot pathogen (*Plasmodiophora brassicae*) and some *forma speciales* of *Fusarium oxysporum* are less sensitive to heat and highly controlled in-vessel composting systems may be required if feedstock material is likely to be contaminated with such pathogens.

Most beneficial microorganisms are also killed during the high-temperature phase of composting. However, some remain in the outer low temperature parts of the compost pile/windrow (Hoitink *et al.*, 1997). The disease-suppressive properties of composts are usually induced during the curing phase, because most biocontrol agents re-colonize the compost after peak heating. A wide range of species has been identified as biocontrol agents in compost-amended substrates. These include *Bacillus* spp., *Enterobacter* spp., *Flavobacterium balustinum* 299, *Pseudomonas* spp., other bacterial genera and *Streptomyces* spp. as well as fungal species including *Penicillium* spp., *Gliocladium virens*, several *Trichoderma* spp. and others (Chung & Hoitink, 1990; Hoitink *et al.*, 1997). Compost moisture content (ideally 40–50% moisture) is critical if the compost is to be successfully colonized by disease-suppressive microorganisms after peak heating.

Compost produced in the open, in an environment which is high in microbial species diversity has been shown to be colonized by a greater variety of microbial species than the same produced in an in-vessel system (Kuter *et al.*, 1983). This may be partly because the survival of a wide range of beneficial microorganism species is less likely in in-vessel systems, because the entire contents of the vessel will reach consistently high temperatures at the same time. Composts made in enclosed systems may require to be cured for longer to improve suppressiveness, incorporated into soils for several months prior to planting or they may require inoculation with specific biological control agents (Hoitink *et al.*, 1997).

Microflora and fauna associated with suppressive composts

The types of microorganisms isolated from disease-suppressive container media are similar to those studied by scientists working on biological control in field soils. It has been demonstrated that fungal populations in composted hardwood bark media suppressive and conducive to rhizoctonia damping-off differ significantly (Kuter *et al.*, 1983). Although no single species dominated in all the media tested, *Trichoderma* and *Gliocladium virens* were abundant in all suppressive media. *Trichoderma* spp. were also identified as important fungal antagonists in composts prepared from larch bark for control of Fusarium brown rot in Chinese yam (Sekiguchi, 1977).

There is little information available on the activity of bacterial antagonists in composts or compost-amended soils. Bacterial antagonists recovered at random by baiting with propagules of *Rhizoctonia solani* or plant roots from suppressive batches of composted hardwood bark include isolates of *Pseudomonas aeruginosa, P. putida, P. stutzeri, Xanthomonas maltophilia, Janthinobacterium lividum, Flavobacterium balustinum, Enterobacter cloacae, E. agglomerans, Bacillus cereus, B. mycoides* and *B. subtilis*. It is not known which of these bacterial antagonists predominate in suppressive composts or what their relative contributions are. However it can be concluded that the bacterial and fungal colonists isolated are generally rapid, primary colonizers of organic matter.

Microarthropods (springtails and mites) may play a role in the suppression of soil-borne plant pathogens in compost-amended substrates, although specific reports on their role in composts are few. It is known that they are most active in soils that contain high levels of organic matter (Brady & Weil, 1999). Long-term organically managed soils tend to have higher organic matter levels and have also been shown to contain higher populations of soil fauna (including collembola, predatory nematodes and mites) than their conventionally managed counterparts (van Bruggen & Termorshuizen, 2003).

Many attempts have been made to develop techniques for predicting the disease-suppressive properties of composts based on the presence of specific microbial antagonists or the level of microbial activity, but prediction of disease suppressiveness in composts is rarely possible, and no one technique can be used to predict the suppressiveness of a compost to all diseases. The presence of specific antagonists, such as *Trichoderma* spp. and/or *Gliocladium virens* in a compost is no guarantee that that compost will be suppressive to diseases caused by *Rhizoctonia solani*, despite the fact that these fungi are known to be important biological control agents of *R. solani*. Several workers have used a simple, rapid enzyme assay to determine microbial activity based on the rate of hydrolysis of fluorescein di-acetate (Boehm & Hoitink, 1992; Ryckeboer, 2001). Results obtained can be reliably used to determine the suppressiveness of potting mixes to Pythium diseases, but it is not possible to predict the level of suppression of diseases caused by other pathogens such as *Rhizoctonia solani* using this technique.

Several approaches have been used to monitor changes in the soil microbial community structure following amendment with compost. These techniques include for example, physiological profiling using Biolog® microplates (Riddech *et al*., 2002; Borrero *et al*., 2006) or analysis of phospholipid fatty acids (PLFAs) (Sundh & Rönn, 2002), plating on selective media, for example, for fatty acid metabolising microorganisms (McKellar & Nelson, 2003), DNA-based techniques such as analysis of terminal restriction fragment-length polymorphism (T-RFLPs) (Michel *et al*., 2002) and denaturing gradient gel electrophoresis (DGGE) (Calvo Bado *et al*., 2002). These approaches may lead to an improved understanding of which changes in microbial communities are associated with increases in disease suppressiveness within composts.

5.7.4.5 *Inoculation of composts with biological control agents*

Compost-amended potting mixes can be produced which provide consistent natural suppression of damping-off and root diseases caused by *Phytophthora* and *Pythium* spp., but variable suppression is obtained for other diseases. It is possible to improve the suppressiveness of potting mixes to other pathogens such as *Rhizoctonia* and *Fusarium* spp. by amending the compost with specific antagonists including *Trichoderma* spp. (Nelson *et al*., 1983; Cotxarrera *et al*., 2002), *Verticillium biguttatum* and non-pathogenic *Fusarium oxysporum* (Postma *et al*., 2003; Cotxarrera *et al*., 2002). The composts produced can be up to 3 times as suppressive as the unamended, naturally suppressive compost, but the compost into which antagonists are introduced must provide an appropriate environment for growth and reproduction of the antagonist if disease suppression is to be enhanced (Postma *et al*., 2003). For example, the degree of decomposition of the compost was important in determining the degree of suppression of Rhizoctonia damping off in container media (Nelson *et al*., 1983).

5.8 Effects of compost extracts/teas on plant disease

Compost extracts or teas have been tested for their ability to suppress a wide range of foliar diseases in glasshouse and field-grown edible and ornamental crops, with variable results. Much of the early work was carried out on plant pathogens cultured *in vitro* or in detached leaf assays, but there are plenty of recent examples where the suppression of diseases on whole plants has been recorded. Examples of diseases controlled in whole plants include grey mould (*Botrytis cinerea;* Weltzien, 1989; Elad & Shtienberg, 1994, Scheuerell & Mahaffee, 2006), powdery mildews (including *Sphaerotheca* and *Uncinula* spp.; Weltzien, 1989; Elad & Shtienberg, 1994, Scheuerell, 2002), downy mildew (*Plasmopara viticola*; Ketterer, 1990), fungal and bacterial blights/leaf spots (caused by a range of pathogens including *Pseudopeziza tracheiphila, Sphaeropsis sapinea, Pseudomonas* spp. and *Xanthomonas* spp.; Weltzien, 1989; Yohalem *et al.*, 1994; Al-Dahmani *et al.*, 2003) and apple scab (*Venturia inaequalis;* Yohalem *et al.*, 1996). Crops that have been studied in this context include mainly edible cereal, vegetable and fruit crops such as maize, barley, sugar beet, potato, bean, lettuce, pepper, tomato, grape, apple and strawberry. The feedstocks used to prepare composts from which the studied teas were prepared, varied, although most included some form of animal manure. The compost production methods used, the type of extract/tea used and the experimental conditions varied greatly and it was difficult to compare results of work carried out on single diseases by different workers. Disease suppression in the work listed above was extremely variable and was not recorded with all pathogens in all tests. Efficacy depended on the crop, the feedstocks used to make the compost, the compost production system, the extract/tea preparation method and the experimental system used to test the extract/tea.

Definitions of the type of extract used in experimental work varied depending on the authors. However, almost all of the studies involved NCTs under the definition adopted in this review and that of Scheuerell & Mahaffee (2002). Several commercial companies in the United States and Europe now promote and sell machinery to make ACTs (www.soilfoodweb.com; www.attra.ncat.org). However, limited work has been published to date to demonstrate the efficacy of ACTs and there is little scientific evidence to show that they are any more effective in controlling disease than NCTs.

Compost teas are also being widely advertised and used on both organic and conventional farms (mainly in the United States) as an inoculant to restore or enhance soil microflora (www.attra.ncat.org). However, very little work has been done to confirm or quantify the benefits from using compost teas in this way. Tränkner (1992) investigated the effect of NCTs on seedling damping-off caused by *Pythium ultimum.* He found that NCTs prepared from either grape marc or cattle manure suppressed mycelial growth of *P. ultimum in vitro.* He also found that application of NCTs significantly increased seed germination, root length and root dry weight when seeds were soaked in NCT and dried prior to being sown in soil inoculated with *P. ultimum.* There has also been work done to show that NCT suppressed the growth of *Rhizoctonia solani in vitro* (Weltzein, 1991), but it is well known that successful disease control *in vitro* does not always translate to field conditions. Recent work has shown that Fusarium wilt of pepper (*F. oxysporum* f. sp. *vasinfectum*) and cucumber (*F. oxysporum* f. sp. *cucumerinum*) was controlled by drenching NCT on to soil under greenhouse conditions (Ma *et al.*, 2001). More comprehensive accounts of diseases which have been fully or partially controlled through the application of compost teas or extracts under experimental conditions are given in Litterick *et al.* (2004) and Scheuerell & Mahaffee (2002).

5.8.1 Factors affecting disease suppression

There is sufficient information to show that in some cases, plant pathogen control has been at least as good with compost extracts or teas as with conventional fungicides

(Ketterer, 1990; Scheuerell & Mahaffee, 2002). However, compost extracts and teas have, in other studies, been shown to have little or no effect on plant pathogens or disease (Al-Dahmani *et al.*, 2003; Al-Mughrabi, 2006). Research also suggests that different composts, preparation and application methods affect the efficacy of the final product (Weltzein, 1989; Urban & Tränkner, 1993; Elad & Shtienberg, 1994; Yohalem *et al.*, 1994; Brinton *et al.*, 1996; Cronin *et al.*, 1996). The main factors affecting disease suppression are discussed below.

5.8.1.1 Aeration

Most commercially available compost tea brewers are designed to produce ACTs rather than NCTs, but there is considerable debate over whether aeration produces a tea which is more effective in suppressing plant pathogens. Ingham (2003) states that NCTs are less effective than ACT's, but the majority of the scientific literature which demonstrates suppression of plant pathogens concerns NCTs. Few studies have directly compared the effects of ACTs and NCTs made from the same compost source, although those that have done so conclude that aeration of the tea has no effect on disease control (Mahafee & Scheuerell, 2006). The popular literature also contains repeated references to the fact that NCTs are often phytotoxic, but again, there appears to be no documented evidence to support these claims.

5.8.1.2 Compost source and age

The nature of the compost used to produce compost teas (including feedstock, production method and age) has been shown to affect the disease-suppressive properties of the teas. It is difficult to draw clear conclusions from past work. For example, some workers found that teas made from composts based on animal manures were more effective than those made from composts based on undigested vegetable matter. However, others demonstrated that those made from composts based on undigested vegetable matter were equally effective. Scheuerell & Mahaffee (2006) examined the effects of compost teas made from 30 different composts and found that disease suppression was associated with the particular batch of compost and not necessarily the feedstocks used to create the compost.

The impact of compost age on the disease-suppressive properties of compost teas has been studied extensively, but it is difficult to draw meaningful conclusions from the results. Several workers have suggested that the ideal time to use a compost to make compost tea depends on its maturity (which has a significant impact on microbial numbers, microbial diversity and the presence of beneficial microorganisms). The time which compost takes to reach maturity in turn depends on the feedstocks used and the composting method. However, definitions of compost maturity vary, which makes interpretation of the literature difficult.

5.8.1.3 Fermentation time

For both NCTs and ACTs, disease-suppressive properties have been found to increase with fermentation time to a maximum, then decline (Ketterer, 1990; Ketterer & Schwager, 1992). It is thought that the optimum fermentation time is likely to depend on the compost source and the fermentation method. For example, Weltzein (1990) showed that late blight lesions (*Phytophthora infestans*) on detached tomato leaves were suppressed to the greatest extent by 7- and 14-day fermentations, as opposed to 1-, 2- and 28-day fermentations. There is a lack of published data to support the statements made in popular literature concerning the effect of fermentation time on the disease-suppressive properties of compost teas. Ingham (1999, 2003) claims that optimum fermentation times coincide with the presence of maximum active microbial biomass in the teas as they ferment, which is usually 18–36 hours. Others have suggested that whilst 24 hour fermentations are best for compost teas which are intended for use primarily as fertilizers, fermentation times of 7–14 days are better when producing compost teas with optimal disease-suppressive properties.

5.8.1.4 Fermentation nutrients

Most compost teas produced for research purposes or for use in gardens or on commercial crops are produced with added nutrients (i.e. nutrients other than those present in the compost). There are many compost tea recipes, the authors of which claim will produce compost teas, which enhance the growth of specific microbial groups, or will result in particular bacteria:fungi ratios. Several brewer manufacturers also sell proprietary formulations of nutrients, which they claim will help produce teas with particular properties. However, there is very little published scientific evidence to support the efficacy claims made for most of the recipes or proprietary nutrient formulations.

A wide range of nutrients have been added to compost extracts and teas prior to fermentation, primarily in attempts to see whether the disease-suppressive properties of the teas were enhanced. Nutrients used to date include sucrose, malt, yeast extract, peptone, starch, nutrient broth, humic acid, kelp, molasses and rock dust (Mahaffee & Scheuerell, 2006). The addition of fermentation nutrients has been shown in some cases to enhance the disease-suppressive properties of the teas (Ketterer, 1990; Sackenheim *et al.*, 1994; Scheuerell & Mahaffee, 2004, 2006), to reduce the disease-suppressive properties (Urban & Trankner, 1993; Scheuerell & Mahaffee, 2004), or in other cases to have no effects on disease-suppressive properties (Elad & Shtienberg, 1994).

5.8.1.5 Dilution

Very few investigations have been carried out to determine whether dilution of compost extracts and teas (with water) affects their ability to suppress disease. Many small-scale growers apply compost tea at the rate of 50 L ha^{-1}, according to instructions in the Compost Tea Brewing Manual (Ingham, 2003). However, the cost and practical problems associated with applying them at this rate on a large scale would likely be prohibitive. Most published studies relate to a single pathogen, and no overall conclusion relating to the effect of dilution has been reached. Some workers have found that the disease-suppressive properties of compost tea could be maintained after dilution dependent on compost source (Elad & Shtienberg, 1994), whereas others found that dilution decreased disease-suppressive properties of the tea (Scheuerell & Mahaffee, 2004).

5.8.1.6 Application frequency

Decisions as to whether compost teas can be part of an economically viable integrated disease control strategy depend on the frequency of applications required to produce disease-suppressive effects. However, the effects of frequency of application of compost teas on disease suppression have not been studied. Most studies carried out to date have used spray intervals similar to those used for synthetic pesticides.

5.8.1.7 Use of adjuvants

Various adjuvants have been studied with the aim of improving the disease-suppressive properties of compost teas. The four main types of adjuvants which have the potential for use include:

(a) *spreaders* – which reduce the surface tension of spray droplets, thus allowing them to spread evenly over leave surfaces rather than lying in beads.
(b) *stickers* – which enhance the ability of compost teas to adhere to plant surfaces.
(c) *protectants* – which protect microbes from stresses due to desiccation, UV light and so forth.
(d) *nutrients* – which can be used as food substrates for the beneficial microorganisms in the extracts/teas.

The majority of researchers working in this field have recorded increased disease suppression where specific adjuvants have been added to compost extracts/teas prior to application. Adjuvants which have been shown to enhance the disease-suppressive properties of compost teas include a range of spreaders and stickers including methyl cellulose (Sackenheim *et al.*, 1994; Scheuerell, 2002; Scheuerell & Mahaffee, 2004) and various nutrients including molasses, bouillon, rape seed oil and casein (Scheuerell & Mahaffee, 2002; Mahaffee & Scheuerell, 2006). Further work is required to determine the best combinations of adjuvants to use for specific situations.

5.8.2 Problems with the use of compost extracts/teas

One of the most controversial issues associated with the use of compost teas at present, is their potential to propagate and spread human pathogens contained in the feedstock. For example, Welke (1999) detected faecal coliform and Salmonella populations in the source compost, the NCT fermentation and on samples of broccoli and leek growing in a field and sprayed with the NCT. Evidence has shown that pathogens can grow during the production of both ACTs and NCTs. However, the indications are that pathogen growth is not supported when ACTs or NCTs are prepared without fermentation nutrients (Ingram & Millner, 2007). Further work is required to ensure that the production and use of compost teas and extracts can be guaranteed not to propagate and spread human pathogens on to food intended for human consumption.

The issue was recently addressed by the United States National Organic Standards Board (US NOSB, 2004). They set out rules for the safe production of compost tea which, if followed, mean that compost teas produced without added nutrients can be applied without restriction. Compost teas made with added nutrients can be used only if the production system has been shown twice to produce teas that meet the United States Environmental Protection Agency guidelines for recreational water, otherwise, a 90-day harvest interval applies when they are to be used on non-grain food crops. In practice, these standards are likely to eliminate the use of compost teas produced with added nutrients on food crops, since few producers are likely to be prepared to pay for the necessary testing. Given the apparent importance of added nutrients to compost tea efficacy (based on recent experimental results), considerable further work is needed to determine whether human pathogen re-growth in compost teas can occur and to what extent. Methods to prevent re-growth are also needed in order to ensure consumer safety and use of compost teas as part of integrated disease control strategies. There is continuing debate in Europe over whether compost extracts and teas require to be registered with national pesticide authorities such as The UK Pesticides Safety Directorate or whether their use should be regulated through national food standards bodies.

5.9 Mechanisms involved in the suppression/control of plant disease using composts and compost extracts/teas

5.9.1 Composts

Four mechanisms have been described for the activity of biocontrol agents present in composts against soil-borne plant pathogens. They are:

(a) successful competition for nutrients (including primarily C, N, Fe), oxygen and infection sites;
(b) antibiosis;
(c) parasitism and predation;
(d) induced systemic resistance.

Most reports of disease suppression suggest that microbiostasis (i.e. competition and/or antibiosis) and hyperparasitism are the principal mechanisms. Competition results when there is a demand by two or more microorganisms for a resource; competition for that resource may lead to disease control, where a non-pathogen successfully outcompetes a plant pathogen. For example, competition seems to be the main mechanism whereby *F. oxysporum* is controlled by *Trichoderma harzianum T-35* (Sivan & Chet, 1989). Siderophores, such as low molecular weight ferric-specific ligands, are produced under iron-limiting conditions by some beneficial microorganisms to ensure that they secure sufficient iron. This in turn can limit iron availability for some pathogens such as *Pythium* spp. and can result in reduced disease incidence.

Antibiosis occurs when the production of specific and/or non-toxic specific metabolites or antibiotics by one organism has a direct effect on another organism. The biological control of several plant pathogens is known to be mediated at least in part through antibiosis. For example, the toxin 'gliotoxin' has been shown to cause the antagonism of *Gliocladium virens* against *Pythium ultimum* (Roberts & Lumsden, 1990). Weller *et al.* (2002) found that take-all decline in barley, caused by *Gaeumannomyces graminis* var *tritici* was related to a build-up of fluorescent pseudomonad bacteria, which produced the antifungal metabolite 2,4-diacetylphloroglucinol.

Parasitism and predation by beneficial microorganisms in composts has been shown by various workers to be an important mechanism in disease suppression. Four stages of the parasitism process can be distinguished, which can be described as: chemotrophic growth, recognition, attachment and degradation of the host cell walls through production of lytic enzymes, chitinases and β-1,333-glucanases (Elad *et al.*, 1983). For example, antagonists such as *Trichoderma harzianum, T. hamatum* and *T. viride* have been shown to parasitize plant pathogens including *Rhizoctonia solani, Sclerotium rolfsii, Phytophthora* spp. and *Pythium* spp. by secreting enzymes which destroy cell walls.

Induced systemic resistance to a broad spectrum of plant pathogens, triggered by beneficial microorganisms has been proven to be effective in many plant species (De Clercq *et al.*, 2004). Induced systemic resistance shows some phenotypic similarities with plant pathogen-induced systemic acquired resistance, but differences between the two resistance mechanisms becomes apparent when they are studied in depth (Walters *et al.,* 2007; see also Chapter 4). Several workers have recently demonstrated that composts can induce systemic responses in plants, leading to reduced disease severity. For example, Hoitink *et al.* (1997) found that compost mulches applied under parts of trees led to improved disease control throughout the root system. Trankner (1992) reported that compost incorporated into soil reduced the severity of powdery mildew on wheat and several workers have shown that composts can induce systemic resistance to pythium root-rot in cucumber when applied to a section of the root system using a split root system (Zhang *et al.*, 1996; Lievens *et al.*, 2001).

Mechanisms of disease suppression can be divided loosely into two general categories described as 'general' and 'specific' (Baker & Cook, 1974). The term 'general' applies where disease suppression can be attributed to the activity of many different types of microorganisms. The propagules of pathogens which are affected by general suppression do not tend to decline rapidly in soil. They are small (<200 μm in diameter), do not store large quantities of nutrients and rely on exogenous C sources such as seed and root exudates for germination and infection (Nelson, 1990). These pathogens are sensitive to the activities of other microorganisms in the soil, that is they are sensitive to microbiostasis.

It is during the curing or maturation phases of the composting process that composts become suppressive to soil-borne plant diseases caused by *Pythium* and *Phytophthora* spp. The exact point during the maturation process at which composts begin to develop suppressive properties depends on various factors (as discussed earlier). Disease suppressiveness in this instance is related to the total microbial biomass present in the compost. The suppressiveness is mediated by diverse mesophilic organisms which recolonize the compost from the outer, low-temperature layer. General suppression tends to result from competition for nutrients and eocological niches by numerous bacterial and fungal species that adversely affect the activity of, or induce microbiostasis of, plant pathogens. For example, potting mixes containing composted hardwood bark and light, sphagnum peat are naturally suppressive to pythium root-rot and damping-off as a result of general disease suppression (Hoitink & Boehm, 1999).

Specific disease suppression occurs where the presence of just one or two microorganisms can explain suppression of a particular pathogen or the disease that it causes. It has been suggested that the specific microorganism or microorganisms responsible for this effect can be transferred from one soil to another in order to confer suppressiveness, whereas those responsible for general suppression cannot be transferred as easily (Baker & Cook, 1974). Examples of pathogens suppressed in this way include *Rhizoctonia solani* and *Sclerotium rolfsii* (Chung & Hoitink, 1990; Gorodecki & Hadar, 1990). These pathogens both produce large propagules known as sclerotia which do not rely on exogenous C sources for germination and infection. During suppression, the sclerotia are colonized by specific hyperparasites (mainly *Trichoderma* spp.) and their inoculum potential is reduced. Suppression of damping-off caused by *Rhizoctonia solani* is variable, mainly due to the random nature of colonization of compost by effective biological control agents after peak heating (Hoitink *et al.*, 1997). The location of the compost pile in relation to naturally occurring biological control agents (e.g. in forests and agricultural systems) is important in this context.

Microorganisms that produce antibiotics and those that induce systemic resistance in plants (to specific pathogens) represent other examples of specific suppression.

5.9.2 Compost extracts and teas

Most workers acknowledge that several modes of activity are involved in disease suppression following application of compost extracts and NCTs. To date, there is no published work that has determined the mechanisms involved with ACTs. The effects of compost extracts/teas appear mainly to be associated with live microorganisms, since the activity of sterilized or micron-filtered compost extracts has in some cases been shown to be reduced against test pathogens (Weltzein & Ketterer, 1986). In a few cases however, activity has been unaffected following sterilization/micron filtration (Yohalem *et al.*, 1994; Cronin *et al.*, 1996).

The main living active agents in compost teas are thought to be bacteria in the genera *Bacillus* and *Serratia* and fungi in the genera *Penicillium* and *Trichoderma*, although other genera are involved (Brinton *et al.*, 1996). Very little is known about importance of total microbial numbers or species diversity in relation to the efficacy of compost extracts and teas. Induced resistance, antibiosis and competition are thought to be the main means by which live microorganisms bring about disease suppression from NCTs. For example, germination of *Sphaerotheca fuliginea* conidia was not inhibited when treated with NCT *in vitro*. However NCT-treated cucumber leaves demonstrated indicators of induced resistance including increased papilla formation, lignification and necrotic reactions when *S. fuliginea* began to infect (Samerski & Weltzein, 1988).

Several studies have demonstrated that antibiosis can be partly responsible for pathogen or disease suppression, based on work which shows no loss of suppressive activity due to NCTs when the teas are sterilized or micron filtered prior to application (Elad & Shtienberg, 1994; Yohalem *et al.*, 1994; Cronin *et al.*, 1996). It is known that many of the microorganisms present in compost extracts and teas can produce compounds that are toxic to other microorganisms. For example, some chemicals produced by *Pseudomonas* spp. (e.g. siderophores) exert a powerful chemical effect against other organisms (Potera, 1994). Antibiotics known to be produced by *Bacillus subtilis* and others, can inhibit growth and germination of many fungal species (Brinton *et al.*, 1996). NCT has also been shown to have an *in vitro* mycolytic effect on microspores and chlamydospores of *Fusarium oxysporum* f. sp. *cucumerinum* (Ma *et al.*, 2001).

5.10 Conclusions and future work

There is increasing pressure on farmers to reduce pesticide use, whilst maintaining crop quality and yields. At the same time, production of composts from organic wastes is rising as most developed countries are attempting to reduce volumes of waste being sent to landfill. There is growing evidence that the use of composts and compost extracts/teas can help suppress plant diseases as part of integrated disease control strategies through improvements in soil health and through direct control of diseases.

In a few documented cases, control of specific diseases using composts or compost extracts/teas in conventional agricultural or horticultural systems has been equal to or better than that achieved with synthetic pesticides. However, for many pests and diseases the level of control, which has been demonstrated in glasshouse and field trials, is lower than that normally considered acceptable for conventional growers. Commercial and domestic produce buyers may find it difficult to accept that the quality and yield of conventional crops treated in this way are often lower. For organic growers, who have no access to synthetic fungicides, composts and compost extracts/teas may provide particularly useful additions to the range of partial disease control solutions to which they have access.

Considerable further work is required to develop protocols that can be used to ensure predictable and reliable disease suppression or control through application of composts and compost extracts/teas on cropping systems and turf in different soil types.

The majority of research that has yielded commercially important results to date has concerned lignin-rich composts, based for example on tree bark. Given that most of the composts now being produced on a large scale in Europe and North America are based on other feedstocks, (mainly organic wastes including green [or yard] and vegetable wastes), further research is required to determine how to optimize their production and use to favour suppression of plant disease in modern crop production systems.

In particular, work is required to determine consistently effective methods to predict disease suppressiveness in composts. Some of the newer techniques based on organic matter characterization or assessment of microbial species diversity or functional diversity may prove helpful in this case.

Several workers have acknowledged that the use of composts of poor or inappropriate quality may have led to low levels of disease suppression. For this reason, the relative importance of several biological, chemical and physical parameters of compost require further study in order to facilitate production of reliably suppressive composts. Several workers have demonstrated that the amendment of composts with specific biological control agents can result in enhanced disease suppression and some composts have been shown to be excellent carriers of such biological control agents. However, further work is required to ensure that the composts into which these biological control agents are introduced have optimal chemical, physical and biological properties to allow rapid colonization by the introduced organisms.

Many of the recent reports of improved plant growth or successful disease control using compost teas are based on anecdotal information or commercially sensitive data held by private companies. There is a need for independent research to demonstrate the effects of compost teas and to elucidate the mechanisms behind reports of disease suppression or improved plant growth. Evidence from recent work suggests that the use of particular nutrient amendments (to ensure the growth of specific groups of beneficial microorganisms) and the use of spray adjuvants can significantly enhance the disease-suppressive properties of compost teas. Further work is required to optimize the use of such additives for use in specific cropping systems. The development of compost tea application systems that are especially suited to deliver live microorganisms (e.g. drenching and spraying equipment) may also help to enhance the disease-suppressive properties of compost teas.

5.11 References

Abawi GS, Widmer TL, 2000. Impact of soil health management practices on soilborne pathogens, nematodes and root diseases of vegetable crops. *Applied Soil Ecology* **15,** 37–47.

Albiach R, Canet R, Pomares F, Ingelmo F, 2000. Microbial biomass content and enzymatic activities after the application of organic amendments to a horticultural soil. *Bioresource Technology* **75,** 43–48.

Al-Dahmani JH, Abbasi PA, Miller SA, Hoitink HAJ, 2003. Suppression of bacterial spot of tomato with foliar sprays of compost extracts under greenhouse and field conditions. *Plant Disease* **87,** 913–919.

Al-Mughrabi K, 2006. Antibiosis ability of aerated compost tea against foliar and tuber potato diseases. *Biotechnology* **5,** 69–74.

Aryantha IP, Cross R, Guest, DI, 2000. Suppression of *Phytophthora cinnamomi* in potting mixes amended with uncomposted and composted animal manures. *Phytopathology* **90,** 775–782.

Baker KF, Cook RJ, 1974. Biological control of plant pathogens. In: Kelman A, Sequira L, eds. *The Biology of Plant Pathogens.* San Francisco: W.H. Freeman and Co, 220–285.

Boehm MJ, Hoitink HAJ, 1992. Sustenance of microbial activity in potting mixes and its impact on severity of Pythium root-rot of poinsettia. *Phytopathology* **82,** 259–264.

Boehm MJ, Wu T, Stone AG, Kraakman B, Iannotti DA, Wilson GE, Madden LV, Hoitink HAJ, 1997. Cross-polarized magic angle spinning 13C nuclear magnetic resonance spectroscopic characterization of soil organic matter relative to culturable bacterial species composition and sustained biological control for pythium root rot. *Applied and Environmental Microbiology* **63,** 162–168.

Borrero C, Ordovás J, Trillas MI, Avilés M, 2006. Tomato Fusarium wilt suppressiveness. The relationship between the organic plant growth media and their microbial communities as characterised by Biolog®. *Soil Biology and Biochemistry* **38,** 1631–1637.

Brady NC, Weil RR, 1999. Organisms and ecology of the soil. In: Brady NC, Weil RR, eds. *The Nature and Properties of Soils.* New Jersey: Prentice-Hall, 404–445.

Brinton WF, Tränkner A, Droffner M, 1996. Investigations into liquid compost extracts. *BioCycle* **37,** 68–70.

Bulluck III LR, Ristaino JB, 2002. Effect of synthetic and organic soil fertility amendments on southern blight, soil microbial communities, and yield of processing tomatoes. *Phytopathology* **92,** 181–189.

Calvo Bado LA, Morgan JAW, Sergent M, Pettit T, Whipps JM, 2002. Molecular characterisation of legionella populations present within slow sand filters used for fungal plant pathogen suppression in horticultural crops. *Applied and Environmental Microbiology* **69,** 533–541.

Chung YR, Hoitink HAJ, 1990. Interactions between thermophilic fungi and *Trichoderma hamatum* in suppression of rhizoctonia damping-off in a bark compost-amended container medium. *Phytopathology* **80,** 73–77.

Cooke DA, 1993. Pests. In: Cooke DA, Scott RK, eds. *The Sugar Beet Crop.* London: Chapman & Hall, 429–484.

Coventry E, Noble R, Mead A, Marin FR, Perez JA, Whipps JM, 2006. Allium white rot suppression with composts and *Trichoderma viride* in relation to sclerotia viability. *Phytopathology* **96,** 1009–1020.

Cotxarrera L, Trillas-Gay MI, Steinberg C, Alabouvette C, 2002. Use of sewage sludge compost and *Trichoderma asperellum* isolates to suppress Fusarium wilt of tomato. *Soil Biology and Biochemistry* **34,** 467–476.

Cronin MJ, Yohalem DS, Harris RF, Andrews JH, 1996. Putative mechanism and dynamics of inhibition of the apple scab pathogen *Venturia inaequalis* by compost extracts. *Soil Biology and Biochemistry* **28,** 1241–1249.

de Brito Alvarez MA, Gagné S, Antoun H, 1995. Effect of compost on rhizosphere microflora of the tomato and on the incidence of plant growth promoting Rhizobacteria. *Applied and Environmental. Microbiology* **61,** 194–199.

De Ceuster TJJ, Hoitink HAJ, 1999. Prospects for composts and biocontrol agents as substitutes for methyl bromide in biological control of plant diseases. *Compost Science and Utilization* **7,** 6–15.

De Clercq D, Vandesteene L, Coosemans J, Ryckeboer J, 2004. Use of compost as suppressor of plant diseases. In: Lens P, Hamelers B, Hoitink HAJ, Bidlingmaier W, eds. *Resource Recovery and Reuse in Organic Solid Waste Management.* London: IWA Publishing, 317–337.

Ditzler CA, Tugel AJ, 2002. Soil quality field tools: Experiences of USDA-NRCS Soil Quality Institute. *Agronomy Journal* **94,** 33–38.

Diver S, 2001. *Notes on Compost Teas: A 2001 Supplement to the ATTRA Publication 'Compost Teas for Plant Disease Control'*. Fayetteville, Arkansas: ATTRA publication.

Doran JW, Sarrantonio M, Liebig MA, 1996. Soil health and sustainability. *Advances in Agronomy* **56,** 2–54.

Elad Y, Chet I, Boyle P, Henis Y, 1983. Parasitism of *Trichoderma* spp. on *Rhizoctonia solani* and *Sclerotium rolfsii* – scanning electron microscopy and fluorescence microscopy. *Phytopathology* **73,** 85–88.

Elad Y, Shtienberg D, 1994. Effect of compost water extracts on grey mould (*Botrytis cinerea*). *Crop Protection* **13,** 109–114.

Epstein E, 1997. *The Science of Composting*. Lancaster, Pennsylvania, USA: Technomic Publishing Company Inc.

Gorodecki B, Hadar Y, 1990. Suppression of *Rhizoctonia solani* and *Sclerotium rolfsii* in container media containing composted separated cattle manure and composted grape marc. *Crop Protection* **9,** 271–274.

Grebus ME, Watson ME, Hoitink HAJ, 1994. Biological, chemical and physical properties of composted yard trimmings as indicators of maturity and plant disease suppression. *Compost Science and Utilization* **2,** 57–71.

Hadar Y, Gorodecki B, 1991. Suppression of germination of sclerotia of *Sclerotium rolfsii* in compost. *Soil Biology and Biochemistry* **23,** 303–306.

Hadar Y, Mandelbaum R, 1986. Suppression of *Pythium aphanidermatum* damping-off in container media containing composted liquorice roots. *Crop Protection* **5,** 88–92.

Hoitink HAJ, 1980. Composted bark, a lightweight growth medium with fungicidal properties. *Plant Disease* **64,** 142–147.

Hoitink HAJ, Boehm MJ, 1999 Biocontrol within the context of soil microbial communities: A substrate-dependent phenomenon. *Annual Review of Phytopathology* **37,** 427–426.

Hoitink HAJ, Boehm MJ, Hadar Y, 1993. Mechanisms of suppression of soilborne plant pathogens in compost-amended substrates. In: Hoitink HAJ, Keener HM, eds. *Science and Engineering of Composting: Design, Environmental, Microbiological and Utilisation Aspects.* Worthington, Ohio: Renaissance Publications, 601–621.

Hoitink HAJ, Fahy PC, 1986. Basis for the control of soilborne plant pathogens with composts. *Annual Review of Phytopathology* **24,** 93–114.

Hoitink HAJ, Stone AG, Han DY, 1997. Suppression of plant disease by composts. *HortScience* **32,** 184–187.

Hoitink HAJ, Watson ME, Faber WR, 1986. Effect of nitrogen concentration in juvenile foliage of rhododendron on phytophthora dieback severity. *Plant Disease* **70,** 292–294.

Howard A, 1943. *An Agricultural Testament.* New York: Oxford University Press.

Inbar Y, Chen Y, Hadar Y, 1989. Solid-state carbon-13 nuclear magnetic resonance and infrared spectroscopy of composted organic matter. *Soil Science Society of America Journal* **53,** 1695–1701.

Ingham ER, 1999. Making a high quality compost tea, Part II, *Biocycle* **40,** 94.

Ingham ER, 2003. *The Compost Tea Brewing Manual.* Corvallis, Oregon, USA: Soil Foodweb Inc.

Ingham ER, Alms M, 1999. *Compost Tea Manual.* Corvallis, Oregon, USA: Soil Food Web Inc.

Ingram DT, Millner PD, 2007. Factors affecting compost tea as a potential source of *Escherichia coli* and *Salmonella* on fresh produce. *Journal of Food Protection* **70,** 828–834.

Kato K, Fukaya M, Tomita I, 1981. Effect of successive applications of various soil amendments on tomato fusarium wilt. *Research Bulletin of the Aichi Agricultural Research Centre* **13,** 199–208.

Ketterer N, 1990. Untersuchungen zur wirkung von kompostextracten auf den blattbefall ser kartoffel und tomate durch *Phytophthora infestans* sowie auf den befall der weinrebe durch *Plasmopara viticola, Pseudopeziza tracheiphila* und *Uncinula necator*. Dissertation, University of Bonn.

Ketterer N, Schwager L, 1992. Effect of compost extracts on the disease incidence and the phyllosphere flora of bush bean and tomato leaves. *Mededelingen van de Faculteit-Landbouwwetenschappen, Universiteit Gent* **57,** 411–421.

Koepf HH, 1992. *Biodynamic Farming: Principles and Practice*. New York: Anthro Press.

Kuter GA, Nelson EB, Hoitink HAJ, Madden LV, 1983. Fungal populations in container media amended with composted hardwood bark suppressive and conducive to rhizoctonia damping-off. *Phytopathology* **73,** 1450–1456.

Lawson D, Brundage J, 2006. Demonstration trials of the utilisation of composted materials in the maintenance of sports and amenity turfgrass. ORG 0027. The Waste and Resources Action Programme, Banbury, UK.

Lewis JA, Lumsden RD, Millner PD, Keinath AP, 1992. Suppression of damping-off of peas and cotton in the field with composted sewage sludge. *Crop Protection* **11,** 260–266.

Lievens B, Vaes K, Coosemans J, Ryckeboer J, 2001. Systemic resistance induced in cucumber against Pythium root rot by source separated household waste and yard trimmings composts. *Compost Science and Utilization* **9,** 221–229.

Litterick AM, Harrier LA, Wallace P, Watson CA, Wood M, 2004. The role of uncomposted materials, composts, manures and compost extracts in reducing pest and disease incidence and severity in sustainable temperate agricultural and horticultural production – a review. *Critical Reviews in Plant Science* **23,** 453–479.

Lumsden RD, Lewis JA, Millner PD, 1983. Effect of composted sewage sludge on several soilborne pathogens and diseases. *Phytopathology* **73,** 1543–1548.

Ma LP, Qiao XW, Gao F, Hao BQ, 2001. Control of sweet pepper fusarium wilt with compost extracts and its mechanism. *Chinese Journal of Applied Environmental Biology* **7,** 84–87.

Mahaffee WF, Scheuerell SJ, 2006. Compost teas: alternative approaches to the biological control of plant diseases. In: Bailey MJ, Lilley AK, Timms-Wilson TM, Spencer-Phillips PTN, eds. *Microbial Ecology of Aerial Plant Surfaces*. Wallingford: CABI International.

Mandelbaum R, Hadar Y, Chen Y, 1988. Composting of agricultural wastes for their use as container media: Effect of heat treatments on suppression of *Pythium aphanidermatum* and microbial activities in substrates containing compost. *Biological Wastes* **26,** 261–274.

McKellar ME, Nelson EB, 2003. Compost-induced suppression of pythium damping-off is mediated by fatty acid-metabolizing seed-colonizing microbial communities. *Applied and Environmental Microbiology* **69,** 452–460.

Michel FC, Marsh TJ, Reddy CA, 2002. Bacterial community structure during yard trimmings composting. In: Insam H, Riddech N, Klammer S, eds. *Microbiology of Composting*. Berlin: Springer Verlag, 25–42.

Nelson EB, 1990. Exudate molecules initiating fungal responses to seeds and roots. *Plant and Soil* **129,** 61–73.

Nelson EB, Boehm MJ, 2002. Compost-induced suppression of turfgrass diseases. *BioCycle* **43,** 51–55.

Nelson EB, Kuter GA, Hoitink HAJ, 1983. Effects of fungal antagonists and compost age on suppression of Rhizoctonia damping-off in container media amended with composted hardwood bark. *Phytopathology* **73,** 1457–1462.

Noble R, Coventry E, 2005. Suppression of soil-borne plant diseases with composts: A review. *Biocontrol Science and Technology* **15,** 3–20.

Ownley BH, Benson DM, 1991. Relationship of matric water potential and air-filled porosity of container media to development of phytophthora root-rot of rhododendron. *Phytopathology* **81,** 936–941.

Postma J, Montanari M, van den Boogert PHJF, 2003. Microbial enrichment to enhance the disease suppressive activity of compost. *European Journal of Soil Biology* **39,** 157–163.

Potera C, 1994. From bacteria: A new weapon against fungal infection. *Science* **265,** 605.

Riddech N, Klammer S, Insam H, 2002. Characterisation of microbial communities during composting of organic wastes. In: Insam H, Riddech N, Klammer S, eds. *Microbiology of Composting*. Berlin: Springer Verlag, 34–52.

Riggle D, 1996. Compost teas in agriculture. *BioCycle* **37,** 65–57.

Roberts DP, Lumsden RD, 1990. Effect of extracellular metabolites from *Gliocladium virens* on germination of sporangia and mycelial growth of *Pythium ultimum*. *Phytopathology* **80,** 461–465.

Roy SC, Newhook FJ, 1970. The influence of cattle excreta on sporulation of *Phytophthora cinnamomi* Rands. *New Zealand Journal of Agricultural Research* **13,** 308–314.

Ryckeboer J, 2001. Biowaste and yard waste composts: Microbiological and hygienic aspects – suppressiveness to plant diseases. PhD thesis. Katholieke Universiteit Leuven, Leuven, Belgium.

Sackenheim R, Weltzien HC, Kast WK, 1994. Effects of microflora composition in the Phyllosphere on biological regulation of grapevine fungal diseases. *Vitis* **33,** 235–240.

Samerski C, Weltzien HC, 1988. Investigations on the mode of action of compost extracts in the host-parasite-system sugar beet – powdery mildew. *Journal of Plant Diseases and Protection* **95,** 176–181.

Scheuerell SJ, 2002. Compost teas and compost-amended container media for plant disease control. PhD Thesis. Oregon State University, Oregon, USA.

Scheuerell SJ, Mahaffee WF, 2002. Compost tea: Principles and prospects for plant disease control. *Compost Science and Utilization* **10,** 313–338.

Scheuerell SJ, Mahaffee WF, 2004. Compost tea as a container medium drench for suppressing seedling damping-off caused by *Pythium ultimum*. *Phytopathology* **94,** 1156–1163.

Scheuerell SJ, Mahaffee WF, 2006. Variability associated with suppression of gray mould (*Botrytis cinerea*) on geranium (*Pelargonium – hortorum*) by non-aerated and aerated compost teas. *Plant Disease* **90,** 1201–1208.

Scheuerell SJ, Sullivan DM, Mahaffee WF, 2005. Suppression of seedling damping-off caused by *Pythium ultimum, P. irregulare* and *Rhizoctonia solani* in container media amended with a diverse range of Pacific Northwest compost sources. *Phytopathology* **95,** 306–315.

Schueler C, Biala J, Vogtmann H, 1989. Antiphytopathogenic properties of biogenic waste compost. *Agriculture, Ecosystems and Environment* **27,** 477–482.

Sekiguchi A, 1977. Control of fusarium wilt on chinese yam. *Annual Report of the Department of Plant Pathology, Entomology, Vegetables and Floriculture Nagano Experimental Station, Japan* **1,** 10–11.

Serra-Wittling C, Houot S, Alabouvette C, 1996. Increased soil suppressiveness to fusarium wilt of flax after addition of municipal solid waste compost. *Soil Biology and Biochemistry* **28,** 1207–1214.

Sivan A, Chet I, 1989. The possible role of competition between *Trichoderma harzianum* and *Fusarium oxysporum* on rhizosphere colonization. *Phytopathology* **79,** 198–203.

Spring DE, Ellis MA, Spotts RA, Hoitink HAJ, Schmitthenner AF, 1980. Suppression of the apple collar rot pathogen in composted hardwood bark. *Phytopathology* **70,** 1209–1212.

Stephens CT, Stebbins TC, 1985. Control of damping-off pathogens in soiless container media *Plant Disease* **69,** 494–496.

Sundh I, Rönn S, 2002 Microbial succession during composting of source-separated urban organic household waste under different initial temperature conditions. In: Insam H, Riddech N, Klammer S, eds. *Microbiology of composting*. Berlin: Springer Verlag, 53–64.

Tilston EL, Pitt D, Fuller MP, Groenhof A, 2005. Compost increases yield and decreases severity of take-all in Winter wheat. *Field Crops Research* **94,** 176–188.

Tränkner A, 1992. Use of agricultural and municipal organic wastes to develop suppressiveness to plant pathogens. In: Tjamos ES, Papavizas GC, Cook RJ, eds. *Biological Control of Plant Diseases*. New York: Plenum Press, 35–42.

Trillas-Gay MI, Hoitink HAJ, Madden LV, 1986. Nature of suppression of *Fusarium* wilt of radish in container medium amendment with composted hardwood bark. *Plant Disease* **70,** 1023–1027.

Tsror [Lakhim] L, Barak R, Sneh B, 2001. Biological control of black scurf on potato under organic management. *Crop Protection* **20,** 145–150.

United States National Organic Standards Board. 2004. Compost Tea Task Force Report April 6 2004. 21 pp. (www.ams.usda.gov/nosb/meetings/CompostTeaTaskForceFinalReport.pdf) Last accessed 20.10.07.

Urban J, Tränkner A, 1993. Control of grey mold (*Botrytis cinerea*) with fermented compost/water extracts. In: Fokkema NJ, Kohl J, Elad Y, eds. *Biological Control of Foliar and Post-harvest Diseases: Proceedings of a Workshop. Bulletin of the International Organization for Biological and Integrated Control of Noxious Animals and Plants, West Palearctic Regional Section*. Montfavet, France, 8–11.

van Bruggen AHC, 1995. Plant disease severity in high-input compared to reduced input and organic farming systems. *Plant Disease* **79,** 976–984.

van Bruggen AHC, Semenov AM, 2000. In search of biological indicators for plant health and disease suppression. *Applied Soil Ecology* **15,** 13–24.

van Bruggen AHC, Termorshuizen AJ, 2003. Integrated approaches to root disease management in organic farming systems. *Australian Journal of Plant Pathology* **32,** 141–156.

Voland RP, Epstein AH, 1994. Development of suppressiveness to diseases caused by *Rhizoctonia solani* in soils amended with composted and noncomposted manure. *Plant Disease* **81,** 461–466.

Walters DR, Bingham IJ, 2007. Influence of nutrition on disease development caused by fungal pathogens: Implications for plant disease control. *Annals of Applied Biology* **151,** 307–324.

Walters D, Newton A, Lyon G, 2007. *Induced Resistance for Plant Disease Control: A Sustainable Approach to Crop Protection*. Oxford: Blackwell Publishing.

Weller DM, Raaijmakers JM, McSpadden Gardner BB, Thomashow LS, 2002. Microbial population responsible for specific soil suppressiveness to plant pathogens. *Annual Review of Plant Pathology* **40,** 309–348.

Welke S. 1999. Effectiveness of compost extracts as disease suppressants in fresh market crops in BC. Organic Farming Research Foundation. Grant Report. Santa Cruz, California, USA, 19–31.

Weltzein HC, 1989. Some effects of composted organic materials on plant health. *Agriculture, Ecosystems and Environment* **27,** 439–446.

Weltzein HC, 1990. The use of composted materials for leaf disease suppression in field crops. In: *Crop Protection in Organic and Low-input Organic Agriculture*. BCPC Monograph No. 45, Farnham, Surrey: British Crop Protection Council, 115–120.

Weltzein HC, 1991. Biocontrol of foliar fungal disease with compost extracts. In: Andrews JH, Hirano SS, eds. *Microbial Ecology of Leaves*. New York, USA: Springer-Verlag, 430–450.

Weltzein HC, Ketterer N, 1986. Control of *Phytophthora infestans* on tomato leaves and potato tubers through water extracts of composted organic wastes. *Phytopathology* **76,** 1104 (abstract 372).

www.attra.ncat.org Website of the National Sustainable Agriculture Information Service. Accessed 18.10.07

www.symbio.co.uk/library/files/Xtractor%20brochure%202%20x%20A3.pdf Website of the commercial company Symbio, Surrey, UK. Accessed 07.11.07

www.composttea.com Website of the commercial company EPM Incorporated. Accessed 18.10.07

www.growingsolutions.com Website of the commercial company Growing Solutions Incorporated. Accessed 18.10.07

www.soilfoodweb.com Website of the commercial company SoilFoodWeb Incorporated. Accessed 18.10.07

www.soilsoup.com Website of the commercial company SoilSoup Incorporated. Accessed 18.10.07

www.solvita.com Website of the commercial company Solvita. Accessed 18.10.07

Yohalem DS, Harris RF, Andrews JH, 1994. Aqueous extracts of spent mushroom substrate for foliar disease control. *Compost Science and Utilization* **2,** 67–74.

Yohalem DS, Nordheim EV, Andrews JH, 1996. The effecf of water extracts of spent mushroom compost on apple scab in the field. *Phytopathology* **86,** 914–922.

Zhang W, Dick WA, Hoitink HAJ, 1996. Compost-induced systemic acquired resistance in cucumber to Pythium root rot and Anthracnose. *Phytopathology* **86,** 1066–1070.

Zibilske LM, 1998. Composting of organic wastes. In: Sylvia DM, Fuhrmann JF, Hartel PG, Zuberer DA, eds. *Principles and Applications of Soil Microbiology.* NJ, USA: Prentice- Hall, Inc., 482–497.

Chapter 6
The use of host plant resistance in disease control

Hugh Wallwork

South Australian Research and Development Institute, GPO Box 397, Adelaide SA 5001, Australia

6.1 Introduction and benefits of resistance

The most effective form of disease control must be for each crop plant to have its own inbuilt defence against infection such that farmers have freedom in other aspects of crop management to pursue maximum economic returns in a sustainable farming system. Resistance naturally occurs against all microbes other than the very few which are pathogenic to a specific crop. For those pathogens, genetic variation exists for resistance at a variety of levels and can be selected in breeding programs. There are, however, a few pathogens where no resistance is currently available, but it is hoped that when the biological basis for resistance to pathogen attack is better understood, new genetic manipulation techniques may provide future varieties with improved resistance mechanisms.

Host resistance provides an easy and cheap control method for farmers to use. The exception may be where higher levels of resistance are only present in lower yielding or poorer quality varieties. This can be remedied where plant breeders put disease resistance as a high priority in their crop development programs although the corollary of this is that yield and quality must take a slightly lower priority.

A second advantage is that in using resistance there is a reduced need to monitor crops or be concerned about the application of fungicides. This will depend on the level of resistance in the variety and severity of the disease in the environment in which it is grown. A third advantage of resistance is that, in lowering the level of disease, less inoculum will be produced and this will in turn reduce the rate of evolution of pathogen virulence providing longer term protection to future crops.

Reliance on resistance has frequently caused a problem when plant breeders have used single, highly effective resistance genes and ignored the genes of smaller effect that provide useful background partial resistance. New virulent pathogen strains have frequently appeared rendering the once resistant varieties susceptible. Where this has occurred repeatedly it has been known as the 'boom and bust' cycle in plant breeding. As these lessons have been learnt, the use of more durable resistances to pathogens with variable virulences has increasingly been sought.

In seeking and using resistance it is critical to have a sound understanding of the genetic basis of host resistance in the varieties and germplasm available to breeding programs and to understand which resistances are likely to be most effective in the long term. This knowledge is also important in determining the best breeding strategy to adopt in developing new varieties.

This chapter will cover the genetics of host resistance, how resistance can be selected and used in a breeding program and some strategies for its deployment in agricultural crops. Some promising new avenues for developing improved resistances are also highlighted.

6.2 Types of resistance

A problem with communicating about disease resistance is the diversity of terms that have been used to describe the different forms of expression of resistance. In most publications, resistance is divided into two types, with one being controlled by single 'strong' resistance genes and the other being controlled by a number of genes of smaller effect. In reality, there is usually a continuum of variation between these two contrasting types of resistance and no one set of contrasting terms is ideal because exceptions are found to any categories that have been developed. In practice, some form of categorisation is helpful in describing variation, so some of the most common terminology that has been used is explained below.

6.2.1 Seedling and adult plant resistance

Seedling resistances are functional from the onset of plant growth and effective throughout the life of the plant. They are generally controlled by single genes and are very effective in the absence of matching virulence in the pathogen. Adult plant resistance (APR) on the other hand, covers a broad range of resistance types distinguished simply by not being effective at the plant's seedling stage. Generally APR is provided by genes of smaller effect, which might operate through a wide variety of mechanisms.

Seedling resistance generally operates in a gene-for-gene manner with the avirulence of the pathogen, as described by Flor (1971). In this model, resistance occurs where a resistance gene product is matched by the presence of an avirulence gene product in the pathogen. This matching leads to an increasingly well-characterised pathway of host responses that leads to a hypersensitive resistance (HR) reaction.

This terminology has been widely used for the rust diseases of cereals where it has been most helpful. It is less useful in describing resistance to some other pathogens such as the necrotrophic pathogens *Pyrenophora tritici-repentis* and *P. teres* f. sp. *maculata* that cause tan spot of wheat and spot form net blotch of barley, respectively. In these instances, seedling resistances are controlled by genes of small effect, some of which are not effective in later growth stages.

6.2.2 Major genes and minor genes

These terms are widely used to describe genes of large and small effect. They roughly match those for seedling/APR and are commonly used and helpful. Historically, major genes have often been associated with race-specific and minor genes with race-non-specific resistance (see below) that has led to some errors and confusion, as many minor genes have been found which are race-specific. The boundary between major and minor genes is not defined and varies across publications.

6.2.3 Multigenic/polygenic resistance

These terms have been widely used to describe resistance controlled by an unknown number of minor genes. The former term avoids the mixing of Latin and Greek derivations. As genetic knowledge of resistance increases, these general terms will tend to be replaced by more precise descriptions.

6.2.4 Race-specific and race non-specific resistance

These terms sought to differentiate between resistance that was subject to loss of effectiveness with the appearance of new virulent strains of a pathogen and resistance that was thought would never be lost because the pathogen was not capable of developing virulence to it. The main problem has been a lack of evidence to suggest that any particular resistance was race-non-specific. The assumption was generally made that genes of large effect (major or seedling genes) were race-specific and genes of small effect (minor genes or APR) were race-non-specific. This has subsequently and comprehensively been shown not to be the case in many host–parasite systems.

6.2.5 Vertical and horizontal resistance

These terms were devised by Van der Plank (1968) to convey the difference between race-specific and race-non-specific resistance. They came from bar graphs used to display resistance to different isolates of the late blight pathogen (*Phytophthora infestans*) in potatoes in South Africa. Where large differences in resistance were noted on the resistance (vertical) axis the resistance was clearly race-specific. Where no significant differences were observed across the different isolates (on the horizontal axis) the resistance was determined to be race-non-specific. These observations caused later confusion when used too rigidly. The principal flaw with these terms was that there was never any good evidence to show that the minor genes that gave a horizontal response were not also susceptible to changes in pathogen virulence.

6.2.6 Qualitative and quantitative resistance

These terms also roughly match those for seedling/APR and vertical/horizontal resistance, with qualitative resistance referring to a high level of resistance controlled by a single gene and quantitative resistance being controlled by several genes of smaller effect. The term 'quantitative' subsequently came to be used to describe locations on chromosomes where genes of small effect were identified through mapping studies, hence 'quantitative trait locus' or QTL. More recently, the use of the term QTL has come to be used for major gene loci as well.

6.2.7 Partial resistance

This term was originally used in studies of the resistance in barley to leaf rust caused by *Puccinia hordei,* to describe resistance that reduced the rate of epidemic development despite having a susceptible reaction type (Parlevliet & Van Ommeren, 1975). In studies on a number of varieties, resistance to leaf rust was found to be governed by up to 6–7 minor genes with additive effects and was correlated with increased latency period and reduced infection frequency, pustule size, infectious period and spore production. The term is now used more generally to describe any resistance that is only partially effective in reducing disease expression and is usually synonymous with minor gene resistance.

6.2.8 Hypersensitive resistance

This refers to highly effective resistance that produces small necrotic spots caused by a strong reaction at the site of infection resulting in localised death of challenged plant cells. This is typically aligned with seedling major gene resistance and has been extensively studied to understand the physiological interaction between host and pathogen. These studies are gradually revealing the complex biochemical interactions that mediate the defence response.

6.2.9 Non-host resistance/immunity

In these responses, no symptoms are seen and the pathogen is not capable of growing on the host.

6.2.10 Durable resistance

This is now one of the most widely used terms in disease resistance breeding, as it describes the resistance that plant breeders and farmers want in their varieties without trying to define the genetic mechanism. The term was suggested for use by Johnson & Law (1975) to avoid problems created by previous terminology and to refer to rust resistance in wheat that in practice had provided stable resistance in varieties that had been grown over a large area for many years. It specifically avoided identifying the resistance with particular phenotypes or suggesting that the resistance would never be lost to a change in pathogen virulence. Specific well-studied examples are the *Sr*2 and *Rpg*1 genes for partial resistance to stem rust in wheat and barley respectively, the genes *Lr*34 and *Yr*18 which provide partial resistance to leaf and stripe rust in wheat and which may in fact be the same gene and the *mlo* locus which provides resistance to powdery mildew in barley.

6.2.11 Other terms used

In some host pathogen systems, minor gene, APR or partial resistance has been observed as resistance that reduces the rate of epidemic spread of a disease. This has been also described as ‘slow rusting’ or ‘slow mildewing’ and so on. Other researchers have noted that some minor gene resistances may be more affected by temperature than major or seedling resistance genes. These have been described as ‘temperature sensitive’ resistances. In reality, resistance provided by genes of small effect are likely to operate by a very wide range of mechanisms including pathogen avoidance, physical barriers, resistance to infection, reduction in pathogen growth and reduced spore production. Reduction in pathogen growth and sporulation in turn are likely to be affected by a wide range of different metabolic pathways, enzymes and other factors. Many of the genes involved may simply modify the expression of resistance rather than affect resistance *per se*.

6.2.12 Tolerance

This is covered in Chapter 7 but is included here as there have been many instances where resistance and tolerance have been confused. Resistance is the ability of a plant to reduce growth and multiplication of a pathogen, whereas tolerance is the ability of a plant to yield despite the presence and growth of a pathogen on the plant. These two traits may be inherited independently of each other, as in the case of cereal cyst nematodes in cereals, which are characterised by having just a single generation in a year. Where growth of a pathogen is continuous or where a nematode has multiple cycles of growth in a season, then resistance and tolerance become less independent of each other.

6.3 Sources of resistance

In developing resistant varieties, plant breeders need access to effective sources of resistance. These may already be present within current varieties and breeding lines in the breeding programs or from similar programs in other regions or countries. Where these are not adequate, the next source of variation comes from national or international germplasm collections where landraces, closely related wild relatives and distantly related species may be found that represent a wider range of useful variation for the trait being sought. Finally, new variation can be generated in a crop using mutation or by genetic engineering using genes identified from other species or newly created versions (alleles) of existing genes or by modified expression of gene activity.

6.3.1 Landraces and wild relatives

Landraces are collections made from fields where the farmer is unlikely to have introduced modern varieties. These landraces are often quite mixed and will have been locally selected over a long period of time and may therefore contain genetic variation which is not present in modern breeding programs, providing useful adaptation to that environment.

Wild relatives, which include crop ancestors, are species that in nature remain largely genetically isolated from the crop species, but which can be hybridised with the crop plant to allow the transfer of a required trait without overwhelming problems of hybrid sterility. Depending on the degree of relatedness, the genetic transfer may involve from two to several rounds of crossing and selection to obtain the required trait in a background suitable for commercial production. Because cultivated wheats will have been selected from a wider population in an earlier age, wild wheat relatives will often contain a much greater degree of genetic variation for most traits.

Where the environments and circumstances in which collections of landraces and wild relatives were made have been carefully documented, it is possible to target accessions that are more likely to reveal useful variation for the trait being sought. Collections may for example be sampled based on the latitude, altitude, soil pH, soil salinity, rainfall zone, farming system or pathogen exposure of the collection site.

International collection centres, which may carry many thousands of accessions of a particular crop species, are increasingly organising their collections such that core groups of germplasm representing the breadth of their collection, but with as little duplication as possible, are identified for first pass evaluation by scientists or breeders who can then return for more targeted sampling. Global information systems (GIS) and computer models are also being used to help identify accessions of particular value for specific purposes.

There are many examples of where wild relatives have provided significant commercial gains to cropping industries. Bioversity International (2008) has listed a few as follows: 'The desirable traits of wild sunflowers (*Helianthus* spp.) are worth an estimated US$267 million to US$384 million annually to the sunflower industry in the United States; one wild tomato variety has contributed to a 2.4 percent increase in solids content worth US$250 million; and three wild peanuts have provided resistance to the root knot nematode, which cost peanut growers around the world US$100 million each year.'

In wheat wild relatives have provided a key source of variation for resistance to stem rust (*Puccinia graminis*) and leaf rust (*P. triticina*), two diseases whose control helps to maintain food supplies to a substantial proportion of the world population. Initially, cultivated tetraploid wheats were used as donors of stem rust resistance for bread wheats (Hayes *et al.*, 1920; McFadden, 1930) and then subsequently wild tetraploid and diploid species were used by others to introduce a range of rust and powdery mildew (*Blumeria graminis* f. sp. *tritici*) resistance genes and other non-biotic stress traits. Possibly the most important such introduction came from deployment of the *Sr*2 gene derived from *Triticum dicoccum* (McFadden, 1930). This gene has subsequently been shown to provide durable partial resistance to stem rust and is being widely used in wheat breeding programs in the twenty-first century to counter this disease.

In rice, resistance to bacterial blight caused by *Xanthomonas oryzae* pv. *oryzae* was identified by researchers in India in 1977 from *Oryza longistaminata*, a wild species of rice from Mali, and transferred into cultivated varieties. The resistance gene, known as *Xa*21, has been transferred and pyramided along with other genes with resistance to bacterial blight into elite breeding lines in India, China, the Philippines and Korea, with some being released for cultivation in China.

6.3.2 Distantly related species

Where closely related species do not provide sufficient variation, then other more distantly related species might be used, providing some means is available for transferring the resistance where straight crosses do not produce viable progeny. This has involved the use of cytogenetic technologies that have facilitated the intercrossing of species that would otherwise remain genetically isolated. In the case of wheat, this has included the use of a 'bridging' species, an example of which was the transfer of rust and eyespot (*Tapesia yallundae*) resistance from *Aegilops ventricosa*. This wild tetraploid species 'V' was crossed to another tetraploid wheat *Triticum persicum* 'P' and the resulting hybrid was then crossed to the French hexaploid wheat variety Marne 'M'. Progeny generated from the hybridisation became known as 'VPM' lines (Doussinault *et al.* 1983) and have been used in the successful development of numerous wheat varieties with resistance to eyespot and rust in the UK, France, United States and Australia (Friebe *et al.* 1996).

A problem with obtaining genes from more distantly related species has been a lack of recombination occurring between the chromosomes of the different species. This has meant that with the required resistance has come many other linked genes, often deleterious to yield and other traits, that were located on the same chromosome. This is known as 'linkage drag'.

Two early methods used for promoting the transfer of smaller segments of chromosome from distantly related species have been spontaneous centric fusion and irradiation. In the former, lines were generated that lacked a chromosome of bread wheat, but which had an additional chromosome from the donor species. Natural, albeit rare, breakage of chromosomes, usually at the centromere, at meiosis is often followed by subsequent fusion of broken chromosome arms leading to rare translocations with a single chromosome carrying segments of both donor and recipient DNA. A very significant example of a centric fusion derived translocation was the transfer of the short arm of chromosome 1R from rye, with the major genes *Sr*31, *Lr*26 and *Yr*9 for resistance to stem, leaf and stripe (yellow) rust respectively, onto the 1B chromosome of wheat (Zeller, 1973). This formed a translocation chromosome known as 1BL/1RS that is now carried in many modern wheat varieties. This segment remains large, however, and also carries an unfavourable quality gene that leads to 'sticky dough' during some industrial processes. A second significant example has been the transfer of the linked genes *Lr*24 and *Sr*24 for leaf and stem rust resistance in wheat on a $3DL/3Ag^e$ translocation from *A. elongatum,* producing a line called Agent (Smith *et al.*, 1968) that has subsequently been used widely in wheat breeding.

Ionizing radiation has been used to generate chromosome breakages that can then lead to subsequent fusion of chromosome segments from donor and recipient species. Sears (1956) used irradiated pollen to transfer the leaf rust resistance gene *Lr*9 from *Ae. umbellulata,* whilst Knott (1961) used irradiated seed to transfer the stem rust resistance gene *Sr*26 from *Agropyrum elongatum* into bread wheat. Ionizing radiation has also been used to transfer resistance to powdery mildew into barley and oats from *Hordeum bulbosum* (Pickering *et al.*, 1995) and *Avena barbata* (Thomas & Aung, 1978), respectively.

The reduction in size of donor chromosome segments to reduce linkage drag has been aided by the discovery and use of a characteristic of some accessions of the wild grass *Triticum speltoides* to suppress a gene *Ph*1 which prevents pairing of non-homologous chromosomes in wheat (Riley *et al.*, 1968). Using genetic stocks with the *Ph* suppressor gene, it became possible to increase recombination between non-homologous (homoeologous) chromosomes. Riley *et al.* (1968) first used this technique to transfer the yellow (stripe) rust resistance gene *Yr*8 from *Aegilops comosa* to produce a line called Compair.

Many examples of resistance transfer have been reported using chromotypes of wheat where the chromosome 5B or 5BL arm carrying the *Ph*1 gene is deleted (nulli-5B), or the segment of chromosome 5BL with the *Ph*1 gene is deleted (the ph1b mutant) (Sears, 1977). In both of these cases, homoeologous chromosome pairing occurs. A notably successful case was the further development of the *Sr*24 and *Lr*24 genes for resistance to stem and leaf rust resistance in wheat, already mentioned above. Sears (1973, 1978) used this method to produce several $3DL/3Ag^e$ translocations from *A. elongatum*, some of which produced white seeded forms through breakage of the linkage to red grain colour in the donor parent. These recombinant lines were used as the source for stem and leaf rust in many past and present Australian wheat cultivars (Friebe *et al.*, 1996) where white grained wheats are required.

More recently, these techniques have been supplemented by the advent of molecular markers, whereby the precise location of recombination events along chromosomes can be identified through the existence of DNA polymorphisms in the region of chromosome translocations. These molecular tags have been used for example to identify crosses made with the ph1b mutant that deliver the smallest possible segments of DNA whilst still retaining the *Sr*26 gene for stem rust resistance in wheat (Dundas and Shepherd, 1998). This has led to breeding lines being developed that should avoid the yield penalty associated with this otherwise successfully used gene in the past (Dundas, 2007).

As well as the crossing of wild species or ancestors with modern wheats, another strategy in which there has been much recent interest has been the use of the diverse diploid and tetraploid ancestors to recreate modern hexaploid wheat. At the International Maize and Wheat Improvement Centre (CIMMYT) in Mexico, an extensive collection of 'synthetic' wheat lines based on crosses between different *Ae. tauschii* (syn. *Ae. squarrosa*) accessions, donor of the D genome in wheat, and tetraploid durums *T. turgidum* ssp. *durum* (AABB genome), have been developed (Mujeeb-Kazi, 1996). Initial crosses were made fertile by chromosome doubling using colchicine and lines with good agronomic adaptation were identified following hybridisation with elite CIMMYT spring bread wheat lines (AABBDD). At the same time, similar synthetic wheats, based in this case on winter wheat crosses, were made in the USA which have led to the identification of bread wheat lines with useful resistances including to Hessian fly (Cox *et al.*, 1995). Synthetic wheats have also been developed in Australia and have been used to transfer resistance to cereal cyst nematode (CCN) caused by *Heterodera avenae* (Eastwood *et al.*, 1991) to adapted wheat lines. Together, these and other projects have released valuable new germplasm for resistance to a wide diversity of pathogens, as well as other useful traits in an easily accessible form for breeding programs (reviewed in van Ginkel & Ogbonnaya, 2006). The first commercial varieties based on the CIMMYT synthetic wheats, Chuanmai42 and Carmona, were released in 2003 to farmers in China and Spain respectively (van Ginkel & Ogbonnaya, 2006). These varieties were not selected specifically for disease resistance, but for their high yield and agronomic adaptation. Future varieties are expected to draw on the impressive range of novel variation for disease resistance.

6.3.3 Transgenic resistance

Genetic manipulation (GM) technologies involving the sequencing and cloning of resistance genes are making the use of distantly related and alien species a much easier option. Many resistance genes have now been isolated from a broad range of hosts (Hulbert *et al.*, 2001), their DNA sequenced and significant similarities in their structure identified. Most have a leucine-rich repeat region (LRR) and a nucleotide-binding site (NBS) region. The LRR regions have been specifically linked to resistance (R) gene specificity, as well as to defence response signalling in the plant, although at this stage generalisations about the specific functions of these regions are likely to hide a great deal of complexity. Another important group of proteins associated with resistance responses are kinases. These probably interact with NBS-LRR proteins in the host resistance pathway. In the case of the rice bacterial blight resistance gene *Xa*21, a kinase is linked with an LRR region and a short transmembrane (TM) region that allows the expressed protein to straddle the cell membrane, with the kinase region located in the cytoplasm (Song *et al.*, 1995). In contrast, the *Rpg*1 gene coding for durable resistance to stem rust in barley has a different type of resistance gene structure, in that it has close homology to receptor kinases, but has no NBS or LRR region (Brueggeman *et al.*, 2002).

*Xa*21 was sequenced and cloned and when transformed into rice found to stably express resistance against a suite of pathogen isolates in 1995 (Wang *et al.*, 1996). It was subsequently transformed into the variety IR72 and the first field tests conducted in 2000. In this particular case, the use of GM technology did not provide an economic breakthrough, as the gene had already been bred into an adapted variety, IRBB21, using non-GM technologies. It has, however, been a pioneering example demonstrating the use of the transgenic technologies in a major crop. At this time, transgenic rice lines with *Xa*21 are still being field tested and their future debated, whilst other transgenic lines are being developed that carry cloned resistance genes for sheath blight and stem borer (Brar & Khush, 2006). As with other major genes, new strains of bacterial blight with virulence on *Xa*21 have been found in Korea and the Philippines, subsequent to the gene being deployed using non-transgenic technologies.

It is envisaged however that once GM rice becomes widely accepted, this technology will be used to pyramid multiple different resistance genes into a single variety providing, it is hoped, more durable resistance. Success for this strategy, as with the use of other major genes, will require agreement and coordination amongst plant breeders and the industry to prevent varieties being grown together that will allow the pathogen to mutate stepwise, one gene at a time, to overcome these gene pyramids.

The successful transfer of major resistance genes may be largely limited to closely related species which have related resistance signalling pathways. Transfer of effective resistance from tomato to tobacco, both *Solanaceous* species, has been clearly demonstrated (Rommens *et al.*, 1995), but to date there has been no reported success of resistance genes being effective when transferred to different families, except where either the corresponding avirulence product was artificially provided or where a race of the pathogen that affected both donor and recipient species was involved (Ayliffe & Lagudah, 2004).

Another possible approach would be to use cloned genes that code for proteins involved in the immune response or non-host resistance. Eukaryotes are resistant to almost all microbes other than a very few which are generally specific to that species or genus. Resistance to all the others is known as non-host resistance or innate immunity and operates through the recognition of pathogen-associated molecular patterns (PAMPs), a term developed by mammalian scientists (Medzhitov & Janeway, 1997). PAMPS, as the name suggests, are conserved molecules that are common across a broad range of potential pathogens and which help to distinguish them as potentially harmful to a host. A more appropriate general term to use perhaps is MAMPs (for microbe- associated molecular patterns), as non-pathogenic microorganisms also elicit the defence response (see review by Bent & Mackey, 2007). Examples of many different known MAMPs are flagellins in bacteria, chitins in fungi and double-stranded RNA in viruses. Recognition of MAMPs occurs through plant surface pattern recognition receptors which elicit the immune or non-host resistance response (MAMP-triggered immunity). Microbes that have become pathogens are those that have evolved the ability to suppress MAMP-triggered immunity by interfering with either MAMP recognition or with the subsequent host response pathway through the production of 'effector' proteins secreted by the pathogen into the host cell (see review by Chisholm *et al.*, 2006). Plants have subsequently evolved to produce proteins that recognise pathogen effector proteins. These resistance or R genes interact with the pathogen effector genes to produce the classic gene-for-gene resistance response (Flor, 1971). Pathogen virulence occurs where the pathogen effector protein is modified in some way such that the resistance gene is no longer as effective in recognising the effector gene.

Improved host resistance to a wide range of pathogens could be developed by transgenic insertion of novel MAMP receptors derived from other species or modified in some way to broaden their effectiveness. It must be borne in mind however that MAMP-activated defences may be switched off by pathogens. Problems may also arise if there is a physiological cost of a plant producing such receptors in the absence of pathogen attack. Natural resistance genes appear to only trigger host defences when required, minimising the cost to the host plant. A similar level of fine control may be needed for transgenic resistance and identifying the appropriate promoters and control mechanisms may not be straightforward.

An option other than inserting novel resistance genes is to modify the expression of existing ones that are involved in the resistance pathway (reviewed by Rommens & Kishore, 2000). A successful demonstration of this possibility involved the *NPR*1 gene in *Arabidopsis* that regulates systemic acquired resistance, overexpression of which increased the plant's resistance to a diverse array of pathogens (Cao *et al.*, 1998; see also Chapter 4).

A specific option for developing improved resistances to pathogenic viruses has been to transform the gene that expresses their coat proteins into the host. The effectiveness of this approach was first demonstrated with tobacco mosaic virus (TMV), where the coat protein delayed symptoms when inserted into the genome of tobacco plants (Powell-Abel *et al.*, 1986).

Papaya ringspot virus (PRSV) is a damaging disease that limits papaya production worldwide. Soon after the TMV demonstration, a PSRV-resistant line was developed on Hawaii using transformed PRSV coat protein incorporated into the host genome (Fitch *et al.*, 1992). Two varieties, Rainbow and SunUp, were subsequently commercialised in 1998 and Rainbow, an F1 hybrid developed from a cross between the homozygous transgenic SunUp and the non-GM variety Kapoho, became widely planted and helped to save the papaya industry on the island from devastation by PRSV (Gonsalves, 2004). Transgenic papaya germplasm was also released for commercial cultivation in China in 2006 and is being developed for other papaya growing areas including South-East Asia, Australia, Brazil and Jamaica. To date, this is the only example of commercial production of a transgenic crop developed for disease resistance. However, with an increasing number of genes associated with the host defence response being identified and with increasing public acceptance of GM technologies, it is expected that varieties with improved disease resistance will be developed using genetic engineering in the coming years.

6.3.4 Mutation

Mutation provides an attractive option for developing alternative disease resistances where insufficient variation for resistance exists in a crop or a related species. The mutation can be generated through point mutation by a chemical, typically ethyl methyl sulphonate (EMS) or sodium azide, or through larger chromosomal changes caused by γ-ray irradiation. For disease resistance work, the former is preferable as the smaller changes generated are less likely to disrupt other genetic processes and traits. Because multiple changes occur with either method, further crossing and selection of mutant lines is usually required in order to produce a variety suitable for commercial release.

A highly successful example is the development of durable resistance to powdery mildew (*Blumeria graminis* f. sp. *hordei*) controlled by resistance at the *mlo* locus in barley. The first mutant at the locus was induced by X-rays in 1942 and since then, a series of new independent mutation events have been developed. Some of these mutants have been widely deployed in high yielding cultivars across Europe. While plenty of naturally generated variation has existed for this trait, there has been a long history of major resistance genes losing their effectiveness through changes in the pathogen population. The *mlo* mutants have provided a source of resistance in adapted backgrounds that appears to be effective against all races, that is it does not conform to the gene-for-gene system identified by Flor (1971). Initially, necrotic leaf spotting that accompanied the resistance and which reduced grain yield, hindered the use of *mlo* mutants. Subsequent breeding work eliminated those problems and from the 1980s onwards many varieties carrying *mlo* resistance have been grown throughout Europe without any virulence to the resistance being detected (Helms Jørgensen, 1992).

It is of interest that the protein coded by the *mlo* gene is quite different from all other resistance genes so far identified, as the wild type allele codes for a cell membrane receptor and it is the non-functional allele of this which provides resistance (Büschges *et al.*, 1997). It is also noteworthy that the mutant *mlo* locus increases susceptibility to the rice blast fungus *Magnaporthe grisea* (Jarosch *et al.*, 1999) and the spot blotch fungus *Bipolaris sorokiniana* (Kumar *et al.*, 2001).

Although *mlo* mutants have been widely used, natural variation at the *mlo* locus has also been found from landrace collections made in Ethiopia that provided similar resistance. Where mutation will have a more profound effect is where it provides resistance to an important pathogen where naturally occurring resistances have not been found and especially where alternative control strategies are unavailable or economically costly. This is the case for many soil-borne diseases, where variation is not present or where screening is very time consuming, costly and/or unreliable. The recent report (Hershey, 2007) of a mutant line with resistance to barepatch disease caused by *Rhizoctonia solani* (AG8) in the wheat variety Scarlet in the USA is therefore of great interest, although it is yet to be seen whether this resistance will operate in the field. The resistance gene will be even more valuable if it is found to operate against *Rhizoctonia* species in other crops.

6.4 Breeding methodology and selection strategies for inbreeding crops

There are three general methods used for breeding for disease resistance: (a) pedigree selection based on simple crosses between two parents, (b) backcrossing and (c) mass selection methods (see Figure 6.1). Using different permutations of these methods in different generations of the breeding process provides a diversity of specific options that can be tailored to the trait, the specific genes involved, the scale and budget for the program and the breeder's personally favoured strategies.

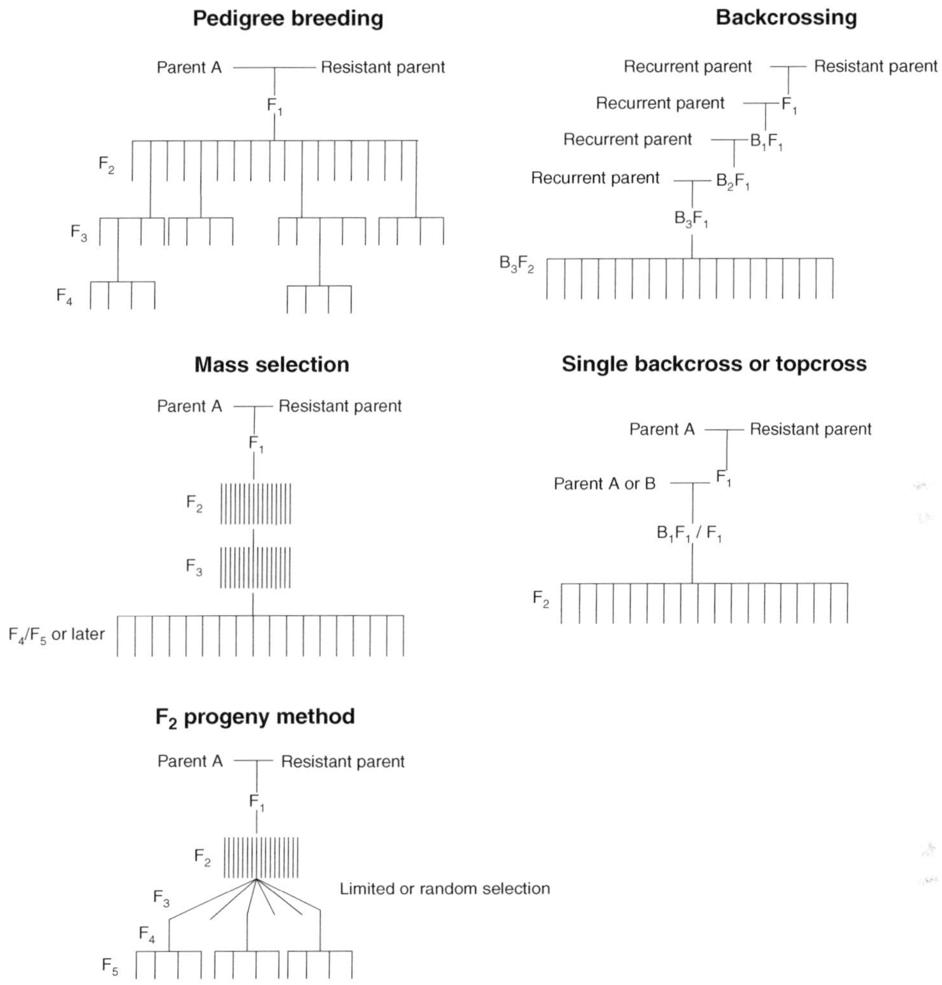

Figure 6.1 Some basic breeding techniques for self-pollinating crops.

Where diseases are generally less important, but have potential for severe damage under specific conditions, then a sensible strategy is for breeding programs to simply avoid the use of highly susceptible parents or to only select against those breeding lines that are highly susceptible.

6.4.1 Phenotypic and marker-assisted selection (MAS)

Until the development of molecular markers, almost all selection for disease resistance was conducted through the phenotypic evaluation of breeding lines in the presence of disease. In a few instances, known linkages with other traits or pleiotropy, as in the case of leaf tip necrosis and the minor genes *Lr*34/*Yr*18 for leaf and stripe resistance, respectively, have been used. Cytogenetic methods and chromosome staining have also been used to identify resistant lines carrying introgressed chromosomal segments from related species.

It is now possible, with the use of molecular markers derived from a variety of technologies, to efficiently select breeding lines based on a plant's genotype rather than its phenotype. Use of this technology is expanding rapidly as the genetic knowledge underpinning it is developed. Molecular markers can only be used where the resistance genes or modifiers required for resistance expression have been linked on the chromosome to markers and where there are a sufficient number of markers polymorphic across the breeder's germplasm available to enable the rapid and accurate identification of specific resistant genotypes. These conditions have only been met for a limited number of resistance genes used in breeding programs and so much development work remains to be done before the technology is mature. Breeding programs are also keen on having 'perfect' markers rather than linked markers. Perfect markers show complete association with the resistance gene and provide unambiguous data on the presence of specific resistance alleles. In many cases, such markers are based on single-nucleotide polymorphisms (SNPs) that underlie variation in the resistance genes. In other cases, insertions or deletions (indels) in the DNA are involved. Few such markers are currently available for most crop species, although progress in their development is being made rapidly.

Once developed, molecular markers allow more precise and rapid selection than phenotypic selection. They also allow a plant breeder to reliably pyramid two or more resistance genes into a single variety more readily, thereby increasing the likelihood that the variety will have durable resistance to multiple strains.

6.4.2 Pedigree breeding

This is standard procedure for many inbreeding crop programs. It involves the selection of individual plants in the F_2 and subsequent generations, so that a precise pedigree of each line can be traced through the program. Generally it is based on simple crosses, but frequently involves topcrosses to a third parent or, less often, more complex crosses.

Pedigree breeding allows for plant selection where maximum diversity is expressed amongst the segregating progeny. The method is most likely to be used where the resistant donor parent is adapted to the region of cultivation or where both adapted parents have some useful resistance and where the disease is reliably expressed in a single plant. Early generation selection for disease resistance ensures that resistance is retained during the selection processes for other traits of economic significance. This is only likely to be achieved, however, where high throughput and low cost screening systems are in place. This is often the case for foliar pathogens that are highly visible and can be inoculated into disease nurseries. It is less likely to be the case for diseases that are more difficult to observe, inoculate uniformly or which are costly to screen.

For pedigree breeding, MAS is most likely to be used in later generations where the numbers of lines are smaller and when a greater proportion of genes are homozygous. This may change as the cost of marker screening is reduced.

6.4.3 Backcrossing

Backcrossing aims to retain a large proportion of the genome of an adapted variety, while introgressing one or a small number of specific genes. It involves the screening of single

F_1 plants for the resistance trait or marker in BC_1F_1, BC_2F_1 and so on generations (or else in BC_xF_2 progeny) to ensure that the resistance is present. The number of backcrosses made will vary depending on the crop species, on the ease of selection in single plants, on the degree of adaptation of the resistance donor and the maximum manageable population size of the subsequent selection generations.

Backcrossing provides a very rapid method for delivering new varieties providing the recurrent parent is suitable and if screening of F_1 plants is effective and reliable. Backcrossing usually requires few resources other than the crossing itself and F_1 screening in the early generations. It is best used where the resistance donors are poorly adapted in the target environment and where several rounds of crossing are required to return to an adapted genotype.

Backcrossing can be speeded up substantially where MAS is used to select for the recurrent parent genotype at the same time as the required resistance from the donor. Such background screening can, for example, result in three backcrosses in maize being the equivalent of 8 backcrosses where only the donor trait is selected (Frisch *et al.*, 1999).

Backcrossing is, however, essentially a conservative approach. It does not allow for much progress in yield or for traits other than the disease resistance gene being sought.

A variation on backcrossing is *partial backcrossing* where, usually, a single backcross is made to the better-adapted parent. This is followed by selection through subsequent generations by the pedigree or other methods. Partial backcrossing provides a compromise between allowing for sufficient variation amongst progeny to make yield and other gains, whilst retaining a high chance of maintaining donor resistance in a sufficiently large proportion of the progeny. Unless very large segregating population sizes can be generated, single backcrosses require at least moderately adapted donor parents.

With partial backcrossing there is an opportunity for screening of single plant BC_1F_1 progeny either by phenotypic response, where the resistance is dominant, or by MAS. Such screening allows a higher proportion of resistant progeny to be present in the subsequent generation or, particularly where molecular markers are used, for a higher proportion of progeny with other desirable traits to be selected. Partial backcrossing is particularly useful where inheritance is multigenic or complex and/or is inherited as a recessive trait and molecular markers are not available, that is where normal backcrossing is not possible. It is also best used where resistance screening is costly, slow and/or has low repeatability such as with root and crown diseases.

6.4.4 Mass selection

This method involves the growing of a large number of early generation progeny in a disease-prone environment with strong natural selection pressure, so that the most resistant lines preferentially survive and form a high proportion of the harvest and thus of later generations. It may involve single or complex crosses and has the advantage of requiring little technology or bookkeeping. Mass selection is rarely used in isolation from other methods and has been even more rarely successful where used as a sole means of selection. The method is inefficient for the production of varieties where quality or other required traits are not naturally selected for in the specific screening environment.

6.4.5 Recurrent selection

This involves the repeated inter-crossing of lines in a mixed population of parents with mass selection of progeny for the required trait occurring at each generation. This method often involves the use of male sterility to reduce the cost of multiple crossing of lines. Recurrent selection has historically been promoted as a means for combining multiple minor genes with minimal resources and thus useful for breeding in developing countries. A significant disadvantage of the technique is that useful gene linkage blocks are often disrupted through the multiple crossing processes. To reduce this, intense selection is required to maintain good agronomic performance and quality. This however, usually requires significant resources that have often not been available. Evidence suggests this method can be successful for some diseases of high heritability such as foliar pathogens, but that it is much less likely to result in well-adapted quality germplasm. It is very unlikely to be successful for root diseases where heritability/repeatability of selection is low.

6.4.6 F_2 progeny method

This is an example of a method that combines components of two of the above general methods. It combines the principal of mass selection for one generation with pedigree breeding over subsequent generations, but with populations rather than single plants being subject to assessment.

The principal aim of the F_2 progeny method (Rathjen & Pederson, 1986) is to postpone heavy selection for a trait such as disease resistance until F_2-derived lines can be replicated in trials over space and/or time. This allows primary selection to occur on the greatest diversity of lines using more objective measurements of adaptation. The F_2 progeny method avoids selection for disease resistance in the F_2, when recessive genes are poorly expressed, but may allow for mass selection based on other traits such as yield. This method therefore has a great advantage for traits of low heritability. It also reduces bookkeeping and work in the early generations.

Many minor disease resistance genes are recessive making this method efficient for selecting lines with multiple numbers of such genes. It is also an appropriate method for selecting for resistance to diseases where screening may require a larger number of plants, perhaps one or more whole plots, for each assessment. The method can also be combined effectively with single backcrosses and top-crosses where resistance enrichment can take place with marker screening in the BCF_1 generation.

6.4.7 Avoiding the most susceptible varieties

This involves ensuring that no variety of a crop is cultivated that is highly susceptible to a pathogen that cannot be easily controlled by other methods. While this is not a breeding method as such, it involves a different principle and is therefore treated separately. It has frequently been observed that some varieties or breeding lines are extremely susceptible to one or more pathogens. They do not just display a susceptible reaction type, but often develop epidemics much faster and may become infected under a wider range of environment conditions than other susceptible varieties. The effect on inoculum production is illustrated in Figure 6.2 (Wallwork, 2007). Whilst only using hypothetical data, this

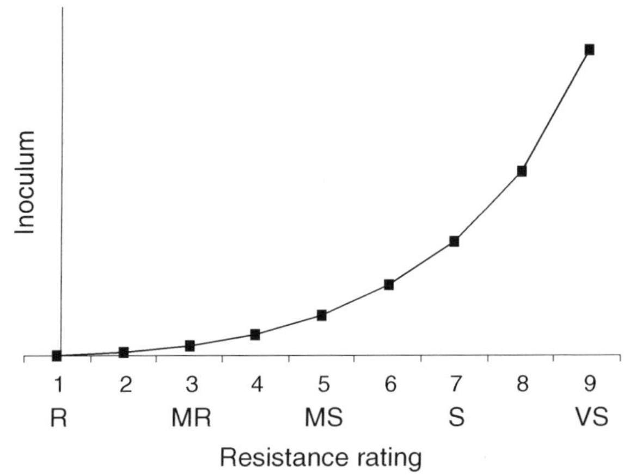

Figure 6.2 Hypothetical relationship between the level of inoculum production and disease resistance rating.

figure illustrates the general relationship between the resistance reaction type of a variety and the amount of inoculum produced by that level of resistance. The relationship is not linear, but exponential, with the most highly susceptible varieties providing a disproportionately higher level of inoculum than other moderately susceptible varieties. Where such a variety susceptible to a pathogen with aerial dispersal is grown over a significant area, it is likely that the large amounts of inoculum produced from it will create a much higher risk environment for all crops in surrounding areas of that species. Indeed, it may only be where such varieties are cultivated that a particular pathogen becomes a damaging disease.

If high levels of inoculum production can be largely averted by removing the most susceptible varieties from a region, a reduced need will exist for high levels of resistance or alternative disease management practices. The advantage of this approach is that selection for resistance in a breeding program can be a lot less stringent and therefore have less impact on yield, quality and other competing traits.

6.4.8 Minimum disease resistance standards

For foliar pathogens that have an aerial dispersal stage, the strategy of reducing disease pressure by avoiding the sowing of the most susceptible varieties across a region, involves obtaining agreement from all other breeding organisations, seed merchants or regulatory authorities, whichever is appropriate to a particular country. The system will only be effective if very few or no crops in a region are producing large quantities of inoculum. This involves the setting of agreed minimum disease resistance standards that are adhered to, most likely on an industry self-regulating basis. Such a system is operating in Australia for control of the wheat rust diseases (Wallwork, 2007).

6.5 Deployment of resistance

A further aspect of disease resistance is the manner in which it is deployed in a country or region. Two options, the deployment of different resistances within a variety mixture or as multilines, are covered in Chapter 8.

Different resistance genes can also be deployed on a local or regional basis. This may be very helpful in slowing the rate of an epidemic locally or restricting it to a region. Such a strategy requires more coordination or regulation than is likely to be possible where many independent farmers are involved but may be feasible where more centrally managed industries are involved.

A critical factor is to avoid deploying single, race-specific resistance genes where pathogen variation is a problem. This practice ensures that the effective working life of a resistance gene is greatly shortened compared to where the gene is deployed in combination with a second or even third resistance gene, so that the pathogen is required to evolve multiple avirulence genes simultaneously. Such pyramided resistances are likely to be quite durable. There is evidence to suggest that the durable resistance of wheat to rye powdery mildew is due to such a gene pyramid (Matsumura & Tosa, 2000). The existence of tightly linked molecular markers and cloned resistance genes is likely to tempt many plant breeders to use these genes preferentially, if doing so results in more rapid development of resistance varieties. The downside of such a situation is that it might lead to a reduction in the diversity of resistance genes being deployed and thus greater vulnerability to losses should one of more of those resistances be overcome.

Another concern is with the deployment of similar resistances across crop species. This will lead to a broader host range being available for some pathotypes and increased economic losses. Using, for example, the same gene to provide resistance to wheat streak mosaic virus in both wheat and maize or stem rust in bread wheat and durum will increase vulnerability in both crops where they are grown together.

Molecular markers can and are being used for the creation of 'gene cassettes', whereby a number of useful genes are being linked together on a single segment of chromosome so that they can be selected in a breeding program as a single Mendelian factor. This is made easier where a range of useful genes has already been located at nearby positions on a chromosome. A good example is the short arm of the 3B chromosome in wheat, where resistance loci for stem rust (*Sr*2), head scab, Septoria *tritici* blotch (*Stb*2) and Septoria *nodorum* blotch have been located in close proximity. Using molecular markers, the alleles responsible for resistance to the first three are being linked so that breeders can select for them as a single unit (Goodwin, 2007).

Useful genes can also be linked onto the same chromosome segments using transgenic approaches. This could involve different genes for resistance to the same pathogen and/ or resistance to different pathogens. This will likely be more cost-effective where the resistances are race-non-specific or durable or else where the gene pyramids are provided some protection from erosion from the separate deployment of individual genes in other varieties.

6.6 Conclusion

In this chapter, the various types of resistance, sources from which new variation can be obtained and breeding methods for inbreeding species have been outlined. While breeding for outcrossing or clonally propagated species or the use of F_1 hybrids in inbreeding species has not been covered in this chapter, similar resistances and breeding principles apply.

It is apparent that genetic resistance is the ideal form of disease control for a farmer, providing it is both durable and that, in its selection, progress in improving other economic traits is not compromised. Two of the main problems with resistance in the past have been a lack of variation for resistance to some pathogens and the loss of resistance to others as the pathogen populations have developed virulence to existing resistance genes. Overcoming the latter problem may be solved in some instances through the development of a better understanding of the genetic control of durable resistance in existing germplasm, while the development of a more detailed knowledge of the biochemical and genetic basis of the interaction between host and pathogen may unlock new opportunities for developing novel forms of resistance that can be introduced through genetic manipulation technologies. Some of these novel resistances are likely to be no more durable than existing resistances, but there is good reason to believe that genes involved in general or non-host resistance mechanisms, less prone to changes in pathogen virulence, can be identified and developed into new varieties.

Existing resistance sources can also be made both more effective and durable if deployed in a more strategic manner. The most effective strategy will be to reduce the level of inoculum in the environment by avoiding the cultivation of the most susceptible varieties and also by adopting chemical and/or agronomic management strategies that will reduce inoculum levels by other means.

6.7 References

Ayliffe MA, Lagudah ES, 2004. Molecular genetics of disease resistance in cereals. *Annals of Botany* **94,** 765–773.

Bent AF, Mackey D, 2007. Elicitors, effectors, and R genes: The new paradigm and a lifetime supply of questions. *Annual Review of Phytopathology* **45,** 399–436.

Bioversity International (2008) (www.bioversityinternational.org) Accessed in January 2008.

Brar DS, Khush GS, 2006. Breeding rice for resistance to biotic stresses: Conventional and molecular approaches. In: Ladha JK, Aggarwal P, Hardy B, eds. *Proceedings of the Second International Rice Congress*, 9–13 October, New Delhi, India http://www.irri.org/science/abstracts/030.asp

Brueggeman R, Rostoks N, Kudrna D, Kilian A, Han F, Chen J, Druka A, Steffenson B, Kleinhofs A, 2002. The barley stem rust-resistance gene Rpg1 is a novel disease-resistance gene with homology to receptor kinases. *Proceedings of the National Academy of Sciences USA* **99,** 9328–9333.

Büschges R, Hollricher K, Panstruga R, Simons G, Wolter M, Frijters A, van Daelen R, van der Lee T, Diergaarde P, Groenendijk J *et al.*, 1997. The barley *Mlo* gene: A novel control element of plant pathogen resistance. *Cell* **88,** 57–63.

Cao H, Li X, Dong X, 1998. Generation of broad-spectrum disease resistance by over-expression of an essential regulatory gene in systemic acquired resistance. *Proceedings of the National Academy of Sciences USA* **95,** 6531–6536.

Chisholm ST, Coaker G, Day B, Staskawicz BJ, 2006. Host–microbe interactions: Shaping the evolution of the plant immune response. *Cell* **124,** 803–814.

Cox TS, Sears RG, Bequette RK, Martin TJ, 1995. Germplasm enhancement in winter wheat x *Triticum tauschii* backcross populations. *Crop Science* **35,** 913–919.

Doussinault G, Delibes A, Sanchez-Monge R, Garcia-Omedo F, 1983. Transfer of a dominant gene for resistance to eyespot disease from a wild grass to hexaploid wheat. *Nature* **303,** 698–700.

Dundas IS, 2007. New sources of rust resistance from alien species: Meliorating linked defects and discovery. *Australian Journal of Agricultural Research* **58,** 545–549.

Dundas IS, Shepherd KW, 1998. Shortening the Agropyron chromosome segment carrying gene *Sr*26 utilising chromosome engineering and molecular markers. In: Slinkard AE, ed. *Proceedings of the 9th International Wheat Genetics Symposium*, Volume 2, Saskatoon, Canada, 2–7 August, University of Saskatchewan: University Extension Press, 35–37.

Eastwood RF, Lagudah ES, Appels R, Hannah M, Kollmorgen JF, 1991. *Triticum tauschii*: A novel source of resistance to cereal cyst nematode (*Heterodera avenae*). *Australian Journal of Agricultural Research* **42,** 69–77.

Fitch MM, Manshardt RM, Gonsalves D, Slightom JL, Sanford JC, 1992. Virus resistant papaya derived from tissues bombarded with the coat protein gene of papaya ringspot virus. *Bio/Technology* **10,** 1466–1472.

Friebe B, Jiang J, Raupp WJ, McIntosh RA, Gill BS, 1996. Characterization of wheat-alien translocations conferring resistance to diseases and pests: Current status. *Euphytica* **91,** 59–87.

Frisch M, Bohn M, Melchinger AE, 1999. Comparison of selection strategies for marker assisted backcrossing of a gene. *Crop Science* **39,** 1295–1301.

Flor HH, 1971. Current status of the gene-for-gene concept. *Annual Review of Phytopathology* **9,** 275–296.

Gonsalves D, 2004. Transgenic papaya in Hawaii and beyond. *AgBioForum*, **7(1&2),** 36–40. http://www.agbioforum.org.

Goodwin, SB, 2007. Back to basics and beyond: increasing the level of resistance to Septoria tritici blotch in wheat. *Australasian Plant Pathology Society* **36,** 532–538.

Hayes HK, Parker JH, Kurtzweil, 1920. Genetics of rust resistance in crosses of varieties of *Triticum vulgare* with varieties of *T. durum* and *T. dicoccum*. *Journal of Agricultural Research* **19,** 523–542.

Helms Jørgensen I, 1992. Discovery, characterization and exploitation of *Mlo* powdery mildew resistance in barley. *Euphytica* **63,** 141–152.

Hershey CH, 2007. WSU seeking patent on root rot-resistant wheat. *Plant Breeding News* **184,** 1–31. www.fao.org/ag/Agp/agpc/doc/services/pbn/pbn-184.htm

Hulbert SC, Webb CA, Smith SM, Sun Q, 2001. Resistance gene complexes: Evolution and utilization. *Annual Review of Phytopathology* **39,** 285–312.

Jarosch B, Kogel KH, Schaffrath U, 1999. The ambivalence of the barley Mlo locus: Mutations conferring resistance against powdery mildew (*Blumeria graminis* f.sp. *hordei*) enhance susceptibility to the rice blast fungus Magnaporthe grisea. *Molecular Plant-Microbe Iinteractions* **12,** 508–514.

Johnson R, Law CN, 1975. Genetic control of durable resistance to yellow rust (*Puccinia striiformis*) in the wheat cultivar Hybride de Bersée. *Annals of Applied Biology* **81,** 385–391.

Knott DR, 1961. The inheritance of rust resistance. VI. The transfer of stem rust resistance from *Agropyron elongatum* to common wheat. *Canadian Journal of Plant Science* **41,** 109–123.

Kumar J, Hückelhoven R, Beckhove U, Nagarajan S, Kogel K-H, 2001. A compromised Mlo pathway affects the response of barley to the necrotrophic fungus *Bipolaris sorokiniana* (Teleomorph: *Cochliobolus sativus*) and its toxins. *Phytopathology* **91,** 127–133.

Matsumura K, Tosa Y, 2000. The rye mildew fungus carries avirulence genes corresponding to wheat genes for resistance to races of the wheat mildew fungus. *Phytopathology* **85,** 753–756.

McFadden ES, 1930. A successful transfer of *emmer* characters to *vulgare* wheat. *Journal of the American Society of Agronomy* **22,** 1020–1034.

Medzhitov R, Janeway C, 1997. Innate immunity: The virtues of a nonclonal system of recognition. *Cell* **91,** 295–298.

Mujeeb-Kazi A, Rosas V, Roldan S, 1996. Conservation of the genetic variation of *Triticum tauschii* (Coss.) Schmalh. (*Aegilops squarrosa* auct. non L.) in synthetic hexaploid wheats (*T. turgidum* L. s.lat. × *T. tauschii*; $2n = 6\times = 42$, AABBDD) and its potential utilization for wheat improvement. *Genetic Resources and Crop Evolution* **43,** 129–134.

Parlevliet JE, Van Ommeren A, 1975. Partial resistance of barley to leaf rust, *Puccinia hordei*. II. Relationship between field trials, micro plot test and latent period. *Euphytica* **24,** 293–303.

Pickering RA, Hill AM, Michel M, Timmerman-Vaughan GM, 1995. The transfer of a powdery mildew resistance gene from *Hordeum bulbosum* L to barley (*H. vulgare* L.) chromosome 2. *Theoretical and Applied Genetics* **91,** 1288–1292.

Powell-Abel P, Nelson RS, De B, Hoffmann N, Rogers SG, Fraley RT, Beachy RN, 1986. Delay of disease development in transgenic plants that express the tobacco mosaic virus coat protein gene. *Science* **232,** 738–742.

Rathjen AJ, Pederson DG, 1986. Selecting for improved grain yield in variable environments. In: Williams TA, Watts GS, eds. *Proceedings of Plant Breeding Symposium*. Agronomy Society of New Zealand. Special Publication, 104–115.

Riley R, Chapman V, Johnson R, 1968. The incorporation of alien disease resistance in wheat by genetic interference with the regulation of meiotic chromosome synapsis. *Genetical Research, Cambridge* **12,** 199–219.

Rommens CM, Kishore GM, 2000. Exploiting the full potential of disease resistance genes for agricultural use. *Current Opinion in Biotechnology* **11,** 120–125.

Rommens CMT, Salmeron JM, Oldroyd GED, Staskawicz BJ, 1995. Intergeneric transfer and functional expression of the tomato resistance gene *Pto*. *Plant Cell* **7,** 1537–1544.

Sears ER, 1956. The transfer of leaf-rust resistance from Aegilops umbellulata to wheat. *Brookhaven Symposia in Biology* **9,** 1–22.

Sears ER, 1973. *Agropyron*-wheat transfers induced by homoeologous pairing. In: Sears ER, Sears LMS, eds. *Proceedings of the Fourth International Wheat Genetics Symposium*, August 6–11, University of Missouri, Columbia, 191–207.

Sears ER, 1977. An induced mutant with homoeologous pairing in common wheat. *Canadian Journal of Genetics and Cytology* **19,** 585–593.

Sears ER, 1978. Analysis of wheat-Agropyron recombinant chromosomes. In: Sanchez-Monge E, Garcia-Olmedo F, eds. *Interspecific Hybridization in Plant Breeding. Proceedings of the 8th EUCARPIA Congress*, Madrid, Spain, 63–72.

Smith EL, Schlehuber AM, Young HC, Edwards LH, 1968. Registration of agent wheat. *Crop Science* **8,** 511–512.

Song WY, Wang GL, Chen LL, Kim HS, Pi LY, 1995. A receptor kinase-like protein encoded by the rice disease resistance gene *Xa21*. *Science* **270,** 1804–1806.

Thomas H, Aung T, 1978. The transfer of mildew resistance from the tetraploid wild oat Avena barbata into the cultivated oat. In: Sanchez-Monge E, Garcia-Olmedo F, eds. *Interspecific Hybridization in Plant Breeding. Proceedings of the 8th EUCARPIA Congress*, Madrid, Spain, 109–112.

Wang GL, Song WY, Ruan DL, Sideris S, Ronald PC, 1996. The cloned gene, *Xa21*, confers resistance to multiple *Xanthomonas oryzae* pv. *oryzae isolates in transgenic plants. Molecular Plant-Microbe Interactions* **9,** 850–855.

Van der Plank JE, 1968. *Disease Resistance in Plants*. New York: Academic Press.

van Ginkel M, Ogbonnaya F, 2006. Using synthetic wheats to breed cultivars better adapted to changing production conditions. In Turner N.C, Acuna T, Johnson, R.C. (2006). 'Ground-breaking stuff'. *Proceedings of the 13th Australian Agronomy Conference*, 10–14 September 2006, Perth, Western Australia. Australian Society of Agronomy. http://www.regional.org.au/au/asa/2006/index.htm

Wallwork H, 2007. The role of minimum disease resistance standards for the control of cereal diseases. *Australian Journal of Agricultural Research* **58,** 588–592.

Zeller FJ, 1973. 1B/1R wheat-rye chromosome substitutions and translocations. *Proceedings of the 4th International Wheat Genetics Symposium*. University of Columbia, Missouri, 209–221.

Chapter 7

Crop tolerance of foliar pathogens: possible mechanisms and potential for exploitation

Ian Bingham[1] *and Adrian Newton*[2]

[1]Scottish Agricultural College, Edinburgh, UK

[2]Scottish Crop Research Institute, Invergowrie, Dundee, UK

7.1 Introduction

Host defence against fungal pathogens encompasses three key processes: disease avoidance (sometimes known as escape), resistance to infection and tolerance of infection (Lovell *et al.*, 2004; Brown & Handley, 2006). Disease avoidance refers to the ability of a crop to restrict spore transfer within the canopy, thereby limiting the spread of the pathogen. Resistance mechanisms, on the other hand, limit the growth of the pathogen within the host following successful transfer and germination of spores. Tolerance refers to the ability of the host to maintain growth and yield in spite of pathogen infection and disease.

In practice it is resistance, and to a lesser extent avoidance, mechanisms that have been exploited most effectively in the management of crop disease. By contrast, tolerance is poorly understood and has not been exploited commercially even though it has potential for reducing the impact of pathogens on yield and the need for fungicides. Recently there has been a revival in interest in tolerance, stimulated by the recognition that crop production must become more sustainable (Brown & Handley, 2006). Unlike most major resistance genes and fungicides, tolerance is a potentially durable form of defence because it places little or no selection pressure on pathogen populations. In this chapter we consider briefly why progress in understanding and improving tolerance of infection in crops has been slow, before discussing which crop traits might confer greater tolerance and the prospects for their exploitation in the future. Successful exploitation will require progress in three areas. Firstly, quantifying intra-specific variation in tolerance and its heritability, secondly, identifying the underlying mechanisms and traits responsible, and thirdly, developing suitable methods for selecting the desirable traits in breeding programmes. We focus on tolerance of foliar pathogens, although similar principles will also apply to tolerance of stem base and root pathogens.

7.2 Concepts and definitions – a historical perspective

It has been recognised for many years that genotypes can differ in their yield response to a given severity of infection and reported examples include some as early as 1894

(Schafer, 1971). Moreover, the differences in response can be large. For example, the yield loss of wheat cultivars under a comparable severity of rust infection ranged from 9.5% to 44.5% (cited by Schafer, 1971). So why, given potential benefits of this scale, is tolerance still so poorly understood? There are several factors that may have contributed to the disappointing progress to date including a lack of consensus on how tolerance should be defined, practical difficulties in quantifying it, and its apparent low heritability.

The term tolerance has been used widely and inconsistently in the literature and for a number of years this hindered the development of a clear conceptual framework for research into the physiological mechanisms underlying tolerance (Gaunt, 1981; Clarke, 1984). Some authors have used the term tolerance to describe partial or incomplete resistance to infection. In this context, 'tolerant' genotypes are distinguished from less 'tolerant' types by exhibiting a lower severity of infection. More commonly, tolerance has been defined in terms of the ability to maintain an acceptable seed yield, or some other measure of plant fitness or productivity, under a given severity of pathogen infection, thus distinguishing it from resistance (Caldwell *et al.*, 1958; Schafer, 1971). A further distinction has been made between tolerance of the pathogen and tolerance of disease (Gaunt, 1981; Clarke, 1984; Newton *et al.*, 1998; Inglese & Paul, 2006). According to Inglese & Paul (2006) tolerance of infection is the relationship between the presence of the pathogen and disruption of normal host physiology, whereas tolerance of disease is the relationship between host growth or fitness and the physiological disruption resulting from infection. For example, the native rust fungus *Coleosporium tussilginis* resulted in smaller reductions in net CO_2 fixation of *Senecio vulgaris* per unit of infection than the alien rust *Puccinia lagenophorae* indicating a greater level of pathogen tolerance in the *S. vulgaris–C. tussilginis* interaction. By contrast there was no difference between the two pathosystems in the reduction in host growth and fitness per unit reduction in photosynthesis, indicating that tolerance of disease in each case was comparable (Inglese & Paul, 2006).

However, attempting to distinguish between tolerance of the pathogen and tolerance of disease can be potentially misleading, because it usually requires singling out specific processes by which to measure physiological dysfunction. In reality, the overall effect of the pathogen on growth and yield will be the net outcome of numerous physiological adjustments stemming from the primary disruption to the tissue. In practice, it can also be difficult to distinguish unequivocally between measures of infection severity and symptoms of disease. For example, to what extent should necrotic tissue within a lesion be considered a measure of pathogen presence or leaf damage and physiological dysfunction? Some authors have suggested that the term tolerance of disease is meaningless or unnecessary and that all tolerance is in reality tolerance of the pathogen (Parbery, 1978; Gaunt, 1981). According to Parbery (1978), if tolerance of the pathogen is defined as the capacity of the host to support pathogen growth with little disturbance to the host metabolism, then complete pathogen tolerance implies no disease. Indeed it is difficult to envisage the converse situation where tolerance of the pathogen may be low (implying significant disruption to host physiology), but tolerance of disease high (little effect on growth and yield), unless the disruption is highly localised and the plant as a whole can compensate for the impaired function. There are examples of this in the insect herbivory literature where damage to apical meristems can lead to the outgrowth of axillary buds (Tiffin, 2000), but little evidence that it occurs in response to pathogen infection.

In spite of the lack of consensus in the literature about the relative importance of disease versus pathogen tolerance to overall tolerance, we consider that useful progress can be made in identifying traits that may contribute to tolerance provided that terminology is defined clearly from the outset. In fact there is a danger in setting criteria for identifying cases of tolerance that are too stringent because it can make quantifying tolerance more difficult, especially in large field experiments, thereby hindering rather than facilitating progress. A further consideration is that for an improved understanding of tolerance to be exploited commercially, it must be defined and measured in terms that relate to disease management practice. At present, management decisions for most foliar pathogens are based on assessments of visible infection or disease symptoms. Understanding and predicting the extent of yield loss for a given severity of visible infection will aid decisions regarding the need for fungicide in a particular crop or variety. Thus, for the purpose of this chapter we define tolerance as less than expected yield loss in response to a given severity of visible disease symptoms or pathogen-induced loss of green area.

7.3 Yield formation

Yield formation in many crop species has been analysed in terms of the quantity of photosynthetically active radiation (PAR) incident upon the crop (Q), the fraction of the radiation intercepted by green tissue (I), the efficiency with which the energy from intercepted PAR is converted into dry matter (radiation use efficiency, RUE) and the partitioning of dry matter into the harvested parts (harvest index, HI). This can be summarised as follows:

$$Y = Q * I * RUE * HI \tag{7.1}$$

Equation (7.1) has provided the conceptual framework for modelling the effects of pathogens and pests on crop growth and yield (Johnson, 1987; Waggoner, 1990; Rossing *et al.*, 1992; Gaunt, 1995; Paveley *et al.*, 2001; Bancal *et al.*, 2007) and is a useful basis for identifying potential tolerance traits. Pathogen infection and subsequent disease can reduce yield through effects on radiation interception, RUE and dry mater partitioning to yield bearing structures, as illustrated in Figure 7.1. A more quantitative treatment of the relationships linking symptom area and yield is given by Gaunt (1995) and Paveley *et al.* (2001). It follows, that tolerant genotypes will be those that possess traits that enable them to maintain high levels of PAR interception, RUE, and the formation of and partitioning of dry matter to yield bearing structures in spite of pathogen infection.

The type and extent of damage inflicted by foliar pathogens depends on the species in question and its mode of nutrition (Walters *et al.*, 2008a). Necrotrophic fungi kill tissue in advance of colonisation by fungal hyphae. Tissue death results in a loss of green area and some shrinkage of the leaf surface. Necrotic regions within leaves may continue to intercept light, but without contributing to photosynthesis. Typically, necrotrophs and hemibiotrophs have relatively limited effects on host photosynthetic metabolism. Thus, yield loss of potatoes following infection by *Phytophthora infestans*, the cause of late blight, has been attributed to the reduction in light interception by green tissue, with no change being found in RUE (van Oijen, 1990, 1991; Rossing *et al.*, 1992). We must sound a word of caution at this point. Care must be taken when interpreting the impact of

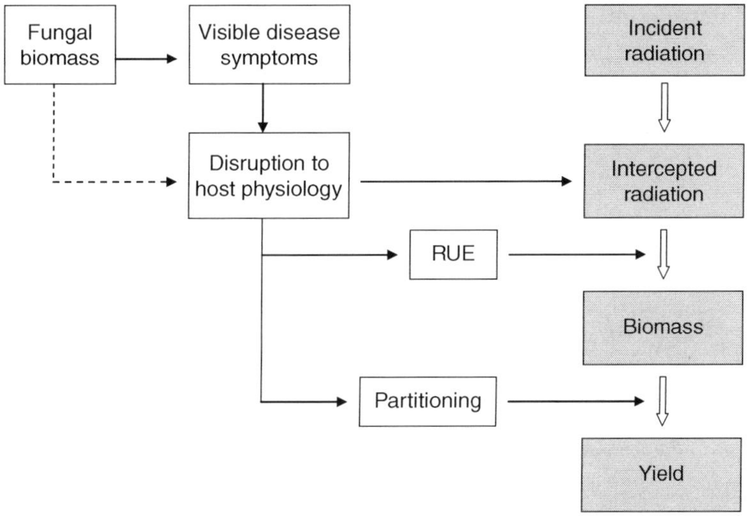

Figure 7.1 Schematic representation of mechanistic links between fungal infection and reduction in crop yield. The broken arrow refers to possible disruption to host physiology prior to formation of, or in the absence of, visible symptoms of infection.

pathogens on radiation interception and RUE, because estimates will depend on whether radiation interception has been assessed on the basis of total leaf area (diseased plus non-diseased) or just green (healthy) area. Thus in contrast to the above work where *P. infestans* was reported to reduce radiation interception and not RUE when assessed on the basis of green tissue, *Mycosphaerella pinodes* of pea reduced RUE but not radiation interception when interception was assessed on the basis of total leaf area (Béasse *et al.*, 2000).

Infection by necrotrophs can result in altered partitioning of carbohydrates and inorganic nutrients, but effects tend to be less pronounced than those observed with biotrophic pathogens (Whipps & Lewis, 1981). Thus export of photosynthates from infected leaves can be reduced, and in some cases, accumulation of nutrients occurs at the infection sites where rates of water loss are high (Ayres & Jones, 1975; Whipps & Lewis, 1981).

In general, host–pathogen interactions with biotrophic fungi are more complex than those with necrotrophs, as they derive their resources for growth and sporulation from living host cells (Walters *et al.*, 2008a). Pathogens, such as rusts and mildews, reduce light interception by masking the leaf surface with pustules and by accelerating leaf senescence (Spitters *et al.*, 1990). Infection may cause a temporary stimulation in the rate of photosynthesis, but then the rate usually declines and chlorophyll is lost (Scholes, 1992). Rates of respiration are also typically increased. Together these effects on photosynthesis and respiration may reduce the RUE of infected leaves (RUE defined as dry matter production per unit of PAR intercepted by green tissue), since net carbon fixation can be reduced in non-symptom expressing (i.e. green) parts of infected leaves (Kral *et al.*, 1993; Scholes & Rolfe, 1995). RUE might also be influenced by effects of pathogens on stomatal regulation and the alteration of normal water relations caused by rupture of the epidermis during sporulation (Ayres, 1981b; Prats *et al.*, 2006). Biotrophic fungi significantly alter the partitioning of carbohydrates and inorganic nutrients within the host plant since the growing fungus in a heavy infection creates a significant 'sink' for diffusion of metabolites to the site of infection (Whipps & Lewis, 1981; Farrar, 1992; Walters *et al.*, 2008a).

The impact on yield of pathogen-induced modifications to radiation interception, RUE and partitioning will depend on the timing of the disease epidemic and its relative effects on 'source' and 'sink' tissues. Here the term 'source' refers to the supply of photosynthates from mature actively photosynthesising leaves or temporary storage reserves. 'Sink' on the other hand refers to the demand for photosynthates by metabolically active non-photosynthesising tissues or tissues that do not have the photosynthetic capacity to meet their own demand. The 'sink' of primary interest in most determinate crops is the developing seed or fruit. In cereals, early disease epidemics that coincide with canopy growth, such as Rynchosporium leaf blotch and powdery mildew of barley, can reduce canopy size and reserves of soluble carbohydrates in the stem at the start of the grain-filling period. Thus, potential assimilate supply for grain filling is limited. However, since the period of canopy growth is also the period during which tillers and spikelets develop, early disease can reduce grain sink capacity by restricting the number of ears produced and to a lesser extent the number of grains per ear (Brooks, 1972; Lim & Gaunt, 1986; Conry & Dunne, 1993). Late epidemics mainly affect the average grain weight (Wright & Hughes, 1987). In other pathosystems, such as the wheat–*Septoria tritci* system, disease tends to develop late after canopy expansion has been completed and the stem carbohydrate reserves deposited. Its main effects are on the duration of the canopy post-flowering and thus photosynthate supply for grain filling rather than the development of grain sink capacity (Paveley *et al.*, 2001).

7.4 How can tolerance be quantified?

Development of techniques for the reliable quantification of tolerance in the field is the first step towards identifying the traits and genes that confer tolerance (Parker *et al.*, 2004). There was a perception, until relatively recently, that tolerance can only be assessed through a comparison of genotypes under identical levels of infection. This presented major practical difficulties in quantifying tolerance because equivalent infection is almost impossible to achieve in field experiments using natural epidemics. More recent studies have adopted a reaction norm approach in which yield is measured over a range of infection or damage severity. Tolerance is given by the slope of the relationship, thereby obviating the need for absolute parity of infection (Parker *et al.*, 2004; Inglese & Paul, 2006). The most common measurement of infection severity is the area under disease progress curve (AUDPC) which provides a measure of lesion area integrated over time (Kramer *et al.*, 1980; Inglese & Paul, 2006). Newton *et al.* (1998) used a variation of this approach to determine the relative tolerance of powdery mildew amongst spring barley genotypes. The yield loss of a genotype in response to disease was plotted against its AUDPC and a common relationship for all genotypes fitted by linear regression (Figure 7.2). Tolerant and intolerant genotypes were identified by their position relative to the regression line; those positioned one or more standard deviations away from the line were designated either tolerant (if below the line) or intolerant (if above the line). Assessments of AUDPC were made from whole canopy scores of infection severity. The approach allows large numbers of genotypes to be evaluated in one experiment, but without supporting measurements provides no information on the mechanisms responsible.

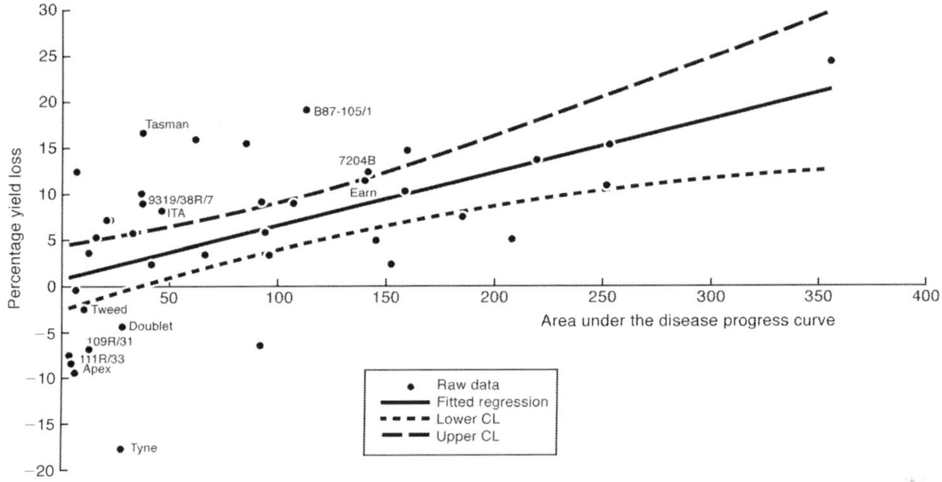

Figure 7.2 Relationship between yield loss and area under the disease progress curve for spring barley cultivars. Data points that fell more than one standard deviation outside the fitted line were considered tolerant or non-tolerant (after Newton *et al.*, 1998; with permission from Elsevier).

Several reports have suggested that tolerance may have low heritability, with its expression depending strongly on the prevailing climatic and edaphic conditions. Thus, the relative tolerance of spring barley cultivars to leaf rust measured in a two-year study differed between years (Kramer *et al.*, 1980). Similarly, spring barley genotypes designated as tolerant of powdery mildew infection varied between years and fertilizer N treatments (Newton *et al.*, 1998, 2000). Part of the variation may arise from the practice of estimating AUDPC in terms of the percentage leaf area occupied by visible disease symptoms scored on a whole plant basis, rather than absolute area, as it provides no indication of the reduction in potential radiation interception resulting from infection. Also, estimates of disease severity on their own do not account for variation between crops in the amount of remaining healthy leaf tissue. The relationship between canopy size and radiation interception is non-linear, thus a given loss of green area through disease may reduce radiation interception more in a small canopy than a large one depending on the disease distribution. Since canopy size is sensitive to soil, climatic and agronomic factors, it is not surprising that relative tolerance also appears to vary with these same factors. To overcome this problem and to provide a more robust estimate of genotypic variation in tolerance, some authors have quantified tolerance from the slope of the relationship between healthy leaf area duration (HAD) and yield (Parker *et al.*, 2004; Foulkes *et al.*, 2006). This approach can be extended by quantifying tolerance from the slope of the relationship between healthy area radiation interception and yield. There are numerous examples of where predictions of disease-induced yield loss based on estimates of HAD or healthy area radiation interception have proved to be more robust across sites and years than those based on percentage AUDPC scores (Johnson, 1987; Waggoner & Berger, 1987; Madden & Nutter, 1995). However, the approach has some disadvantages; firstly, it is time-consuming, thereby limiting the number of genotypes that can be examined, and secondly, some potentially useful tolerance traits may not be identified, such as the maintenance of healthy area through compensatory adjustments in leaf growth (see below). The nature of the 'no disease' control is also important as fungicides may have direct growth regulatory effects on the crop in addition to those resulting from the control or prevention of disease. For example, greening effects and anti-gibberellin activity has been reported for some triazoles (Rademacher, 2000) and yield enhancement with the early generation strobilurins has been associated with modifications to auxin and ethylene metabolism and a delay in leaf senescence (Grossman & Retzlaff, 1997; Grossman *et al.*, 1999). Whether the fungicides are systemic or not may also be influential as systemic fungicides have the potential to control asymptomatic systemic infection such as *Ramularia collo-cygni* of barley (Walters *et al.*, 2008b). Control of asymptomatic infection by fungicides could lead to a yield increase with no associated change in visible symptoms giving the impression that the genotype is non-tolerant if tolerance is defined as yield loss per unit of visible disease.

Exclusion of pathogen challenge may have different effects from protection using fungicides, because challenge of resistant varieties by fungal inoculum can induce defence reactions that are metabolically expensive. Thus, resistant varieties exposed to inoculum may appear non-tolerant as growth may be reduced with few or no visible symptoms of disease because assimilates are diverted from growth to defence. Newton & Thomas (1994) found that powdery mildew inoculum challenge had a greater effect than disease on yield loss of spring barley in glasshouse experiments, but fungicide protection compared with no protection modified this relationship. It is unclear how inoculum challenge and fungicide interactions might influence measurements of tolerance in the field where inoculum challenge is almost invariably present. Newton & Thomas (1994) found no correlation between field and glasshouse experiments in tolerance designations indicating that it is difficult to extrapolate conclusions from glasshouse experiments to field situations.

In recent years, advances have been made in the development of techniques for quantifying fungal presence within plant tissues. These include enzyme-linked immunosorbent assays (ELISA) of fungal biomass and polymerase chain reaction (PCR) methodologies for quantifying fungal DNA (Newton & Reglinski, 1993; Foroughi-Wehr *et al.*, 1996; Fountaine *et al.*, 2007). These techniques have enabled the presence of pathogens to be identified in otherwise symptom-less tissue (Fountaine *et al.*, 2007).

In principle, tolerance can be quantified from the relationship between fungal biomass (or a surrogate measure such as the quantity of DNA) and yield (Newton *et al.*, 1998). However, tolerant genotypes identified from estimates of fungal biomass do not necessarily correspond to those identified from measurements of visible symptom area (Newton *et al.*, 1998), which may reflect genotypic variation in the relationship between the amount of fungal biomass in the tissue, the extent of symptom development and the disruption caused to host physiology. Consequently, when selecting a technique for quantifying tolerance it is important to be aware of what is actually being measured and the mechanisms through which the measured variable might impact on yield. Quantification of fungal presence using ELISA or PCR methodologies is more expensive than assessment of visible symptoms and the improvement in ability to predict the effects of infection on yield, if any, may not justify the extra cost. For example, ELISA assessments of powdery mildew severity accounted for less of the variation in yield between spring barley varieties than AUDPC scores based on visible symptoms (Newton *et al.*, 1998).

Assessments of fungal biomass or DNA also do not measure pathogen challenge. For that, gene expression would have to be measured, particularly that of defence genes, and would require better knowledge than is currently available as to the costs and trade-offs associated with such gene induction (Walters & Heil, 2007). However, we do know that under field conditions there may be a general high-level expression of defence genes compared with glasshouse-grown plants (Pasquer *et al.*, 2005) which will have an energetic and therefore possible yield cost to the plant (Smedegaard-Petersen & Tolstrup, 1985). The ability of the plant to recognise potential pathogens and prime rather than fully express defence pathways would reduce the metabolic cost and may be important for maximising tolerance.

7.5 Potential crop traits conferring tolerance

Based on theoretical considerations of radiation interception, RUE and partitioning, several plant and crop traits can be identified that might influence the tolerance of pathogen infections. However, for many of these traits, experimental evidence supporting their role in tolerance is insubstantial or absent. There are some parallels between the physiological responses of plants to leaf loss caused by herbivory and that resulting from pathogen infection, although plant–pathogen interactions are generally more complex (Ayres, 1992; Gaunt, 1995). As such, the more extensive literature on tolerance of herbivory can, if interpreted with care, provide useful insights to guide investigations on pathogen tolerance. Historically, however, studies of both herbivory and pathogen tolerance have suffered from the same short-coming; relatively few have measured tolerance and expression of candidate mechanisms in the same experiment (Tiffin, 2000).

7.5.1 Canopy size and structure

The fraction of incident PAR that is intercepted by the crop depends on the size and structure of its canopy. It can be estimated from the canopy size (GAI, green area index) and the extinction coefficient for PAR transmission (k) according to an analogy of Beer's law (Bingham *et al.*, 2007). Large canopies will intercept a larger proportion of the incident radiation than small ones, but because the relationship between size and interception is non-linear, each increment of GAI results in a progressively smaller increase in radiation interception (Figure 7.3). The interaction between canopy size and impact of disease-induced 'loss' of GAI on radiation interception will be highly dependent on the vertical distribution of disease (Bastiaans & Kropff, 1993; Bancal *et al.*, 2007). If disease is located in the lower leaf layers the loss of green area will have a negligible effect on total radiation interception by healthy tissue in a large canopy, but a significant effect in a small canopy. This is because in the former, relatively little of the radiation penetrates to the lower leaves. However, if disease is distributed more evenly throughout the canopy, including the upper leaf layers, a reduction in green area through disease will reduce radiation interception by healthy tissue in both large and small canopies. Disease tends to be located in the lower leaf layers in tall crops infected with splash-dispersed pathogens. It also occurs during canopy expansion in pathosystems (e.g. powdery mildew of cereals) where successive leaves emerge more rapidly than pathogens can infect and lead to symptom development on the emerged leaves. In these cases there may be some scope for increasing the

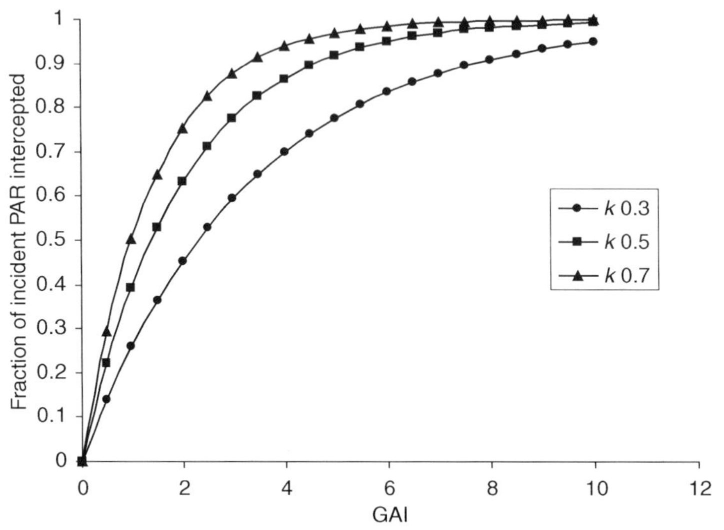

Figure 7.3 Theoretical relationship between the canopy size (GAI) and fraction of radiation intercepted for canopies with different extinction coefficients.

tolerance of the crop through careful management of canopy size. However, any benefit in terms of greater tolerance must be balanced against the risk of increased disease severity. The latter may occur if nitrogen fertiliser is used to manage canopy size (Walters & Bingham, 2007).

The extinction coefficient for transmission of PAR through the canopy is influenced by several factors including leaf inclination, leaf shape, leaf size distribution through the canopy, elevation of the sun and the relative proportions of direct and diffuse radiation (Kramer *et al.*, 1980; Campbell, 1986). Crop species and varieties with planophile leaf habits tend to have a larger extinction coefficient than those with erect leaves and a higher proportion of the incident radiation is intercepted with a small canopy size (Figure 7.2) (Angus *et al.*, 1972). As canopies with planophile leaves intercept a larger fraction of the incident radiation in the upper canopy (Angus *et al.*, 1972) we would expect them to be more tolerant of disease in the lower leaf layers than those with more erect leaves. At present there is relatively little experimental evidence to support this hypothesis. However, results from experiments on spring barley genotypes differing in canopy structure suggest that tolerance of Rhynchosporium leaf blotch is indeed greater in those with planophile habits (Bingham *et al.*, unpublished results).

7.5.2 Photosynthetic activity and compensatory adjustments

The impact of fungal pathogens on net photosynthetic activity within an infected leaf depends on the location within the leaf, as well as the host species and race of pathogen. Some studies based on measurements of gas exchange have suggested that an increase in the rate of photosynthesis can occur in symptom-less areas of infected leaves (Last, 1963; Habeshaw, 1984), but more commonly a decrease has been reported (Martin, 1986; Bastiaans, 1991). Chlorophyll-fluorescence imaging, which provides a greater spatial resolution than can be achieved with gas exchange techniques, has confirmed that in some pathosystems photosynthesis can be impaired in green tissue beyond the region of the lesion and before a loss of chlorophyll occurs (Scholes & Rolfe, 1995). Bastiaans (1991) developed the concept of the virtual lesion to account for the relationship between visible lesion size and the photosynthetic rate of infected leaves. The virtual lesion is the area of the leaf within which the photosynthetic rate is negligible. A ratio of virtual lesion area to visible lesion area (β in Bastiaans' model) greater than 1.0 implies that photosynthesis is inhibited outwith the visible lesion. Values of β ranging from 1.3 to 11 have been reported for rust and mildew infected cereals (Rossing *et al.*, 1992; Robert *et al.*, 2005, 2006), but the value can vary depending on the stage of lesion development and the nature of the symptom assessed. Values were greatest when only the sporulating area of rust-infected wheat was included in the visible lesion area and least when sporulating, necrotic and chlorotic tissue was included (Robert *et al.*, 2005, 2006). The virtual lesion concept has been widely used to model the effects of disease on canopy photosynthesis (Bastiaans & Kropff, 1993; Robert *et al.*, 2004).

Genotypic variation in the extent to which photosynthesis and respiration are modified by infection in symptom-less areas of leaves will lead to apparent variation in disease tolerance if tolerance is expressed as AUDPC, HAD, or healthy area PAR interception and yield. Some inter-specific variation has been found in cereals. Powdery mildew-infected leaves of a wild oat genotype showed a smaller reduction in net photosynthetic rate than two cultivated oat genotypes, even though the severity of infection was greater in the wild oat (Sabri *et al.*, 1997). The wild oat leaves also showed a slower rate of disease-associated senescence. Similar, but less striking effects have been found in a comparison of wild and cultivated barley genotypes (Akhkha *et al.*, 2003). However, to date, the effect of this variation on tolerance at the scale of the crop canopy has not been determined. In a comparison of winter wheat genotypes infected with rust (*Puccinia recondita* Rob. f. sp. *tritici*), no difference was found in the rate of photosynthesis expressed per unit of remaining green area, but genotypes did differ in the rate of rust-associated leaf senescence (Spitters *et al.*, 1990). There is also some limited evidence of intra-specific variation in the response of spring wheat leaves to infection by septoria leaf blotch. Zuckerman *et al.* (1997) observed a 3.5-fold increase in the rate of photosynthesis in the remaining green tissue of the upper three leaves of the variety Miriam compared to Barkai and this was associated with a smaller reduction in average grain weight in Miriam.

Increased photosynthetic activity of non-infected leaves in response to infection elsewhere on the plant has been observed in several pathosystems (Ayres, 1981a; Williams & Ayres, 1981; Roberts & Walters, 1986; Rooney & Hoad, 1989; Murray & Walters, 1992). The scale of response reported differs between studies, and is often greater for dicotyledonous plant pathosystems than monocots. Similar responses have been reported for plants subjected to herbivory or mechanical defoliation (Prins & Verkaar, 1992; Tiffin, 2000; Macedo *et al.*, 2006). However, increased photosynthetic activity is not a universal response to partial defoliation. In some studies the rate of photosynthesis was unaffected or even temporarily decreased by defoliation (Prins & Verkaar, 1992; Zangerl *et al.*, 1997). It is not clear what mechanisms are underlying these responses and why such a range of responses has been observed. Some of the variation may be associated with the growth stage at which treatments are imposed and the relative capacities of source and sink tissues. Inoculations of wheat with *Septoria nodorum* at the young vegetative stage (three-leaf stage) resulted in a significant increase in rate of photosynthesis of non-inoculated leaves, but no increase was found when plants were inoculated at a later growth stage (six-leaf stage) (Rooney & Hoad, 1989). It is tempting to speculate that the lack of response at the later growth stage resulted from a smaller sink limitation of photosynthesis due to the presence of the developing ear so that photosynthesis was operating at a rate close to its full capacity.

An increase in the photosynthetic rate of healthy leaves following defoliation or infection elsewhere is often interpreted as compensating for the reduction in photosynthetic activity of the damaged leaves. If this is the case, those genotypes that are better able to increase their rate would be expected to display a greater tolerance of disease or herbivory in the field. However, the increase in rate may not be 'compensatory' as such and thus may not confer tolerance. For example, it could be associated with an enhanced biosynthesis of defence compounds induced by pathogen infection or wounding rather than production of biomass that is utilised in the formation of yield (Tiffin, 2000). It would also be expected to be a resource-dependent response, modified by nutrient availability or other environmental variables and few studies have attempted to distinguish between these possibilities.

7.5.3 Compensatory adjustments in growth

Morphological adjustments following herbivory or pathogen infection may help ameliorate the effects of leaf damage on net assimilation (Prins & Vekaar, 1992; Tiffin, 2000). A reduction in the proportion of biomass allocated to the root system and an increase in leaf area ratio (ratio of leaf area to total plant biomass) is a common response to defoliation and serves to re-establish the photosynthetic surface as rapidly as possible (Brouwer, 1983; Prins & Verkaar, 1992). An increase in shoot:root biomass ratio has been observed in several species following infection with foliar pathogens (Walters & Ayres, 1981; Rooney & Hoad, 1989; Farrar, 1992). Adjustments in shoot morphology made in response to foliar pathogens have been reported less frequently, but increases in leaf area ratio have been found in some pathosystems (Paul & Ayres, 1986).

Those genotypes that show the greatest morphological plasticity may be able to maintain high levels of reproductive output following defoliation and hence display the greatest tolerance. For example, the grazing tolerant grass *Agropyron desertorum* reduced its root elongation following defoliation, whilst the grazing sensitive *A. spicatum* did not (Richards, 1984). Maintaining canopy expansion at the expense of root growth could potentially predispose the crop to water and nutrient limitations later in the season (Walters & Ayres, 1981). However, soil conditions can also modify root morphology (Bingham, 2001) and negate some of the potentially deleterious effects of infection. As the soil profile dries from the surface downwards root growth may be enhanced in moist soil layers. Powdery mildew infection of barley increased the shoot:root ratio of plants in moist soil, but not those in dry soil (Ayres, 1981b).

The timing of disease epidemics will have a significant influence on the potential for morphological adjustments to contribute to tolerance. In some pathosystems (e.g. barley–*Ramularia collo-cygni*) epidemics tend to occur late in the season at a time when canopy growth is at or nearing completion. Thus, there is little opportunity for compensatory adjustments to be made in leaf or tiller growth. By contrast, epidemics of Rhynchosporium leaf blotch and powdery mildew of barley can occur during early canopy growth and thus there may be potential for selecting genotypes that have a greater capacity for morphological adjustment. However, the extent of any genotypic variation in this response has not yet been quantified.

Ramularia collo-cygni is interesting in another respect, as it is present in the plant systemically long before it becomes symptomatic, that is as an endophyte (Walters *et al.* 2008b). Other endophytes have been found to modify defence reactions and the tolerance of some abiotic stress factors. For example, *Piriformospora indica* when it invades barley does not transition to a pathogenic state, rather it confers greater disease resistance and salt tolerance and results in higher yield (Waller *et al.*, 2005). It effectively 'primes' the defence mechanisms with elevated anti-oxidative capacity due to an activation of the glutathione–ascorbate cycle. It is thought to achieve the mutualism of the endophytic interaction through interference with host cell death mechanisms (Deshmukh *et al.*, 2006). Since interactions with non-pathogenic organisms can modify the response to pathogens, crop tolerance should be defined in its functional context with all the relevant community of interacting organisms in its normal environment. This highlights further some of the potential difficulties when applying results derived from glasshouse experiments to the field situation.

7.5.4 Source–sink relations and storage reserves

The seed or grain yield potential of annual crops is often analysed in terms of source versus sink limitation (Borrás *et al.*, 2004). It has been hypothesised that crops with a large potential assimilate supply relative to the storage capacity of grains will be relatively tolerant of post-flowering loss of green leaf area to disease (Kramer *et al.*, 1980; Gaunt, 1995). This could arise from the possession of a large canopy size and/or water soluble carbohydrate (WSC) storage reserve relative to the number of grains. The relative balance between source and sink during the post-flowering period appears to differ between crop species and production regions. On reviewing a large number of experiments on wheat, soybean and maize grown in high light environments, Borrás *et al.* (2004) concluded that in most cases yield was limited by the potential grain size and not by the supply of assimilate for grain filling. By contrast, under the relatively dull conditions of the United Kingdom, post-flowering source and sink capacities of wheat seem to be in close balance (Beed *et al.*, 2007). The effects of foliar disease on grain sink capacity are relatively small, whilst those on post-anthesis assimilation are large and thus for UK crops disease yield–loss relationships can be explained well in terms of post-anthesis assimilate availability (Paveley *et al.*, 2001). The tolerance of these crops might therefore be improved by selection of traits that enhance post-anthesis radiation interception, RUE, or the utilisation of storage reserves in the presence of disease.

There is evidence that potential assimilate supply generally exceeds the capacity for storage in barley, even in low light environments, and that the extent of the imbalance can differ widely between sites and years (Bingham *et al.*, 2007). This suggests that tolerance of late season disease could be greater in UK barley crops than wheat. Disease earlier in the season (pre-flowering) can reduce the development of potential grain sites in barley as well as healthy leaf area, so that radiation interception per unit grain number during grain filling may not be altered substantially (Bingham *et al.*, unpublished data).

There is evidence of intra-specific variation in the concentration of stem WSC reserves in wheat and it has been suggested that genotypes with large reserves may be more tolerant of post-anthesis loss of leaf area and photosynthetic activity (Foulkes *et al.*, 2002; Ehdaie *et al.*, 2006). However, there is no direct evidence to indicate that selection for large reserve deposition will improve tolerance of disease. Tolerance of septoria leaf blotch was negatively correlated with the amount of soluble carbohydrate reserves in wheat (Foulkes *et al.*, 2006). As yet the reason for the negative correlation has not been established. Barley genotypes also differ in their concentration of stem WSC (Gay *et al.*, 1999). In spring barley, disease reduced the amount of storage reserves deposited, but increased the proportion subsequently used for grain filling (Gaunt & Wright, 1992). However, contrary to the hypothesis that crops with large storage reserves and low yield potential (small grain number) would be more tolerant of disease (Gaunt & Wright, 1992), no relationship was found between yield potential and yield loss caused by leaf rust (Whelan *et al.*, 1997).

7.6 Is there a physiological or ecological cost to tolerance?

Several reports have highlighted a negative correlation between the tolerance of a genotype and its yield potential in the absence of disease (Kramer *et al.*, 1980; Lim & Gaunt, 1986; Parker *et al.*, 2004), which suggests that there may be a physiological cost associated with tolerance. If tolerance is to be exploited as part of a disease management programme it will be important to identify possible tolerance traits that are not negatively associated with yield potential.

Although crop canopies with planophile leaf habits and high extinction coefficients may be more tolerant of disease in the lower canopy, intrinsically, they have a lower RUE under conditions of high solar radiation in the absence of disease (Angus *et al.*, 1972). This is because photosynthesis of the upper leaves is light saturated, whilst the lower leaves are shaded. In canopies with more erect upper leaves, there is a greater transmission of radiation to the lower canopy, but at the same time high rates of photosynthesis can be maintained by the upper leaves with a smaller degree of light saturation. The same arguments may also apply to a constitutive increase in size of the flag leaf of cereals accounting for the negative correlation observed between flag leaf size and yield of healthy wheat genotypes (Foulkes *et al.*, 2007). An ability to make compensatory adjustments in leaf growth in response to disease would not be expected to be deleterious to the yield of healthy crops. However, as discussed above, if this is associated with a reduction in partitioning of biomass to the root system it could make the diseased crop more susceptible to drought.

Upregulation of photosynthesis in response to defoliation or pathogen infection is often interpreted in terms of the relief of sink limitation of photosynthesis (Williams & Ayres, 1981; Zuckerman *et al.*, 1997). If this mechanism is correct, we would expect tolerance to be greatest in genotypes with the greatest sink limitation. Such sink limitation, however, represents a reduction in yield potential in the absence of disease. Indeed, increasing grain numbers and grain size of cereals has been suggested as a possible breeding target to reduce sink limitation of photosynthesis, thereby increasing RUE and yield (Reynolds *et al.*, 2007). This could lead to further reductions in tolerance of pathogens. One trait that may confer tolerance without impairing yield in the absence of disease is delayed disease-induced leaf senescence (Sabri *et al.*, 1997). Until the mechanism of the delayed senescence is understood we cannot determine what its metabolic cost is. However, there is no reason to suspect that it involves a cost in the absence of disease.

In species of natural vegetation there is some evidence of a trade-off between the evolution of resistance to herbivory and evolution of tolerance (Meijden *et al.*, 1988; Fineblum & Rausher, 1995). This may arise because the biosynthesis of defence compounds (resistance) and regrowth following defoliation (tolerance) both represent a demand on a limited supply of carbon skeletons and energy reserves. It is conceivable, therefore, that a similar trade-off may occur between resistance to and tolerance of pathogen infection. If genetic improvement in induced resistance leads to a greater biosynthesis of defence compounds, it could be at the expense of assimilates used for yield formation in diseased crops.

7.7 Role of modelling

Crop growth models are useful tools for predicting the effects of pests and pathogens on crop growth and for investigating the interactions between specific mechanisms of damage and the prevailing climatic conditions (Rossing *et al.*, 1992). Such an approach has been used in a range of pathosystems to determine which component of damage contributes most to the yield response to disease (Rabbinge *et al.*, 1985; Rossing *et al.*, 1992; Bastiaans & Kropff, 1993). For example, modelling the effects of rice blast on canopy photosynthesis indicated that the reduction in rate was the result of adverse effects of lesions on leaf photosynthetic rate and to shading of healthy leaf area by dead leaf tissue (Bastiaans & Kropff, 1993). Other uses of modelling include estimating the effects on crop growth of infection by two or more pathogens at the same time (Robert *et al.*, 2004). In the same way, modelling can be used to estimate the relative effects of different

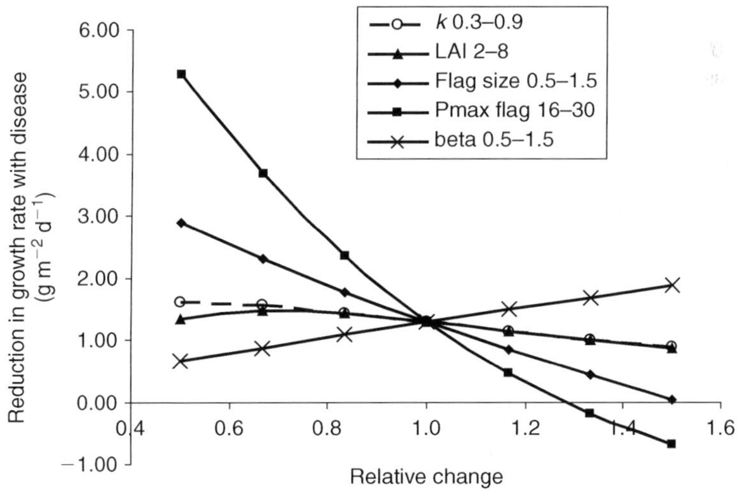

Figure 7.4 Output from a sensitivity analysis of the response of above ground growth rate (dry matter gain) to foliar disease in spring barley just prior to booting. Graph shows predicted effects of changing the canopy light extinction, the light extinction coefficient (k), canopy size (LAI, of leaf plus stem), the ratio of virtual lesion to actual lesion size (β), compensatory changes in size of the flag leaf (flag size) and light saturated rate of photosynthesis of the flag leaf (Pmax flag). Values in the key indicate the range over which the variable was altered; the default value was the mid-point in the range (relative change = 1.0). When a particular variable was altered, others were held at their default value. Flag leaf size is in units of LAI, and Pmax in units of μmol CO_2 m^{-2} s^{-1}; other variables are dimensionless (Bingham *et al.*, unpublished data).

traits on tolerance, and for quantifying the potential costs of these traits in the absence of disease. Figure 7.4 shows the results of a sensitivity analysis in which the effect of disease on above ground dry matter production of spring barley was estimated after varying several leaf and canopy traits. The traits varied were canopy size (GAI), extinction coefficient (k), the virtual lesion size (β; Bastiaans, 1991), and changes in photosynthetic rate and size of the flag leaf in response to disease on the lower leaves. The analysis demonstrates that for spring barley, changes in canopy morphology had a relatively small effect on tolerance compared to compensatory adjustments in leaf growth and photosynthetic rate or virtual lesion size. This type of approach can be used to target research at those traits likely to have the greatest impact on tolerance.

7.8 Strategy for improving tolerance

Although pathogen tolerance has the potential to contribute to disease management, for most crop species we are not yet in a position to be able to include tolerance as an objective in plant breeding programmes. Further research effort is needed before this can be achieved. The necessary steps involved are outlined below.

(a) Tolerance traits must be identified for individual pathosystems. A number of candidate traits have been highlighted above, but traits appropriate for one pathosystem may not be suitable for another because of differences, for example, in host phenology, timing of epidemic development, the nature of the infection and the type of damage induced.
(b) The extent of intra-specific variation in the traits identified must be determined. For some traits, for example the light extinction coefficient, there is evidence of variation within some species that could be exploited. However, for most traits little is known about the extent of any variation, and also, importantly, how the trait of interest influences tolerance in the field. A major shortcoming of much of the research on tolerance to date is that few field studies have quantified both tolerance and the expression of putative tolerance traits simultaneously. If insufficient variation is found within the existing germplasm, it may be necessary to look to wild relatives of crops for a source of tolerance (Sabri *et al.* 1997; Akhkha *et al.*, 2003). However, incorporating genes from wild relatives into the adapted background of elite germplasm requires a considerable breeding effort (Ellis *et al.*, 2000).
(c) Determine the heritability of the trait in field crops. Ideally the expression of tolerance traits should be relatively insensitive to environmental conditions, so that tolerance can be used as a predictable and robust component of disease management. An improved mechanistic understanding of tolerance and the underlying traits that confer it will provide greater predictability should expression prove to be sensitive to some environmental conditions (Tiffin, 2000). Tolerance could then still be of practical value in particular crop situations.
(d) Screening and selection of tolerance. Depending on the nature of the tolerance trait, specific high-throughput screening or marker-assisted selection techniques may need to be developed to enable the trait to be selected within a breeding programme.
(e) The final stage of the process will be to incorporate disease tolerance into an integrated disease management programme. Tolerance is unlikely to give sufficient protection of yield on its own, but in situations where resistance is incomplete, or

disease pressure on resistance is high, it could be a valuable component of disease management enabling the dependence on fungicides to be reduced.

The above strategy is based on first understanding the mechanisms of tolerance and the role of particular traits and then exploring the genetics of the traits to aid selection in breeding. The reverse, but complementary, approach is to make use of the existing data bases and genotyping platforms to identify possible associations between tolerance and genetic markers in a large proportion of the current germplasm. The underlying mechanisms can then be investigated on a subset of genotypes to provide information on how best to deploy and manage the tolerance. Genotyping platforms are now available for high-throughput SNP (single nucleotide polymorphism)-genotyping and these genotyping resources are being associated with detailed phenotype data in public databases (i.e. association genetics). By calculating putative tolerance indices from robust and reliable multi-site field trial data such as that used to compile the UK's CEL (Crop Evaluation Limited) Recommended List, genomic regions and linked markers can be associated with tolerance measures. Co-location of other phenotypic traits will indicate the possible sub-traits that confer the tolerance for further evaluation. If the variance in tolerance accounted for is sufficiently great, then such markers may be used for selection in breeding programmes. The advantage of this approach is that it represents an efficient use of resources since it harnesses information from a vast number of genotypes in extant databases. Its weakness, however, is that tolerance is a complex character and it may not be possible to generate a reliable index of tolerance from information on yield responses to fungicide treatment in variety trials. Therefore, a combination of both approaches would seem to offer the best prospects for understanding and improving the disease tolerance of important crop species in the future.

7.9 Acknowledgements

Scottish Agricultural College and Scottish Crop Research Institute receive financial support from the Rural and Environment Research Analysis Directorate of the Scottish Executive.

7.10 References

Akhkha A, Clarke DD, Dominy PJ, 2003. Relative tolerances of wild and cultivated barley to infection by *Blumeria graminis* f. sp. *hordei* (Syn. *Erysiphe graminis* f. sp. *hordei*). II – the effects of infection on photosynthesis and respiration. *Physiological and Molecular Plant Pathology* **62,** 347–354.

Angus JF, Jones R, Wilson JH, 1972. A comparison of barley cultivars with different leaf inclinations. *Australian Journal of Agricultural Research* **23,** 945–957.

Ayres PG, 1981a. Powdery mildew stimulates photosynthesis in uninfected leaves of pea plants. *Phytopathologische Zeitschrift* **100,** 312–318.

Ayres PG, 1981b. Effects of disease on plant water relations. In: Ayres PG, ed. *Effects of Disease on the Physiology of the Growing Plant*. Cambridge: Cambridge University Press, 131–148.

Ayres PG, 1992. Plants versus pests and pathogens: An old story but the same story? In: Ayre PG, ed. *Pests and Pathogens. Plant Responses to Foliar Attack*. Oxford: Bios Scientific Publishers, 1–12.

Ayres PG, Jones P, 1975. Increased transpiration and the accumulation of root absorbed ^{86}Rb in barley leaves infected by *Rhynchosporium secalis* (leaf blotch) *Physiological Plant Pathology* **7,** 49–58.

Bancal MO, Robert C, Ney B, 2007. Modelling wheat growth and yield losses from late epidemics of foliar diseases using loss of green leaf area per layer and pre-anthesis reserves. *Annals of Botany* **100,** 777–789.

Bastiaans L, 1991. The ratio between virtual and visual lesion size as a measure to describe reduction in leaf photosynthesis of rice due to leaf blast. *Phytopathology* **81,** 611–615.

Bastiaans L, Kropff MJ, 1993. Effects of leaf blast on photosynthesis of rice. 2. Canopy photosynthesis. *Netherlands Journal of Plant Pathology* **99,** 205–217.

Béasse C, Ney B, Tivoli B, 2000. A simple model of pea (*Pisum sativum*) growth affected by *Mycosphaerella pinodes*. *Plant Pathology* **49,** 187–200.

Beed FD, Paveley ND, Sylvester-Bradley R, 2007. Predictability of wheat growth and yield in light-limited conditions. *Journal of Agricultural Science* **145,** 63–79.

Bingham IJ, 2001. Soil–root–canopy interactions. *Annals of Applied Biology* **138,** 243–251.

Bingham IJ, Blake J, Foulkes MJ, Spink J, 2007. Is barley yield in the UK sink limited? I. Post-anthesis radiation interception, radiation use efficiency and source–sink balance. *Field Crops Research* **101,** 198–211.

Borrás L, Slafer GA, Otegui ME, 2004. Seed dry weight response to source–sink manipulations in wheat, maize and soybean: A quantitative reappraisal. *Field Crops Research* **86,** 131–146.

Brooks DH, 1972. Observations on the effects of mildew *Erysiphe graminis* on growth of spring and winter barley. *Annals of Applied Biology* **70,** 149–156.

Brouwer R, 1983. Functional equilibrium: Sense or nonsense? *Netherlands Journal of Agricultural Science* **31,** 335–348.

Brown JKM, Handley RJ, 2006. Fight or flinch? *Heredity* **96,** 3–4.

Caldwell RM, Schafer JF, Compton LE, Patterson FL, 1958. Tolerance to cereal leaf rusts. *Science* **128,** 714–715.

Campbell GS, 1986. Extinction coefficients for radiation in plant canopies using an ellipsoidal inclination angle distribution. *Agriculture and Forest Meteorology* **36,** 317–321.

Clarke DD, 1984. Tolerance of parasitic infection in plants. In: Wood RKS, Jellis GJ, eds. *Plant Diseases: Infection Damage and Loss*. Oxford: Blackwell Scientific Publications, 119–127.

Conry MJ, Dunne B, 1993. Effects of nitrogen timing and frequency of fungicide applications on grain yields of winter barley in Ireland. *Journal of Agricultural Science* **120,** 181–186.

Deshmukh S, Hueckelhoven R, Schaefer P, Imani J, Sharma M, Weiss M, Waller F, Kogel KH, 2006. The root endophytic fungus *Piriformospora indica* requires host cell death for proliferation during mutualistic symbiosis with barley. *Proceedings of the National Academy of Sciences of the United States of America* **103,** 18450–18457.

Ehdaie B, Alloush GA, Madore MA, 2006. Genotypic variation for stem reserves and mobilization in wheat: II. Postanthesis changes in internode water-soluble carbohydrates. *Crop Science* **46,** 2093–2103.

Ellis RP, Forster BP, Robinson D, Handley LL, Gordon DC, Russell JR, Powell W, 2000. Wild barley: A source of genes for crop improvement in the 21st century? *Journal of Experimental Botany* **51,** 9–17.

Farrar JF, 1992. Beyond photosynthesis: The translocation and respiration of diseased leaves. In: Ayres PG, ed. *Pests and Pathogens. Plant Responses to Foliar Attack.* Oxford: Bios Scientific Publishers, 107–127.

Fineblum WL, Rausher MD, 1995. Tradeoff between resistance and tolerance to herbivore damage in a morning glory. *Nature* **377,** 517–520.

Foroughi-Wehr B, Zuchner S, Rabenstein F, 1996. Enzyme-linked immunosorbent assay for detection of *Rhynchosporium secalis* (Oud) J. J. Davis in winter barley. *Zeitschrift für Pflanzenkrankheiten und Pflanzenschutz* **103,** 267–271.

Foulkes MJ, Scott RK, Sylvester-Bradley R, 2002. The ability of wheat cultivars to withstand drought in UK conditions: Formation of grain yield. *Journal of Agricultural Science* **138,** 153–169.

Foulkes MJ, Paveley ND, Worland A, Welham SJ, Thomas J, Snape JW, 2006. Major genetic changes in wheat with potential to affect disease tolerance. *Phytopathology* **96,** 680–688.

Foulkes MJ, Snape JW, Shearman VJ, Reynolds MP, Gaju G, Sylvester-Bradley R, 2007. Genetic progress in yield potential in wheat: Recent progress and future prospects. *Journal of Agricultural Science* **145,** 17–29.

Fountaine JM, Shaw MW, Napier B, Ward E, Fraaije BA, 2007. Application of real-time and multiplex polymerase chain reaction assays to study leaf blotch epidemics in barley. *Phytopathology* **97,** 297–303.

Gaunt RE, 1981. Disease tolerance – an indicator of thresholds? *Phytopathology* **71,** 915–916.

Gaunt RE, 1995. The relationship between plant disease severity and yield. *Annual Review of Phytopathology* **33,** 119–144.

Gaunt RE, Wright AC, 1992. Disease–yield relationship in barley. II. Contribution of stored stem reserves to grain filling. *Plant Pathology* **41,** 688–701.

Gay AP, Spink JH, Foulkes MJ, 1999. Preliminary assessment of the potential for variety typing in winter barley: Stem water soluble carbohydrate measurements. *Home-Grown Cereals Authority Project Report No. 186*. London: HGCA, 1–19.

Grossmann K, Retzlaff G, 1997. Bioregulatory effects of the fungicidal strobilurin kresoxim-methyl in wheat (*Triticum aestivum*). *Pesticide Science* **50,** 11–20.

Grossmann K, Kwiatkowski J, Casper G, 1999. Regulation of phytohormone levels, leaf senescence and transpiration by the strobilurin kresoxim-methyl in wheat (*Triticum aestivum*). *Journal of Plant Physiology* **154,** 805–808.

Habeshaw D, 1984. Effects of pathogens on photosynthesis. In: Wood RKS, Jellis GJ, eds. *Plant Diseases: Infection Damage and Loss*. Oxford: Blackwell Scientific Publications, 63–72.

Inglese SJ, Paul ND, 2006. Tolerance of *Senecio vulgaris* to infection and disease caused by native and alien rust fungi. *Phytopathology* **96,** 718–726.

Kral G, Oerke EC, Schonbeck F, 1993. Effect of powdery mildew on the water balance and photosynthesis of spring barley cultivars differing in tolerance. *Zeitschrift für Pflanzenkrankheiten und Pflanzenschutz* **100,** 1–10.

Kramer T, Gildemacher BH, van der Ster M, Parlevliet JE, 1980. Tolerance of spring barley cultivars to leaf rust, *Puccinia hordei*. *Euphytica* **29,** 209–216.

Johnson KB, 1987. Defoliation, disease and growth: A reply. *Phytopathology* **77,** 1495–1497.

Last FT, 1963. Metabolism of barley leaves inoculated with *Erysiphe graminis* Marchal. *Annals of Botany* **27,** 685–690.

Lim LG, Gaunt RE, 1986. The effect of powdery mildew (*Erysiphe graminis* f. sp. *hordei*) and leaf rust (*Puccinia hordei*) on spring barley in New Zealand. I. Epidemic development, green leaf area and yield. *Plant Pathology* **35,** 44–53.

Lovell DJ, Parker SR, Hunter T, Welham SJ, Nichols AR, 2004. Position of inoculum in the canopy affects the risk of *Septoria tritici* blotch epidemics in winter wheat. *Plant Pathology* **53,** 11–21.

Macedo TB, Peterson RKD, Weaver DK, 2006. Photosynthetic responses of wheat, *Tritcum aestivum* L., plants to simulated insect defoliation during vegetative growth and at grain fill. *Environmental Entomology* **35,** 1702–1709.

Madden LV, Nutter FW, 1995. Modelling crop losses at the field scale. *Canadian Journal of Plant Pathology* **17,** 124–137.

Martin PJ, 1986. Gaseous exchange studies of barley leaves infected with *Rhynchosporium secalis* (Oudem) J.J. Davis. *Physiological and Molecular Plant Pathology* **28,** 3–14.

Meijden E, van der Wijn M, Verkaar HJ, 1988. Defence and regrowth, alternative plant strategies in the struggle against herbivores. *Oikos* **51,** 355–363.

Murray D, Walters DR, 1992. Increased photosynthesis and resistance to rust infection in upper, uninfected leaves of rusted broad bean (*Vicia faba* L.). *New Phytologist* **120,** 235–242.

Newton AC, Reglinski T, 1993. An enzyme-linked immunosorbent assay for quantifying mildew biomass. *Journal of Plant Disease and Protection* **100,** 176–179.

Newton AC, Thomas WTB, 1994. Detection of tolerance of barley cultivars to infection by powdery mildew (*Erysiphe graminis* f. sp. *hordei*). *Euphytica* **75,** 179–187.

Newton AC, Thomas WTB, Guy DC, Gaunt RE, 1998. The interaction of fertiliser treatment with tolerance to powdery mildew in spring barley. *Field Crops Research* **55,** 45–56.

Newton AC, Guy DC, Gaunt RE, Thomas WTB, 2000. The effect of powdery mildew inoculum pressure and fertilizer levels on disease tolerance in spring barley. *Journal of Plant Diseases and Protection* **107,** 67–73.

Parbery DG, 1978. A consideration of tolerance to parasites in plants. *Annals of Applied Biology* **89,** 373–378.

Parker SR, Welham S, Paveley ND, Foulkes J, Scott RK, 2004. Tolerance of septoria leaf blotch in winter wheat. *Plant Pathology* **53,** 1–10.

Pasquer F, Isidore E, Zarn J, Keller B, 2005. Specific patterns of changes in wheat gene expression after treatment with three antifungal compounds. *Plant Molecular Biology* **57,** 693–707.

Paul ND, Ayres PG, 1986. The effects of infection by rust (*Puccinia lagenophorae* Cooke) on the growth of groundsel (*Senecio vulgaris* L.) cultivated under a range of nutrient concentrations. *Annals of Botany* **58,** 321–331.

Paveley ND, Sylvester-Bradley R, Scott RK, Craigon J, Day W, 2001. Steps in predicting the relationship of yield on fungicide dose. *Phytopathology* **91,** 708–716.

Prats E, Gay AP, Mur LAJ, Thomas BJ, Carver TLW, 2006. Stomatal lock-open, a consequence of epidermal cell death, follows transient suppression of stomatal opening in barley attacked by *Blumeria graminis*. *Journal of Experimental Botany* **57,** 2211–2226.

Prins AH, Verkaar HJ, 1992. Defoliation: Do physiological and morphological responses lead to (over) compensation? In: Ayres PG, ed. *Pests and Pathogens. Plant Responses to Foliar Attack*. Oxford: Bios Scientific Publishers, 13–31.

Rabbinge R, Jorritsma ITM, Schans J, 1985. Damage components of powdery mildew in winter wheat. *Netherlands Journal of Plant Pathology* **91,** 235–247.

Rademacher W, 2000. Growth retardants: Effects on gibberellin biosynthesis and other metabolic pathways. *Annual Review of Plant Physiology and Plant Molecular Biology* **51,** 501–531.

Reynolds M, Calderini D, Ciondon A, Vargas M, 2007. Association of source/sink traits with yield, biomass and radiation use efficiency among random sister lines from three wheat crosses in a high-yield environment. *Journal of Agricultural Science* **145,** 3–16.

Richards JH, 1984. Root growth response to defoliation in two *Agropyron* bunchgrasses: Field observations with an improved root periscope. *Oecologia* **64,** 21–25.

Robert C, Bancal MO, Nicolas P, Lannou C, Ney B, 2004. Analysis and modelling of effects of leaf rust and *Septoria tritici* blotch on wheat growth. *Journal of Experimental Botany* **55,** 1079–1094.

Robert C, Bancal MO, Lannou C, Ney B, 2006. Quantification of the effects of *Septoria tritici* blotch on wheat leaf gas exchange with respect to lesion age, leaf number, and leaf nitrogen status. *Journal of Experimental Botany* **57,** 225–234.

Robert C, Bancal MO, Ney B, Lannou C, 2005. Wheat leaf photosynthesis loss due to leaf rust, with respect to lesion development and leaf nitrogen status. *New Phytologist* **165,** 227–241.

Roberts AM, Walters DR, 1986. Stimulation of photosynthesis in uninfected leaves of rust-infected leeks. *Annals of Botany* **56,** 893–896.

Rooney JM, Hoad GV, 1989. Compensation in growth and photosynthesis of wheat (*Triticum aestivum* L.) following early inoculations with *Septoria nodurum* (Berk.) Berk. *New Phytologist* **113,** 513–521.

Rossing WAH, van Oijen M, van der Werf W, Bastiaans L, Rabbinge R, 1992. Modelling the effects of foliar pests and pathogens on light interception, photosynthesis, growth rate and yield of field crops. In: Ayres PG, ed. *Pests and Pathogens. Plant Responses to Foliar Attack*. Oxford: Bios Scientific Publishers, 161–180.

Sabri N, Dominy PJ, Clarke DD, 1997. The relative tolerances of wild and cultivated oats to infection by *Erysiphe graminis* f. sp. *avenae*: II. The effects of infection on photosynthesis and respiration. *Physiological and Molecular Plant Pathology* **50,** 321–335.

Schafer J, 1971. Tolerance to plant disease. *Annual Review of Phytopathology* **9,** 235–252.

Scholes JD, 1992. Photosynthesis: Cellular and tissue aspects in diseased leaves. In: Ayres PG, ed. *Pests and Pathogens. Plant Responses to Foliar Attack*. Oxford: Bios Scientific Publishers, 85–106.

Scholes JD, Rolfe SA, 1995. How do biotrophic pathogens affect the photosynthetic metabolism of their hosts? *Aspects of Applied Biology*, **42,** *Physiological Responses of Plants to Pathogens*, 91–99.

Smedegaard-Petersen V, Tolstrup K, 1985. The limiting effect of disease resistance on yield. *Annual Review of Phytopathology* **23,** 475–490.

Spitters CJ, van Roermund HJW, van Nassau HGMG, Scheper J, Mesdag J, 1990. Genetic variation in partial resistance to leaf rust in winter wheat: Disease progress, foliage senescence and yield reduction. *Netherlands Journal of Plant Pathology* **96,** 3–15.

Tiffin P, 2000. Mechanisms of tolerance to herbivore damage: What do we know? *Evolutionary Ecology* **14,** 523–536.

van Oijen M, 1990. Photosynthesis is not impaired in healthy tissue of blighted potato plants. *Netherlands Journal of Plant Pathology* **96,** 55–63.

van Oijen M, 1991. Light use efficiencies of potato cultivars with late blight (*Phytophthora infestans*). *Potato Research* **34,** 123–132.

Waggoner PE, 1990. Defoliation, disease and growth. In: Rabinge R, Goudriaan J, van Keulen H, Penning de Vries FWT, van Laar HH, eds. *Theoretical Production Ecology: Reflections and Prospects*. Wageningen: Pudoc, 149–180.

Waggoner PE, Berger RD, 1987. Defoliation, disease and growth. *Phytopathology* **77,** 393–398.

Waller F, Achatz B, Baltruschat H, Fodor J, Becker K, Fischer M, Heier T, Huckelhoven R, Neumann C, von Wettstein D, Franken P, Kogel KH, 2005. The endophytic fungus *Piriformospora indica* reprograms barley to salt-stress tolerance, disease resistance, and higher yield. *Proceedings of the National Academy of Sciences of the United States of America* **102,** 13386–13391.

Walters DR, Ayres PG, 1981. Growth and branching pattern of roots of powdery mildew infected barley. *Annals of Botany* **47,** 159–163.

Walters DR, Bingham IJ, 2007. Influence of plant nutrition on disease development caused by fungal pathogens: Implications for disease control. *Annals of Applied Biology* **151,** 307–324.

Walters DR, Havis ND, Oxley SJP, 2008a. Ramularia cygni-collo: The biology of an emerging pathogen of barley. *FEMS Microbiology Letters* **279,** 1–7.

Walters DR, Heil M, 2007. Costs and trade-offs associated with induced resistance. *Physiological and Molecular Plant Pathology* **71,** 3–17.

Walters DR, McRoberts N, Fitt BDL, 2008b. Are green islands red herrings? Significance of green islands in plant interactions with pathogens and pests. *Biological Reviews* **83,** 79–102.

Whelan HG, Gaunt RE, Scott WR, 1997. The effect of leaf rust (*Puccinia hordei*) on yield response in barley (*Hordeum vulgare* L.) crops with different yield potentials. *Plant Pathology* **47,** 397–406.

Whipps JM, Lewis DH, 1981. Patterns of translocation, storage and interconversion of carbohydrates. In: Ayres PG, ed. *Effects of Disease on the Physiology of the Growing Plant.* Cambridge: Cambridge University Press, 47–83.

Williams GM, Ayres PG, 1981. Effects of powdery mildew and water stress on CO_2 exchanges in uninfected leaves of barley plants. *Plant Physiology* **68,** 527–530.

Wright D, Hughes LG, 1987. Effects of fungicides and variety on development, grain growth and yield of spring barley. *Annals of Applied Biology* **111,** 89–102.

Zangerl AR, Arntz AM, Berenbaum MR, 1997. Physiological price of an induced chemical defence: Photosynthesis, respiration, biosynthesis, and growth. *Oecologia* **109,** 433–441.

Zuckerman E, Eshel A, Eyal Z, 1997. Physiological aspects related to tolerance of spring wheat cultivars to *Septoria tritici* blotch. *Phytopathology* **87,** 60–65.

Chapter 8
Plant disease control through the use of variety mixtures

Adrian Newton

Scottish Crop Research Institute, Dundee, UK

8.1 Introduction

The principles driving use of variety mixtures for disease control are soundly based in ecology. Epidemics are the exception in natural and semi-natural ecosystems, reflecting the balance derived from the co-evolution of hosts and pathogens. However, in modern agriculture in particular, this balance is far from equilibrium and epidemics would be frequent were it not for highly effective pesticides and a plant breeding industry which introduces new cultivars to the market with new or different resistance genes. Such a situation is generally profitable when commodity prices are high, but it is costly and rates very poorly on sustainability and ecological or environmental parameter scales.

It may not be overstating the case to say that a major reason why use of variety mixtures is not common in more intensive agriculture is that they have been promoted mainly for their disease control attributes. The problems with this are: (a) disease control levels in practice vary considerably from almost complete to small increases in disease, (b) there are alternative, highly effective disease control methods, principally fungicides, (c) there is a perception that there are quality problems for end users with mixtures, (d) the other advantages of mixtures have not been demonstrated or 'championed' to the same degree. So, whilst this chapter concerns 'plant disease control through the use of variety mixtures', it should be emphasised that there other advantages of growing mixtures even where there is no disease or full fungicide programmes are used (Finckh *et al.*, 2000; Newton *et al.*, 2009).

The non-disease control aspects of variety mixtures can be summarised as better resource utilisation. For example, a mixture component not able to occupy a canopy niche will be compensated for by one which can. This needs to be highlighted in the context of disease control, and stress in general, as components of the canopy will be disadvantaged by the damage caused by both biotic and abiotic stress, thus compensated for by undamaged mixture components. Thus, yield response in variety mixtures is rarely highly correlated with disease reduction as it comprises the whole crop's compensatory and competitive response to all the environmental factors to which it is exposed. Promotion of variety mixtures outwith this context for disease control is likely to fail to achieve success except where other options are limited, such as some organic and subsistence systems.

8.2 Trial demonstrations of mixtures

Focusing on yield response, much experimental work has been carried out with cereals, mainly wheat and barley. In spring barley, the level of disease control or yield increase demonstrated has often been only moderate (Newton & Thomas, 1992; 1993; Newton *et al.*, 2002; Mercer *et al.*, 2006), but in winter barley yield increases of up to 17% have been achieved on both small (Day, 1984; Newton *et al.*, 1998) and large plot scales (Newton & Guy, 2008). The difference between the spring and winter crops of barley may be attributable to lower disease levels and the shorter and less stressed season in the former case. It may also reflect the morphological similarities of the recent recommended cultivars in the UK where many of the reported trials were grown and where fewer complementary traits were available to interact beneficially. More complementary traits may have been present in elite germplasm some years ago as spring barley trials carried out between 1978 and 1982 gave yield increases up to 16% and a mean of 5.69% (Day, 1984). In contrast, Paynter & Hills (2007) generally found no overall yield increases or disease control using current Australian winter barley germplasm. Using non-elite germplasm such as doubled-haploid lines from a mapping population segregating for two dwarfing genes grown in mixtures, yield enhancement of even two-component mixtures of lines differing in dwarfing genes was 6.6% and 9.9% for three components (Newton *et al.*, 2004).

The strategy of complementing cultivars with contrasting traits was demonstrated using Canadian barley cultivars differing in height and maturity date, illustrating not only the potential for yield enhancement with appropriate combinations, but also some which can lead to negative interactions (Essah & Stoskopf, 2002), and similar contrasts have been demonstrated in current wheat and barley mixture trials in Scotland (Newton AC & Hoad S, unpublished data) and wheat mixtures in Canada (Knott & Mundt, 1990). Combining varieties with contrasting traits does not result in the problems one might expect due to trait convergence. Convergence of many characters has been reported anecdotally, but there is also experimental data (Newton *et al.*, 2002). Using mixtures in which the components were physically distinct, it was possible to show some convergence for malting quality attributes, as well as agronomic characters (Swanston *et al.*, 2005, 2006), but it has not been demonstrated whether a similar mechanism operates in mixture components with similar morphology and genetic background.

Finckh & Mundt (1992) clearly illustrated why genetically heterogeneous populations should be treated in a holistic way, as in their wheat experiments between 52% and 58% of the yield variation was attributable to disease in monocultures whereas in mixtures this dropped to between 10% and 31%, illustrating some of the potential different plant–plant interaction effects. Day (1984) recorded powdery mildew reduction of around 35% unrelated to growth stage or absolute level of infection, and Newton *et al.* (2002) found that high inoculum pressure reduced mixtures efficacy, all factors which confound the causal relationships between yield and stresses.

The mechanism whereby mixtures achieve disease control can be attributed to dilution of the density of susceptible plants, introduction of resistant plants as physical barriers limiting pathogen spread, and induced resistance from enhanced frequency of avirulence factors on adjacent susceptible plants (Chin & Wolfe, 1984). In the absence of exposure to multiple hosts, races of pathogens tend to simple virulence patterns suggesting there is a cost in maintaining complex virulence mechanisms (Vanderplank, 1968). Despite this, on exposure to multiple host resistance genotypes, though not necessarily in mixtures, virulence patterns increase in complexity as each season progresses (Newton *et al.*, 1998). Thus, exposure to multiple host genotypes is the basis on which mixtures both inhibit epidemics of pathogens with simple virulence characteristics and select for races with complex virulence thereby avoiding triggering resistance mechanisms. This suggests there is a trade-off in the extent to which the pathogens are exposed to multiple host genotypes with an optimum level of exposure to maximise epidemic control but minimise selection for complex virulence.

Pathologists have often proposed multilines rather than variety mixtures to control disease. These are isogenic or near-isogenic lines differing only in specific resistance genes (Browning & Fey, 1969; Wolfe, 1985; Mundt, 2002a). Whilst potentially useful for optimising control of pathogens with specific gene-for-gene interactions, they target only such specific pathogens and will not capitalise on other heterogeneous interactions. Furthermore, considerable investment is made in developing such lines in a variety which may be superseded and for which therefore there is no market demand.

Even for disease control, the efficacy of mixtures is not just dependent on specific interactions between host and pathogen populations. The canopy structure can be crucial in a number of ways, which can be well illustrated by the pathogen *Rhynchosporium secalis*, causal agent of 'rhynchosporium', 'scald' or 'leaf blotch' on barley and common in many regions of the world. It is splash-dispersed, such as from the soil, where spore inoculum may survive on crop debris, then from leaf to leaf as it forms necrotic lesions or asymptomatic infections and sporulates (Zhan *et al.*, 2008). An open canopy will enable rapid transmission by rain splash up the plant. Epidemic progress can be reduced by molecular and morphological mechanisms. Genetic resistance results when specific genotypes (races) of the pathogen are recognised by plant defence genes causing induction of resistance expression mechanisms. If the host plant has only a single recognition gene then the pathogen will mutate to produce new races not recognised by the pathogen, and multiple host plant genotypes with many different recognition genes could be deployed, which may lead to stability as discussed above. However, morphological differences between component varieties in the mixture can be manipulated to reduce the epidemic progress too. Plants with different height and leaf angle can be deployed together, providing a complex canopy structure and disrupting vertical splash pathways for pathogen dispersal. Multiple contrasting morphological types are particularly effective in this, particularly those with contrasting dwarfing gene (Newton *et al.*, 2004). Mixtures are effective against other splash-dispersed pathogens which have even less host specificity expressed, such as Septoria Leaf Blotch (*Mycosphaerella graminicola*) on wheat (Cowger & Mundt, 2002) and glume blotch (*Phaeosphaera nodorum*) on wheat (Jeger *et al.*, 1981a) and again in *R. secalis* on barley (Jeger *et al.*, 1981b).

Mixtures have been reported to have efficacy against pathogens in the root environment. For example, eyespot was reduced in one study on wheat (Mundt *et al.*, 1995), but not on barley (Gieffers & Hesselbach, 1988). Perhaps more importantly when considering the crop system rather than just disease control, yield benefits have been reported even when disease is not always reduced for several soil pathogens such as *Phytophthora sojae* in soyabean (Wilcox & St. Martin, 1988) and *Cephalosporium graminearum* in wheat (Mundt, 2002b). *Rhizoctonia solani* was reduced in sugar beet (Halloin & Johnson, 2000) and *Helminthosporium victoriae* in oats (Ayanru & Browning, 1977). Virus diseases such as wheat soil-borne mosaic are also reduced through control of their vector, *Polymixa graminis* (Hariri *et al.*, 2001).

8.3 Mixtures used in practice

There is a clear relationship between increased number of components in a mixture and increased disease control (Nitzsche & Hesselbach, 1983; Mundt, 1994; Newton *et al.*, 1997) (Figure 8.1), but even two-component mixtures can achieve disease control, although this may be more effective against some diseases than others. Cox *et al.* (2004), for example, found a proportion range of two-component wheat mixtures to be more effective at reducing leaf rust than tan spot. The optimum structure will depend upon the 'objective' and, for example, for disease control it will differ depending upon the dispersal characteristics of the pathogen and its population structure. Multiple diseases with different life styles and particularly dispersal characteristics can be controlled by the same mixture but at any given scale, component composition it is likely to be more effective against one pathogen than another.

The disease reduction successes in practice are often scale and pathogen dispersal mechanism dependent in terms of both the total area and individual genotype unit area (Holt & Chancellor, 1999). In simulation work, the general conclusion is that the smaller the homogeneous genotype area or genotype unit area (GUA), the greater the mixture efficacy (Goleniewski & Newton, 1994; Xu & Ridout, 2000) and this has been demonstrated in the field in a number of host–pathogen interactions. In work with oat crown rust, Mundt & Leonard (1985) found that mixtures based on clumps of 200 seeds were not effective at reducing disease but random mixtures were, and that with bean rust, reducing GUAs from 0.84 to 0.023 m^2 resulted in progressive disease reduction (Mundt & Leonard,

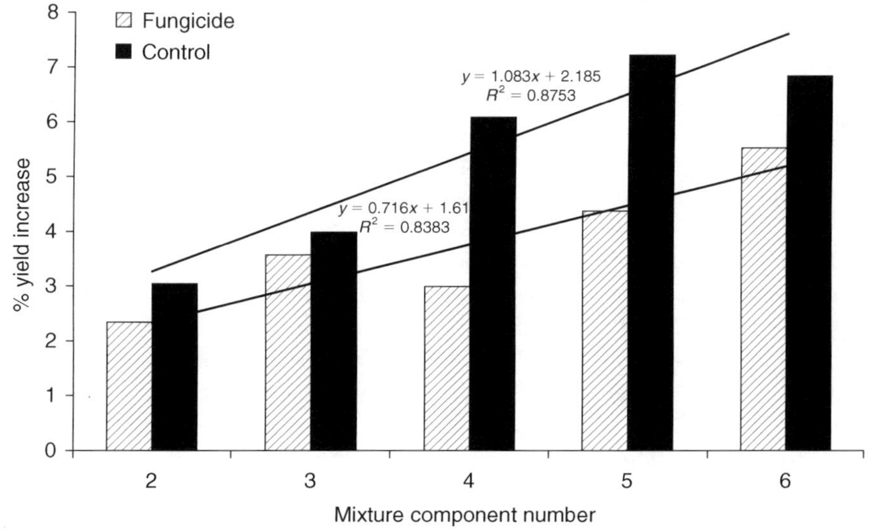

Figure 8.1 Change in *Rhynchosporium secalis* infection of mixtures of winter barley cultivars compared with the mean of their components with different numbers of component cultivars.

1986). Using much more contrasting GUAs of 0.003 m^2 and 11 m^2 with wheat-yellow rust and wheat-leaf rust, Brophy & Mundt (1991) again determined that the smaller GUA was more effective at reducing disease in mixtures. However, this was not always the case, as Mundt & Browning (1985) found that increasing the scale at which component cultivars were deployed from a GUA of 0.003–0.84 or 0.58 m^2, had no effect on the crown rust infection of oats. Similarly, Mundt & Leonard (1986) found no differences in maize rust infection between GUAs ranging from 0.21 to 1.88 m^2. Differences may be due to particular pathogens and spore dispersal conditions and models often do not take into account the dispersal mechanisms of the pathogen. For rhynchosporium (*R. secalis*) on winter barley, the optimum GUA or patch size was determined at about 4 m^2 in one study (Newton & Begg, 2008), which lead to experiments to determine whether a patchy arrangement of component cultivars might be better than homogeneous mixing even within fields. Patchy sowing was found to both reduce disease and increase yield more than homogeneous sowings compared with the mean of monocultures sown alongside, even in the absence of disease (Newton & Guy 2008). Controlling the spatial nature of the 'patchiness' at a field scale could be problematic but this work clearly shows that crude patchiness was easy to deploy in practice, as no pre-mixing of seed was necessary, and whilst it didn't always result in big yield increases, it did deliver disease reduction and the same or more yield than homogeneously pre-mixed seed. Experimental data also indicate that for powdery mildew, in patches (rows) the selection for complex races was much less intense than in random mixtures (which have smaller GUAs) (Huang *et al.*, 1994), so a patchy field sowing should reduce selection for complexity also.

One of the biggest mixture experiments was on rice, where a diversification programme in Yunnan province demonstrated effective control of rice blast. In the first year, the area was 812 ha, expanding to 3342 ha in the second year and comprising row mixtures of susceptible and resistant cultivars. This achieved 94% less severe rice blast than when grown as monocultures and increased the yield by 89% (Zhu *et al.*, 2000). Clearly scale was crucial to this experiment, echoing the experience in the former German Democratic Republic, where up to 92% of the spring barley crop was grown as mixtures (Wolfe, 1997). Mildew declined from over 50% to less than 10%, thereby reducing the fungicide requirement substantially. Numerous cultivars were used, but most used the same resistance and yet it was still a success. Other notable large scale successful uses of mixtures are Poland, where over 90 000 ha of cereal mixtures are reported (Gacek, 1997), Denmark with 62 000 ha in 1996 (Munck, 1997), the USA where, in Oregon, 10% of the soft white winter wheat and 76% of club wheat (Mundt, 1994), and in Kansas 7% of wheat (Bowden *et al.*, 2001) are grown as mixtures. On a smaller scale, mixtures are supported in the Swiss low input 'Extenso' cereal production programme (Merz & Valenghi, 1997) and in Scotland, winter barley mixtures have been grown for some years and winter wheat mixtures are increasing (2007) for distilling (Newton AC, unpublished data). It is also common practice to grow cereals as mixtures in many developing countries where higher total yield, better yield stability, better food quality, animal feed and resistance to pests over time compared with monocultures are quoted as the rationale (Smithson & Lenne, 1996; Woldeamlak *et al.*, 1998a, 1998b), and the stability of wild crop relatives from which landraces were derived is cited as not only sources of resistance, but also indicators of good ecological practice to be re-discovered in modern agriculture (Akem *et al.*, 2000; Wood & Lenne, 2001).

Other crops have been grown as mixtures to control disease. Still in cereals, oats have been trialed in mixtures to reduce the impact of barley yellow dwarf disease (Peltonensainio & Karjalainen 1991; Karjalainen & Peltonensainio, 1993). Efficacy has been very variable in terms of yield, depending on the maturity type (Helland & Holland, 2001). Mundt & Browning (1985) reported considerable reduction in crown rust infection in oat isogenic multiline mixtures, but further work showed that mixtures based on clumps of seeds were not effective at reducing disease, although random mixtures were (Mundt & Leonard, 1985). Yield stability is still always likely to be beneficial even in the absence of significant yield gains (Pfahler & Linskens, 1979).

Late blight (*Phytophthora infestans*) is normally the most economically important and epidemiologically devastating disease of potato, requiring multiple pesticide applications. One might expect that, although polycyclic, the epidemic is normally so fast that the slowing effect of heterogeneity would be unlikely to have much impact, but Garrett & Mundt (2000) achieved 36–37% reductions in the AUDPC in resistant-susceptible mixtures compared with the mean of the components grown in monoculture. Under higher disease pressure Phillips *et al.* (2005) and Pilet *et al.* (2006) found little effect, and similarly, high pathogen diversity levels also reduce the impact of mixtures (Garrett *et al.*, 2001). Susceptible cultivars benefit most in mixtures, whilst partial and fully resistant cultivars are little affected (Andrivon *et al.*, 2003; Phillips *et al.*, 2005).

To reduce anthracnose disease (*Colletotrichum gloeosporiodes*), the tropical pasture legume *Stylosanthes scabra* is frequently marketed and grown as a mixture, but reductions were small, mixture-dependent and very sensitive to the proportion of susceptible components (Chakraborty *et al.*, 1991; Davis *et al.*, 1994). Lucerne (*Medicago sativa*) is also grown as very complex mixtures, or rather as populations with resistance sources for controlling Phytophthora root rot (*Phytophthora medicaginis*) and colletotrichum crown rot (*Colletotrichum medicaginis*) (Mackie & Irwin, 1998; Musial *et al.*, 2002).

In red kidney bean crops, mixtures have been used to reduce anthracnose (*Collecotrichum lindemuthianum*) (Ntahimpera *et al.*, 1996) and in sorghum, they are used to control leaf blight (*Exserohilum turcicum*), the benefits being in proportion to the resistant cultivar proportion included in the mixture (Barrera & Frederikses, 1994). Limited experimentation has demonstrated disease reduction in oilseed rape in two-component mixtures (Walker KC, personal communication). Coffee is grown in mixtures to control its rusts, where mixtures have been planted on a large scale in Colombia (Moreno-Ruiz *et al.*, 1990), and similarly willow (*Salix* spp.) is grown for biomass as mixtures of clones, largely to control rusts (McCracken *et al.*, 2001; Hunter *et al.*, 2002) and reduce beetle damage (Peacock *et al.*, 2001).

8.4 Conclusion

Clearly there are good reasons to use variety mixtures to control disease and they are more practical than most people envisage. They are appropriate for intensive and high input systems as well extensive, low input, subsistence and organic systems where disease control options are more limited. Optimisation of mixture composition is needed for particular disease threats, but complexity is a general rule for increasing efficacy. Variety mixtures have the advantage that the best of the elite germplasm can be exploited, and further benefits derived simply from the enhanced resilience effects of heterogeneous deployment, but also specifically compensating for any weaknesses by other components of the mixture. In intensive agriculture, region markets drive variety choice, not disease resistance traits. Variety mixtures offer an opportunity to utilise market-leading varieties whilst incorporating disease-reducing principles of simply deploying heterogeneity, and specifically including varieties with traits which maximise the effect.

8.5 References

Akem C, Ceccarelli S, Erskine W, Lenne J, 2000. Using genetic diversity for disease resistance in agricultural production. *Outlook on Agriculture* **29**, 25–30.

Andrivon D, Lucas JM, Ellisseche D, 2003. Development of natural late blight epidemics in pure and mixed plots of potato cultivars with different levels of partial resistance. *Plant Pathology* **52**, 586–594.

Ayanru DKG, Browning JA, 1977. Effect of heterogeneous oat populations on the epiphytotic development of Victoria blight. *New Phytologist* **79**, 613–623.

Barrera JAS, Frederikses RA, 1994. Evaluation of sorghum hybrid mixtures for controlling sorghum leaf-blight. *Plant Disease* **78**, 499–503.

Bowden RJ, Shoyer J, Roozeboom K, Claasen M, Evans P, Gordon B, Heer B, Janssen K, Long J, Martin J, Schlegel A, Sears R, Witt M, 2001. Performance of wheat variety blends. *Kansas State University Agricultural Extension Bulletin* **128**. (www.oznet.ksu.edu/library/crps12/SRL128.pdf)

Brophy LS, Mundt CC, 1991. Influence of plant spatial patterns on disease dynamics, plant competition and grain yield in genetically diverse wheat populations. *Agricultural Ecosystems and Environment* **35**, 1–12.

Browning JA, Fey KJ, 1969. Multiline cultivars as a means of disease control. *Annual Review of Phytopathology* **14,** 355–382.

Chakraborty S, Pettitt AN, Cameron DF, Irwin JAG, Davis RD, 1991. Anthracnose development in pure and mixed stands of the pasture legume *Stylosanthes scabra*. *Phytopathology* **81**, 788–793.

Chin KM, Wolfe MS, 1984. The spread of *Erysiphe graminis* f. sp. *hordei* in mixtures of barley varieties. *Plant Pathology* **33,** 89–100.

Cowger C, Mundt CC, 2002. Effects of wheat cultivar mixtures on epidemic progression of *Septoria tritici* blotch and pathogenicity. *Phytopathology* **92,** 617–623.

Cox CM, Garrett KA, Bowden RL, Fritz AK, Dendy SP, Heer WF, 2004. Cultivar mixtures for simultaneous management of multiple diseases: Tan spot and leaf rust of wheat. *Phytopathology* **94,** 961–969.

Davis RD, Chakraborty S, Cameron DL, Irwin JAG, Boland RM, 1994. The influence of mixtures of *Stylosanthes* spp. accessions on the occurrence of anthracnose caused by *Colletotrichum gloeosporioides*. *Australian Journal of Agricultural Research* **45,** 203–210.

Day KL, 1984. The effect of cultivar mixtures on foliar bdisease and yield in barley and wheat. MSc thesis, University of Newcastle Upon Tyne.

Essah SYC, Stoskopf NC, 2002. Mixture performance of phenotypicaly contrasting barley cultivars. *Canadian Journal of Plant Science* **82**, 1–6.

Finckh MR, Mundt CC, 1992. Plant competition and disease in genetically diverse wheat populations. *Oecologia* **91,** 82–92.

Finckh MR, Gacek ES, Goyeau H, Lannou C, Merz U, Mundt CC, Munk L, Nadziak J, Newton AC, de Vallavieille-Pope C, Wolfe MS, 2000. Cereal variety and species mixtures in practice, with emphasis on disease resistance. *Agronomie: Plant Genetics and Breeding* **20,** 813–837.

Gacek E, 1997. Sumarised variety mixture information given to polish farmers. In: Wolfe MS, ed. *Variety Mixtures in Theory and Practice*. European Union Variety and Species Mixture working group of COST Action 817. Online at: http://www.scri.ac.uk/research/pp/pestanddisease/rhynchosporiumonbarley/otherwork/cropmixtures/varietymixtures.

Garrett KA, Mundt CC, 2000. Host diversity can reduce potato late blight severity for focal and general patterns of primary inoculum. *Phyopathology* **90**, 1307–1312.

Garrett KA, Nelson RJ, Mundt CC, Chacon G, Jaramillo RE, Forbes GA, 2001. The effects of host diversity and other management components on epidemics of potato late blight in the humid highland tropics. *Phytopathology* **91,** 993–1000.

Gieffers W, Hesselbach J, 1988. Disease incidence and yield of different cereal cultivars in pure stands and mixtures. II. Winter wheat (*Hordeum vulgare* L.). *Zeitschrift fur Pflanzenschutz* **95,** 63–69.

Goleniewski G, Newton AC, 1994. Modelling the spread of fungal diseases using a nearest neighbour approach: The effect of geometrical arrangement. *Plant Pathology* **43**, 631–643.

Halloin JM, Johnson DJ, 2000. Reduction of sugarbeet losses from Rhysoctonia crown and root rot by use of mixtures of resistant and susceptible varieties. *Phytopathology* **107**, S33.

Hariri D, Fouchard M, Prud'homme H, 2001. Incidence of *Soil-borne wheat mosaic virus* in mixtures of susceptible and resistant wheat cultivars. *European Journal of Plant Pathology* **107**, 625–631.

Helland SJ, Holland JB, 2001. Blend responses and stability and cultivar blending ability in oat. *Crop Science* **41**, 1689–1696.

Holt J, Chancellor TCB, 1999. Modelling the spatio-temporal deployment of resistant varieties to reduce the incidence of rice tungro disease in a dynamic cropping system. *Plant Pathology* **48**, 453–461.

Huang R, Kranz J, Welz HG, 1994. Selection of pathotypes of *Erysiphe graminis* f. sp. *hordei* in pure and mixed stands of spring barley. *Plant Pathology* **43**, 458–470.

Hunter T, Peacock L, Turner H, Brain P, 2002. Effect of plantation design on stem-infecting form of rust in willow biomass coppice. *Forest Pathology* **32**, 87–97.

Jeger MJ, Griffiths E, Jones DG, 1981a. Effects of cereal cultivar mixtures n disease epidemics caused by splash-dispersed pathogens. In: Jenkyn JF & Plumb RT, eds. *Strategies for Control of Cereal Disease*. Oxford: Blackwell Science Publications, 81–88.

Jeger MJ, Jones DG, Griffiths E, 1981b. Disease progress of non-specialised fungal pathogens in intraspecific mixed stands of cereal cultivars. II. Experiments. *Annals of Applied Biology* **98**, 187–198.

Knott EA, Mundt CC, 1990. Mixing ability analysis of wheat cultivar mixtures under diseased and non diseased conditions. *Theoretical and Applied Genetics* **80**, 313–320.

Karjalainen R, Peltonensainio P, 1993. Effect of oat cultivar mixtures on disease progress and yield reduction caused by barley yellow dwarf virus. *Journal of Plant Disease and Protection* **100,** 58–68.

Mackie JM, Irwin JAG, 1998. The proportion of individual plants resistant to *Phytophthora medicaginis* and *Colletotrichum trifolii* in Australian cultivars. *Australian Journal of Experimental Agriculture* **38**, 41–44.

McCracken AR, Dawson WM, Bowden G, 2001. Yield responses of willow (Salix) grown in mixtures in short rotation coppice (SRC). *Biomass & Energy* **21**, 311–319.

Mercer P, Ruddock A, Reavey C, 2006. Problems facing the growing of organic cereals in N. Ireland. *Proceedings Crop Protection in Northern Britain* 2006, 163–168.

Merz Un Valenghi D, 1997. "Extenso" production of cereal mixtures in Switzerland. In: Wolfe MS, ed. *Variety Mixtures in Theory and Practice*. European Union Variety and Species Mixture working group of COST Action 817. Online at: http://www.scri.ac.uk/research/pp/pestanddisease/rhynchosporiumonbarley/otherwork/cropmixtures/varietymixtures

Moreno-Ruiz G, Castillo-Zapata J, Sifuentes J, Browning JA, 1990. The variety Colombia: A variety of coffee with resistance to rust (*Hemileia vastatrix* Berk. & Br.). CENICAFE (Centro Nacional de Invesigaciones de Café).

Munck L, 1997. Variety mixtures: 19 years of experience in Denmark. In: Wolfe MS, ed. *Variety Mixtures in Theory and Practice*. European Union Variety and Species Mixture working group of COST Action 817. Online at: http://www.scri.ac.uk/research/pp/pestanddisease/rhynchosporiumonbarley/otherwork/cropmixtures/varietymixtures

Mundt CC, 2002a. Use of multiline cultivars and cultivar mixtures for disease management. *Annual Review of Phytopathology* **40,** 381–410.

Mundt CC, 2002b. Performance of wheat cultivars and cultivar mixtures in the presence of *Cephalosporium* stripe. *Crop Protection* **21,** 93–99.

Mundt CC, 1994. Use of host genetic diversity to control cereal diseases: Implications for rice blast. In: Zeigler RS, Leong SA, Teng PS, eds. *Rice Blast Disease*. London: CAB International, 293–308.

Mundt CC, Browning JA, 1985. Development of crown rust epidemics in genetically diverse oat populations: Effect of genotype unit area. *Phytopathology* **75,** 607–610.

Mundt CC, Leonard KJ, 1985. Effect of host genotype unit area on epidemic development of crown rust following focal and general inoculations of mixtures of immune and susceptible oat plants. *Phytopathology* **75,** 1141–1145.

Mundt CC, Leonard KJ, 1986. Effect of host genotype unit area on development of focal epidemics of bean rust and common maize rust in mixtures of resistant and susceptible plants. *Phytopathology* **76,** 895–900.

Mundt CC, Brophy LS, Schmitt MS, 1995. Disease severity and yield of pure-line wheat cultivars and mixtures in the presence of eyespot, yellow rust, and their combination. *Plant Pathology* **44,** 173–182.

Musial JM, Basford KE, Irwin JAG, 2002. Analysis of genetic diversity within Australian lucerne cultivars and implications for future genetic improvement. *Australian Journal of Agricultural Research* **53**, 629–636.

Newton AC, Begg G, 2008. Host diversity, scale and the epidemics of plant pathogens. *Journal of Plant Pathology* (in press).

Newton AC, Guy DC, 2008. The effect of uneven, patchy cultivar mixtures on disease control and yield in winter barley. *Field Crops Research* (in press; doi:10.1016/j.fcr.2008.09.002).

Newton AC, Thomas WTB, 1992. The effect of specific and non-specific resistance in mixtures of barley genotypes on infection by mildew (*Erysiphe graminis* f. sp. *hordei*) and on yield. *Euphytica* **59**, 73–81.

Newton AC, Thomas WTB, 1993. The interaction of either an effective or a defeated major gene with non-specific resistance on mildew infection (*Erysiphe graminis* f. sp. *hordei*) and yield in mixtures of barley. *Journal of Phytopathology* **139**, 268–274.

Newton AC, Begg G, Swanston JS, 2009. Deployment of diversity for enhanced crop function. *Annals of Applied Biology* **154** (in press; doi:10.1111/j.1744-7348.2008.00303x).

Newton AC, Ellis RP, Hackett CA, Guy DC, 1997. The effect of component number on *Rhynchosporium secalis* infection and yield in mixtures of winter barley cultivars. *Plant Pathology* **46**, 930–938.

Newton AC, Swanston JS, Guy DC, Ellis RP, 1998. The effect of cultivar mixtures on malting quality in winter barley. *Journal of the Institute of Brewing* **104,** 41–45.

Newton AC, Guy DC, Nadziak J, Gacek E, 2002. The effect of inoculum pressure, germplasm selection and environment on spring barley cultivar mixtures efficacy. *Euphytica* **125,** 325–335.

Newton AC, Swanston JS, Guy D, 2004. Enhanced durability and utility of genes for resistance by deployment in cultivar mixtures. *Proceedings Molecular Plant-Microbe Interactions XI,* St. Petersburg, 18–26 July, 240–243.

Nitzsche W, Hasselbach J, 1983. Sortenmischungen statt Viellinien-Sorten. 1. Somergerste (*Hordeum vulgare* L.) *Zeitschrift für Pflanzenzuchtung* **90**, 68–74.

Ntahimpera N, Dillard HR, Cobb AC, Seem RC, 1996. Anthracnose development in mixtures of resistant and susceptible dry bean cultivars. *Phytopathology* **86,** 668–673.

Paynter B, Hills A, 2007. Mixing feed barley cultivars to decrease leaf diseases and increase grain yield. In: *13th Australian Barley Technical Symposium*, Perth, Western Australia.

Peacock L., Hunter T., Turner H, Brain P, 2001. Does host genotype diversity affect the distribution of insect and disease damage in willow cropping systems? *Journal of Applied Ecology* **8,** 1070–1081.

Peltonensainio P, Karjalainen R, 1991. Agronomic evaluation of growing oat cultivar mixtures under various stress conditions in Finland. *Agriculturae Scandinavica* **41**, 47–53.

Pfahler PL, Linskens HF, 1979. Yield stability and population diversity in oats (*Avena* sp.). *Theoretical and Applied Genetics* **54**, 1–5.

Phillips SL, Shaw MW, Wolfe MS, 2005. The effect of potato variety mixtures on epidemics of late blight in relation to plot size and level of resistance. *Annals of Applied Biology* **147,** 245–252.

Pilet F, Chacon G, Forbes GA, Andrivon D, 2006. Protection of susceptible potato cultivars against late blight in mixtures increases with decreasing disease pressure. *Phytopathology* **96**, 777–783.

Smithson JB, Lenne JM, 1996. Varietal mixtures: A viable strategy for sustainable productivity in subsistence agriculture. *Annals of Applied Biology* **128**, 127–158.

Swanston JS, Newton AC, Brosnan JM, Broadhead A, 2005. Determining the spirit yield of wheat varieties and variety mixtures. *Journal of Cereal Science* **42,** 127–134.

Swanston JS, Newton AC, Hoad S, Spoor W, 2006. Variation across environments in patterns of water uptake and endosperm modification in barley varieties and variety mixtures. *Journal of the Science of Food and Agriculture* **86,** 826–833.

Vanderplank JE, 1968. *Disease Resistance in Plants*. New York: Academic Press.

Wilcox JR, St Martin SK, 1988. Soyabean genotypes resistant to *Phytophthora sojae* and compensation for yield losses of susceptible isolines. *Plant Disease* **82,** 303–306.

Woldeamlak A, Struick PC, Gagnew G, 1998a. Yield advantages of mixtures of wheat and barley land races in Eritrea. In: Zima M, Barstova ML, eds. *Proceedings of the European Society of Agronomy,* June 28–July 3, Nitra, 67–68.

Woldeamlak A, Struick PC, Gagnew G, 1998b. Density and component crop proportion on wheat and barley mixtures in Eritrea. In: Zima M, Barstova ML, eds. *Proceedings of the European Society of Agronomy,* June 28–July 3, Nitra, 65–66.

Wolfe MS, 1985. The current status and prospects of multiline cultivars and variety mixtures for disease resistance. *Annual Review of Phytopathology* **23,** 251–273.

Wolfe MS, 1997. Variety mixtures: Concept and value. In: Wolfe MS, ed. *Variety Mixtures in Theory and Practice*. European Union Variety and Species Mixture working group of COST Action 817. Online at: http://www.scri.ac.uk/research/pp/pestanddisease/rhynchosporiumonbarley/otherwork/cropmixtures/varietymixtures

Wood D, Lenne J, 2001. Nature's fields: A neglected model for increasing food production. *Outlook on Agriculture* **30,** 161–170.

Xu XM, Ridout MS, 2000. Stochastic simulation of the spread of race-specific and race-nonspecific aerial fungal pathogens in cultivar mixtures. *Phytopathology* **49,** 207–218.

Zhan J, Fitt BDL, Pinnschmidt HO, Oxley SJP, Newton AC, 2008. Resistance, epidemiology and sustainable management of *Rhynchosporium secalis* populations on barley. *Plant Pathology* **57**, 1–14.

Zhu Y, Chen H, Fan J, Wang Y, Li Y, Chen J, Fan JX, Yang S, Hu L, Leung H, Mew TW, Teng PS, Wang Z, Mundt CC, 2000. Genetic diversity and disease control in rice. *Nature* **406**, 718–722.

Chapter 9
Biofumigation for plant disease control – from the fundamentals to the farming system

John Kirkegaard

CSIRO Plant Industry, Canberra, ACT 2601, Australia

9.1 Introduction

Biofumigation is a term originally coined to describe the suppression of plant pests and diseases by biocidal hydrolysis products, notably the isothiocyanates (ITCs), released by glucosinolate (GSL)-containing plants in soil (Kirkegaard *et al.*, 1993). The term has since been popularized to now encompass a range of benefits derived from these and other organic amendments. Although the primary focus in this chapter is on recent developments to maximize the ITC-related biofumigation benefits as originally defined, a more complete understanding of the variable disease suppression reported demands a consideration of the other mechanisms operating in amended soils. Increasingly, field-based biofumigation studies are revealing a significant contribution of mechanisms other than those related to GSL hydrolysis products to the observed disease control, as well as other system benefits such as improved soil structure, increased organic matter and erosion control. Ellenby's early research (Ellenby, 1945) on the effects of *Brassica* root exudates on potato cyst nematode (*Globodera* spp.) foreshadowed considerable contemporary interest in the pest and disease control potential of GSL-containing plants. The phenomenon has been observed for centuries, but has received renewed interest recently fuelled by the need to seek disease control alternatives to the widely used soil fumigant methyl bromide (Martin, 2003), and a desire to reduce dependence on other synthetic pesticides. Advances in soil and plant analytical techniques have underpinned an improved understanding of the fundamental processes on which to develop sound disease control strategies using biofumigation and these have been extensively reviewed (Fenwick *et al.*, 1983; Chew, 1988; Brown & Morra, 1997; Rosa *et al.*, 1997; Mithen, 2001; Matthiessen & Kirkegaard, 2006). Yet a mostly empirical approach to much subsequent work and an apparent disconnect between fundamental and applied research has limited the development of the concept to the stage of commercial adoption (Kirkegaard & Matthiessen, 1999). Indeed in a recent review of biofumigation field studies published between 1992 and 2006, only 1 out of the 18 studies reported the GSL content of the incorporated tissue (Matthiessen & Kirkegaard, 2006). The level of commercial adoption has, until recently, belied the heavily documented potential. A more recent systematic approach to biofumigation and advances in species selection and plant incorporation techniques have made it possible to generate ITC concentrations in soil more comparable to those detected during commercial use of the synthetic-ITC generating pesticide metham sodium (Matthiessen *et al.*, 2004; Gimsing & Kirkegaard, 2006). A broader IPM (Integrated Pest Management) approach to biofumigation as a component of disease control strategies, assisted by new soil microbial ecology techniques is providing insights to optimize the ITC and non-ITC related benefits of biofumigants, although success is by no means universal (Hartz *et al.*, 2005; Njoroge *et al.*, 2008). This chapter reviews the fundamental processes underlying the biofumigation concept, considers recent progress in refining approaches for successful commercial adoption and provides some case studies to highlight opportunities and challenges for further development of the concept.

9.2 The glucosinolate–myrosinase system

GSLs are a class of sulfur compounds occurring as secondary metabolites almost exclusively in plants belonging to the Order Capparales, which includes the Brassicaceae (Fenwick *et al.*, 1983; Fahey *et al.*, 2001). GSLs are characterized by a common chemical entity (β-thioglucoside with a sulphonated oxime moiety) with a variable chemical side-chain [R] that distinguishes individual types (Mithen, 2001) (Figure 9.1). The 120

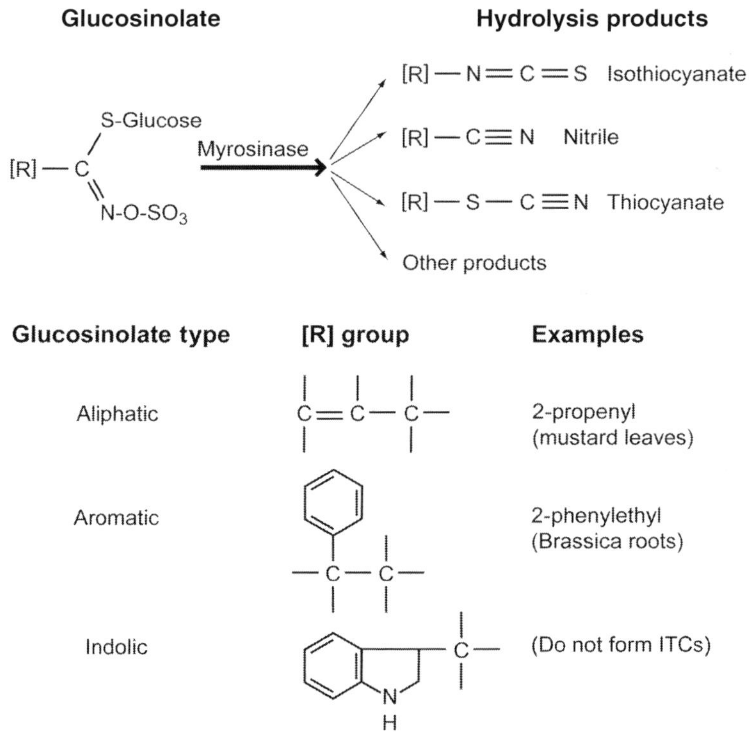

Figure 9.1 Glucosinolates are hydrolyzed by the myrosinase enzyme when intact tissues of glucosinolate-containing plants are disrupted to release a range of hydrolysis products. The isothiocyanates are considered the most bioactive. The structure of the organic side-chain [R] varies in different glucosinolates and is retained in the isothiocyanates and influences their biological activity. The [R] group structure of some common aliphatic and aromatic isothiocyanates are shown to illustrate the diversity.

GSLs currently identified are divided into three groups comprising aromatic, aliphatic or indolyl side chains, which influence the type and biological activity of the various hydrolysis products. All GSL-containing plants also produce a hydrolytic enzyme (thioglucosidase hydrolase) commonly known as myrosinase, which is physically separated from the GSLs in the intact plant tissue (Rosa *et al.*, 1997). On tissue disruption the myrosinase hydrolyzes the GSLs to form a number of hydrolysis products, a dynamic evolutionary link that has led to the term 'glucosinolate–myrosinase system' (Bones & Rossiter, 1996; Rask *et al.*, 2000) (Figure 9.1). The system is thought to have evolved in plants as a defense against generalist herbivores although complex biological interactions between specialist pests and pathogens have since emerged (Benderoth *et al.*, 2006). GSLs themselves have limited biological activity, but the various hydrolysis products are responsible for the biofumigant properties as well as the flavor, anti-nutritional and therapeutic characteristics of *Brassica* vegetables and spices (Fahey *et al.*, 2001; Holst & Williamson, 2004).

The biologically active hydrolysis products include ITCs, organic cyanides, oxazolidinethiones and ionic thiocyanates (Brown & Morra, 1997). Among the degradation products, most focus related to disease control has centered on the ITCs, which are liberated from aliphatic and aromatic GSLs (Mithen, 2001) (Figure 9.1). They have been shown to be the most bioactive of the hydrolysis products, and have been recognized since early in the twentieth century as broad-spectrum biocides (Walker *et al.*, 1937). The use of the synthetic compound methyl ITC (metham sodium) as a soil fumigant replacement for methyl bromide also generated interest in the idea of utilizing ITCs of natural origin for disease control (Matthiessen & Kirkegaard, 2006). The variation in the side-chain structure [R] of some ITCs commonly released from plant tissues is shown in Figure 9.1.

9.3 Modes of utilization

Biofumigation can involve GSL-containing plants as rotation crops, or intercrops (Kirkegaard *et al.*, 2000), by incorporating fresh plant material as green manure (Matthiessen & Kirkegaard, 2006), or utilizing processed plant products high in GSLs such as seed meals (Borek *et al.*, 1997), or dried plant material treated to preserve ITC activity (Lazzeri *et al.*, 2004). Formulations of extracted pure compounds as by-products of oilseed extraction have also been developed (Palmieri, 2004) and while there may be a niche for such products, they are likely to be regarded as pesticides by regulatory authorities and may face significant hurdles in implementation compared with utilization of rotation crops and green manures (Askew, 2004). The application of high concentrations of specific compounds whether 'natural' or synthetic, can bring with it all of the ecological problems that IPM strategies seek to avoid. Biofumigation implementation processes all share common features related to the particular chemistry and behavior of GSL-hydrolysis products in soil discussed later in this chapter. The central focus in this Chapter is on the utilization of GSL-containing plants as incorporated green manures, but a brief consideration of the specific aspects of the other modes of implementation is warranted.

9.3.1 Rotation or intercrops – a role for root GSLs

Biofumigation by rotation crops or intercrops, where above-ground material is harvested or left to mature above-ground, relies on root exudates of growing plants throughout the season, leaf washings or root and stubble residues. In these circumstances it is important to distinguish between active disease suppression through biofumigation and the well established non-host or break-crop effect (Kirkegaard *et al.*, 2000). Several studies have detected both GSLs and ITCs in the rhizospheres of intact plants, and these have been implicated in the suppression of pests and pathogens in both natural and managed ecosystems (van Dam *et al.*, 2008). An extensively studied instance in broad-acre agricultural systems involves the possible role of biofumigation in the widely observed rotational benefits of the *Brassica* break crops canola (*Brassica napus*) and Indian mustard (*Brassica juncea*) to subsequent cereals in the rotation (Angus *et al.*, 1994; Kirkegaard *et al.*, 2000). This was thought to be associated with the release of 2-phenylethyl ITC from the roots of *Brassica* rotation crops, a compound shown to be highly toxic to cereal pathogens *in vitro* (Sarwar *et al.*, 1998; Smith & Kirkegaard, 2002). The roots of canola varieties with high levels of the precursor 2-phenylethyl GSL reduced the inoculum levels of the take-all fungus (*Gaeumannomyces graminins* var. *tritici*) in the soil (Kirkegaard *et al.*, 2000), and also reduced the hosting of *Pratylenchus neglectus* nematodes on canola, which could otherwise multiply to infect subsequent wheat crops (Potter *et al.*, 1999). Rumberger & Marschner (2003, 2004) demonstrated ITC-induced changes in rhizosphere bacteria of canola, although the low levels of ITC detected (1–2 nmol g^{-1} soil) were two orders of magnitude lower than those detected under commercial fumigation with methyl ITC, and cast some doubt as to a direct suppression of disease related to ITC release. Indeed, Smith *et al.* (2004) failed to demonstrate yield improvements in wheat associated with reduced disease in the field using biofumigation and considered the limitation in ITC release from intact, growing roots would reduce disease suppression. Recent studies demonstrating the localization of 2-phenylethyl GSL in the outer tissues of secondary thickened canola roots suggest an evolutionary role in protection of the larger root system from soil-dwelling herbivores and a mechanism to ensure a continuous localized source of biotoxic hydrosolates into the rhizosphere (McCully *et al.*, 2008). Such mechanisms may account for the pest control properties of mustard intercrops grown in mixtures with crops such as potatoes (Akhtar & Alam, 1991). Generally, there is considerably more scope to utilize biofumigation strategies where purposely selected green manures from a range of species can be selected, grown and the whole plant incorporated to improve the timing and amount of ITC release for pest control. As root GSLs can constitute 24% (range 2–81%) of total plant GSL content during the vegetative stages (Kirkegaard & Sarwar, 1998), they should not be overlooked as a source of biofumigation when whole plants are incorporated as green manure.

9.3.2 Seed meals and other processed plant products

The seed meal or 'oilcake' byproducts which remain after pressing rapeseed or mustard seed for oil constitute a convenient, high GSL material suitable for soil amendment for high-value horticultural crops. These products contain sufficient intact myrosinase to ensure effective hydrolysis of the GSLs upon wetting. These have been investigated as a biopesticide (Brown & Morra, 1997) and biofertilizer (Balesh *et al.*, 2005) for some time, and interest has grown recently as a high-value end use for the seed meal arising from high-GSL biodiesel oilseeds is sought. Brassicaceous seed meals have demonstrated significant suppressive activity to a range of insect (Borek *et al.*, 1997), nematode (Mazzola *et al.*, 2001; Rahman & Somers, 2005), fungi (Smolinska *et al.*, 1997) and weeds (Brown & Morra, 1995), although at the high rates of amendment sometimes used (~8 t ha^{-1}), non-GSL related impacts on plant diseases have also been demonstrated (Mazzola *et al.*, 2001; Cohen *et al.*, 2005) (see Section 9.5). In addition to products derived from seed meals, other commercial products comprising the dry pellets of dehydrated high-GSL plants have also been developed for use as biofumigants (Lazzeri *et al.*, 2004). In both cases, these products provide an opportunity to maximize the amount plant tissues with high GSL concentration applied to the soil, and a great deal of flexibility in the timing and incorporation management of the biofumigant material. However, the high cost of transport is likely to limit the use to high-value horticultural industries such as fruit, vegetables and cut flowers or in potting media for glasshouse production. In general, the disease control principles associated with biofumigation using these products align with those of incorporated green manures, but without the sacrifice of time and space required to grow the material on-site.

9.3.3 Green manuring and biofumigation

Incorporated biofumigant green manures or plow-downs can potentially combine the beneficial elements of rotation crops (Section 9.3.1) with a more concentrated release of biocidal GSL-hydrolysis products at the time of incorporation. Green manures also provide benefits to subsequent crops and farming systems through maintenance of soil cover, soil sanitization, reduced erosion, increased soil organic matter and soil structural improvements (Bailey & Lazarovts, 2003; Thorup-Kristensen *et al.*, 2003). Brassicaceous green manures are no exception, and improvements in soil structure (Chan & Heenan, 1996), erosion control (McGuire, 2004) and nutrient cycling (Thorup-Kristensen *et al.*, 2003) have been reported for various Brassicaceous green manures in different parts of the world. Kirkegaard & Matthiessen (2004) recognized the difficulty in isolating the ITC-related disease control effects of green manures from these other effects in many biofumigation field studies and summarized this diagrammatically (Figure 9.2). This diagram provides a framework to reconsider the widely varying impacts of biofumigant green manures previously reported, and highlights the need for caution in ascribing these to ITC-related biofumigation alone.

9.4 Separating GSL-related suppression from other effects of biofumigants

In a practical sense, the mechanism of disease control from Brassicaceous soil amendments may be of little interest to farmers provided it is effective, predictable and repeatable. However the strategies required to improve disease control will vary greatly depending on the principal mechanism/s operating within particular systems (Figure 9.2). In that regard it is important to briefly consider some of the notable non-ITC related mechanisms of disease suppression shown in Figure 9.2 before a more detailed consideration of the ways in which ITC-related biofumigation can be maximized for disease control.

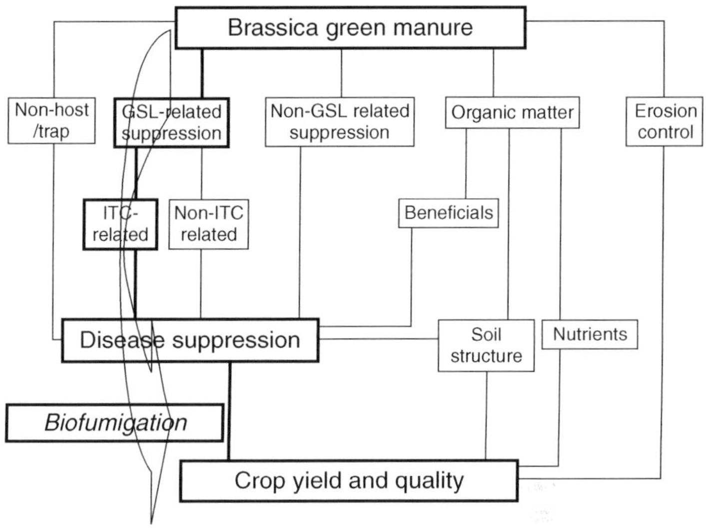

Figure 9.2 Mechanisms by which *Brassica* green manures can influence disease levels, and yield and quality of subsequent crops. The pathway by which ITC-related biofumigation influences disease suppression and crop yield is shown by the large arrow. Disease suppression in amended soil can result from other GSL hydrolysis products, non-GSL-related products, changes in microbial populations associated with the organic matter addition. Crop yield can be influenced by other system benefits of the green manures (from Kirkegaard & Matthiessen, 2004, with permission from Istituto Sperimentale per le Colture Industriali).

9.4.1 Trap crops and non-hosting

Perhaps the most well-documented case involving the use of Brassicaceous green manures as trap or catch crops is for the control of sugar-beet nematodes (*Heterodera schactii*) in northern Europe (Müller, 1999; Schlathoelter, 2004) Fodder radish (*Raphanus sativus*) and white mustard (*Sinapis alba*) are grown as green manures preceding sugar-beet crops. These crops are invaded by the nematodes, which develop within the roots, but have their sexual differentiation disrupted. This results in very low numbers of females in the subsequent generation, causing a significant decline in the population and reducing the infection of subsequent sugar-beet crops. It is a unique example related to specific nematode-resistant cover crops rather than a general non-hosting effect, and a role for GSLs or their hydrolysis products has not been demonstrated. McCleod *et al.* (2001) found little evidence that GSLs were involved in the reduced host status (invasion, egg laying or development) of a range of Brassicaceous crop plants to root knot nematodes (*Meloidogyne javanica*). Potter *et al.* (1999) demonstrated reduced hosting of *Pratylenchus neglectus* on canola plants with higher levels of 2-phenylethyl GSL in the roots, but clearly demonstrated that other non-GSL mechanisms were also involved. For a green manure to be effective in disease control, it is generally desirable that it does not host the pathogen in question so that a decline in population or inoculum occurs during its growth. Indeed, the fact that *Brassica* species are generally moderate hosts to some other important plant parasitic nematodes (e.g. *Meloidogyne* spp.), has reduced their applicability as biofumigant green manures as very careful management is required to avoid population increases on the biofumigant crop, particularly in warmer climates (McLeod & Steele, 1999; Stirling & Stirling, 2003). In such cases it is possible that the suppressive effects of the incorporated tissues could still reduce the population to manageable levels, but it is generally desirable to select biofumigants that are poor hosts, or to grow them at times of the year when pathogens do not build up. In strawberry production systems rotations with *Brassica* vegetables such as broccoli have been shown to provide effective control for pathogens such as Verticillium wilt (*Verticillium dahliae*), but not for *Pythium* spp. (Subbarao *et al.*, 2007). The greater reduction of *V. dahliae* microsclerotia and higher vigor and yield of strawberry following broccoli or brussel sprout rotation crops compared with lettuce occurred at both infested and non-infested sites indicating benefits other than suppression of those diseases were also involved.

9.4.2 Non-glucosinolate or ITC-related effects

Several studies have demonstrated significant suppression by incorporated *Brassica* amendments that are not associated with the GSL concentration of the tissue. A study by Potter *et al.* (1998) showed clearly that the significant suppressive effects (60–95%) of leaf tissues from six diverse *Brassica* species to the nematode *Pratylenchus negelectus* were unrelated to either the total or specific GSL content of the tissues. Similar results have also been shown for root knot nematodes (*Meloidogyne javanica*) by McLeod & Steele (1999). Mazzola *et al.* (2001) and Cohen *et al.* (2005) have also demonstrated that the suppression of the microbial complex associated with apple replant disease which includes the fungal pathogen *Rhizoctonia solani* using *Brassica* seed meal was unrelated to the GSL levels. There are usually two explanations proposed for these observations, but these may occur simultaneously. Firstly, the incorporation of organic matter itself can increase the populations of antagonistic organisms in the soil, as was demonstrated by the involvement of *Streptomyces* spp. and nitric-oxide producing bacteria in the suppression of *Rhizoctonia solani* in seed meal amended soil (Cohen & Mazzola, 2006). The stimulation of soil microbial activity or alteration of the communities to suppress specific disease in potato crops has also been demonstrated for *Rhizoctonia* (Larkin & Honeycut, 2006) and *Verticillium* wilt (Davis *et al.*, 1996), and in other pathosystems by Smolinska (2000). These mechanisms are summarized in Figure 9.2 via the 'organic matter' and 'beneficials' pathways. Secondly, many potentially biologically active, non-GSL compounds such as methanethiol are released from *Brassica* amendments (Bending & Lincoln, 1999), and other products of microbial decomposition of tissues including fatty acids or ammonia can also be biologically active (Bailey & Lazarovts, 2003). These are summarized in Figure 9.2 under 'non-GSL suppression'. Irrespective of the mechanisms responsible, these disease control mechanisms that are unrelated to GSLs, and can occur using non-*Brassica* amendments significantly confound the interpretation of biofumigation studies if appropriate controls are not included in the experiments.

Although ITCs are generally considered the most toxic of the GSL hydrolysis products, a range of other potentially toxic compounds including nitriles, epinitriles, and ionic thiocyanates can also be released (Brown & Morra, 1997; Morra, 2004; Rollin & Palmieri, 2004; Palmieri, 2004) (Figure 9.1). Generally they are less toxic, although Morra (2004) has shown that much of the weed suppression noted following incorporation of Brassicaceous seed meals is likely to arise from the ionic thiocyanate rather than the ITCs. To resolve the question of ITC-related suppression it is generally desirable to correlate the level of pest suppression with measured levels of GSLs in the tissues or ITC released in soil, a task that is not trivial given the relatively rapid loss from soils as a result of many different processes (Brown & Morra, 1997).

9.5 Maximizing biofumigation potential

Disease control using biofumigants can be maximized by selecting biofumigants which can generate high quantities of the GSL precursors to the ITCs which have the greatest toxicity to the target organism. Growing conditions and incorporation strategies can also influence these compounds in plant tissue and thus the potential ITC available for pest suppression.

9.5.1 GSL profiles

The type, concentration and distribution of GSLs in different plant parts vary between *Brassica* species and cultivars (Josefsson, 1967; Sang *et al.*, 1984) and so the capacity to generate ITCs varies accordingly. Kirkegaard & Sarwar (1998) found that differences in total GSL production on a ground area basis among 80 autumn-sown brassicas sampled at mid-flowering varied from 8 to 453 moles ha^{-1}. The variation derived equally from variations in both biomass and GSL concentration which were not correlated in either root or shoot tissues. The proportion of total GSLs which were ITC-liberating also varied considerably from close to 100% in the various mustard species (*B. juncea, B. carinata, B. nigra*) to 50% or less in rapeseed (*B. napus*) and other *Brassica* vegetables (*B. oleracea*). Thus, it is possible to select for biofumigants which produce high biomass, with high concentrations of ITC-liberating GSLs. In a related study (Sarwar & Kirkegaard, 1998), GSL concentrations were also shown to vary 3–10-fold depending on the growing environment, and tended to increase under warmer conditions and longer days in spring with maximum GSL production reaching 1100 moles ha^{-1}. Increases in tissue concentrations can also be induced by various biotic and abiotic plant stresses such as insect attack, drought and sulfur nutrition (Rosa *et al.*, 1997). GSL concentrations also change with plant development, and generally increase in concentration during vegetative growth, decline with the onset of flowering and are lowest in mature tissues. As a result of these developmental changes, there is usually an optimum period when the balance of biomass and GSL concentration in the tissues maximizes total plant GSL production on a ground area basis. Fortuitously, this appears to coincide with early flowering period (Clossais-Besnard & Larher, 1991; Sarwar & Kirkegaard, 1998), a convenient time for the ease of mechanical incorporation of the vegetative material and its rapid decomposition. Thus, plant species and varietal selection, environment, growing conditions and plant ontogeny can all influence the plant-related components of biofumigation potential.

9.5.2 Toxicity of GSL hydrolysis products

The biocidal properties of ITCs have been recognized for some time (Walker *et al.*, 1937), and their activity results from their non-specific and irreversible reaction with sulfhydral groups, disulfide bonds and amines in proteins and amino acids (Brown & Morra, 1997). Variation in the side-chain structure of different ITCs (as determined by the precursor GSL; Figure 9.1) can influence important characteristics related to their biocidal activity such as volatility, solubility and hydrophobicity. The range in the sensitivity of various disease organisms to different ITCs may differ by an order of magnitude (Brown &

Morra, 1997; Smith & Kirkegaard, 2002), which provides opportunities to select *Brassica* species that produce large quantities of ITCs most toxic to the target organism. For example, screening of various pure ITCs *in vitro* demonstrated the sensitivity of cereal fungal pathogens to the aromatic-ITC 2-phenylethyl ITC contained in canola roots (Sarwar *et al.*, 1998), while sulfur-substituted aliphatic ITCs present in radish types were most toxic to *Fusarium* (Manici *et al.*, 1997) and to *Pythium* and *Rhizoctonia* (Manici *et al.*, 2000). *In vitro* screens using pure ITCs or hydrated, freeze-dried tissues must be interpreted carefully because the apparent toxicity of individual ITCs can vary depending on whether a vapor phase or contact toxicity is established in the experimental protocol due to the variation in volatility of the different ITCs (Matthiessen & Kirkegaard, 2006). For example, the short-chain aliphatic ITCs such as 2-propenyl (found in mustard leaves) were more toxic to cereal pathogens than the aromatic 2-phenylethyl ITC (found in canola roots) in headspace experiments, but the reverse was true when the fungi were grown on agar containing the ITCs. Notwithstanding the obvious simplifications in laboratory assays compared with the complex interaction in soils (see later), matching the most potent brassicas with target organisms shown to be sensitive to particular ITCs provides the best chance of achieving successful suppression in the field.

9.5.3 Purposeful selection and development of biofumigants

The variability in the reported suppression of different pathogens in previous field experiments is perhaps not surprising given the failure to monitor the GSL levels of the varieties used, and the lack of purposeful selection of cultivars high in the most appropriate GSL precursors (Matthiessen & Kirkegaard, 2006). Recently, more systematic wide screening of several GSL-containing plant species has led to the identification of several biofumigants that produce high quantities of the most potent GSLs and these have progressed to the stage of commercialization in Europe (Lazzeri *et al.*, 2004), USA (Gies, 2004) and Australia (Kirkegaard & Matthiessen, 2004). In most cases these products are marketed for their general soil health and soil structural benefits together with disease control attributes (Gies, 2004; Patalano, 2004). A current example of significant adoption (16 000 ha in 2003) is in the wheat–potato rotations on light-textured soils of the Pacific Northwest in USA, where mustard green manures (*Sinapis alba* and *B. juncea*) could replace conventional applications of metham sodium with no penalty in disease, yield or quality of potato tubers, demonstrable improvements in soil structure and erosion control and a substantial economic saving (McGuire, 2004). Disease suppression is often not explicitly separated from these other benefits at the stage of commercial adoption because farmers are focussed on the overall systems benefits. However, the wide range in GSL profiles and the differential toxicity of the ITCs evolved to different disease organisms provides scope to select or breed crops with enhanced biofumigation potential for particular target organisms.

9.6 Release efficiency, fate and activity of hydrolysis products in soil

The purposefully selected biofumigants discussed in the previous section contain enough GSL to potentially produce levels of ITC equal to those applied using commercial methyl-ITC application (Kirkegaard & Sarwar, 1998). However, it is the efficiency with which the GSLs can be converted into isothiocyanates, and their fate in the soil that ultimately influences their effectiveness in disease control.

9.6.1 ITC release efficiency

The separation of GSLs and myrosinase in intact *Brassica* tissues prevents the release of the toxic hydrolysis products *in planta*, and the tissue disruption generated by generalist herbivores is understood to underpin the evolution of the glucosinolate–myrosinase system. Typically, the types of incorporation techniques utilized for biofumigant green manures involve disc ploughs or rotary tillers designed to chop and bury the plant material in one operation leaving large plant fragments intact (Matthiessen & Kirkegaard, 2006). Few attempts were originally made to measure the levels of ITCs in soil resulting from these operations, but in laboratory simulations of these processes, ITC formation from incorporated fresh tissues represented less than 5% of that potentially available in the plant tissues (Gardiner *et al.*, 1999; Bending & Lincoln, 1999; Morra & Kirkegaard, 2002). The need for significant tissue disruption at the cellular levels to maximize ITC release from tissue was demonstrated by Morra & Kirkegaard (2002), who showed that by freeze–thawing mustard leaf tissue ITC release into soil could be increased by two orders of magnitude (from <1 to 75 nmol g^{-1} soil). Moreover, the addition of sufficient water to the soil samples to waterlog them doubled the level of ITC detected in the soil, achieving 30% ITC release efficiency. Matthiessen *et al.* (2004) subsequently demonstrated the value of thorough pulverizing of the above-ground tissues followed by thorough watering achieving ITC concentrations in the field of around 100 nmol g^{-1} soil compared with <4 nmol g^{-1} when the material was chopped. Using a similar macerate-incorporate-irrigate strategy at a different site, Gimsing & Kirkegaard (2006) were able to achieve an ITC release efficiency from a high GSL mustard green manure of around 56% (91 nmol g^{-1}soil), although 13% of the original plant GSL remained un-hydrolyzed in soil immediately following incorporation. These results suggest there may be further scope to increase ITC release efficiency. Failure to utilize efficient incorporation techniques thus negates efforts to select high GSL biofumigants (Section 9.5), as the release efficiencies of traditional methods are so low that minor ITC release occurs, and this has undoubtedly contributed to the variability in field responses and the failure of some investigators to detect disease suppression.

9.6.2 Fate and activity of ITCs in soil

The processes influencing ITC activity in soils have been comprehensively reviewed (Brown & Morra, 1997; Morra, 2004; Gimsing & Kirkegaard, 2008), and several key points in relation to their effectiveness in disease control emerge. ITCs and other GSL hydrolysis products are generally short-lived in soil, with rapid decline in concentration within the first few days and persistence for up to 14 days (Figure 9.3). The disease control properties result from the reactivity with proteins and amino acids, but they are also subject to various loss processes including volatilization, sorption by soil organic matter, microbial degradation and leaching (Figure 9.3).

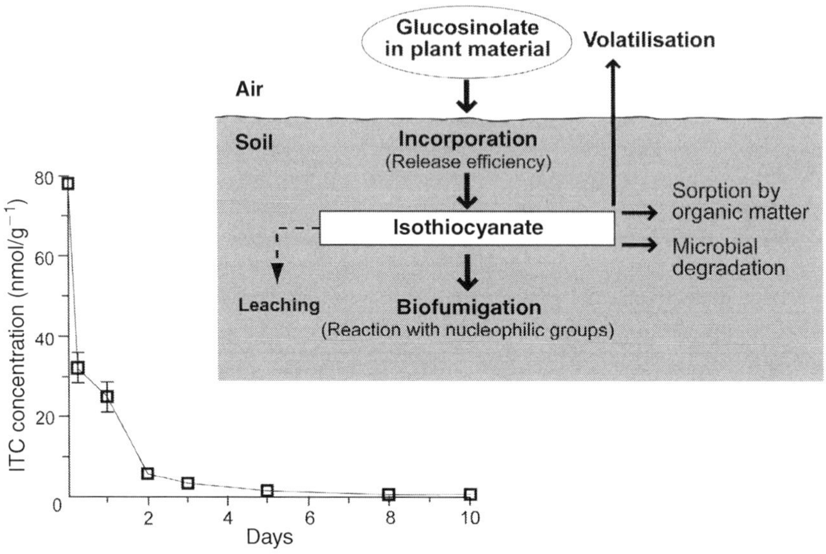

Figure 9.3 Rapid decline in concentrations of 2-propenyl isothiocyanate in field soil following incorporation of pulverized green manure of high-GSL mustard biofumigant (from Gimsing & Kirkegaard, 2006). Isothiocyanates are subject to various loss processes in soils which may reduce the amounts available for biofumigation via reaction with the nucleophilic groups in soil-borne disease organisms (modified from Gimsing & Kirkegaard, 2008, with kind permission of Springer Science and Business Media).

The rates and relative significance of these processes are affected by the specific side-chain structure of the ITC and soil properties such as organic matter content, water content and temperature. ITCs are not as volatile as other synthetic soil fumigants, but volatile losses, particularly for short-chained aliphatic ITCs with higher vapour pressure (e.g. 2-propenyl ITC) can represent a significant source of loss from soil. Rapid incorporation, soil covering techniques and surface watering can serve to reduce this loss, while higher soil temperature will increase it (Price *et al.*, 2005). As a result of their hydrophobic nature, ITCs are rapidly sorbed to organic matter in soil and not to clay (Gimsing *et al.*, 2008) and their pest control potential has been shown to be significantly reduced in soils with higher levels of organic matter (Matthiessen & Shackleton, 2005). Sorption is likely to be greater for longer-chained aliphatic or aromatic ITCs which can reduce their pesticidal effectiveness in headspace toxicity tests. (Matthiessen & Shackleton, 2005). Studies comparing loss from sterile and non-sterile soils demonstrate that ITCs are also degraded by microbes (Warton *et al.*, 2003; Rumberger & Marschner, 2003; Price *et al.*, 2005; Gimsing *et al.*, 2008) and persistence in sterile soil is often increased from hours to days. Enhanced biodegradation by specialized organisms selected for by repeated use of synthetic or plant-derived ITCs provides further evidence of microbial degradation (Warton *et al.*, 2003). Increasing soil water content generally increases ITC concentration and longevity in amended soil, presumably by facilitating more rapid GSL hydrolysis and by increasing ITC in solution thereby reducing volatile losses. Temperature, pH and soil texture tend to have less impact on ITC behavior in soil at the ranges normally experienced in field soils (Brown & Morra, 1997; Gimsing & Kirkegaard, 2008).

9.7 Ecological considerations

The effectiveness of biofumigation for disease suppression will not only be influenced by the differences in the inherent sensitivity of the pathogen to various ITCs, but the particular growth stage of the pathogen which is targeted, the capacity for survival and re-infection in the soil as well as other interactions with non-target organisms.

9.7.1 Pathogen sensitivity and ecology

Inherent sensitivity to ITCs assessed using *in vitro* studies can vary by orders of magnitude for different pathogens (Brown & Morra, 1997), among isolates of the same pathogens (Smith & Kirkegaard, 2002), or for different life-cycle stages such as spores, conidia or mycelia (Mari *et al.*, 1993). Thick-walled oospores or chlamydospores of various oomyces, or sclerotia of *Rhizoctonia* and *Sclerotinia* may be less sensitive to ITCs than germinating propagules of *Aphanomyces euteiches* or *Fusarium oxysporum* (Smolinska & Horbowicz, 1999). Recent studies also demonstrated that hyphae of *Rhizoctonia solani* arising from sclerotia were more resistant to ITCs than those arising from agar plugs (Yulianti *et al.*, 2006). Disease control arising from direct ITC-related suppression may be expected to be more effective for obligate parasites such as that reported for the take-all fungus (*Gaeumannomyces tritici*) and powdery scab in potato (*Spongospora subterranea*) where inoculum recovery requires a host, and more variable for saprophytes such as *Rhizoctonia solani*. Indeed, pathogens such as some *Pythium* spp. which are less sensitive to ITCs and can utilize the incorporated *Brassica* amendment as a food source, can increase after biofumigation (Stephens *et al.*, 1999). The non-specific nature of biofumigation means that non-target organisms may also be affected either directly by the ITCs and other hydrolysis products or via the pathways related to the general organic matter addition described in Figure 9.2. These non-target organisms may include non-pathogenic but functionally important mediators of nutrient transformations, but also pathogen antagonists or suppressors which can influence disease control. Reported changes in non-target soil microbial communities include shorter term ITC-related changes (e.g. Rumberger & Marschner, 2003) but also longer term changes which persist for many weeks and are unrelated to GSL-hydrolysis products (Mazzola *et al.*, 2001; Larkin & Honeycutt, 2006). In some cases these changes involve organisms known to have disease-suppressive capability such as *Streptomyces* spp. and flourescent *Pseudomonas* (Cohen *et al.*, 2005), or *Trichoderma* spp. (Kirkegaard *et al.*, 2004), but also include many as-yet unidentified mechanisms. The complexity of these interactions has been exemplified recently in studies investigating control of apple replant complex by *Brassica* seed meals (see Section 9.8.2.2). In a review of biofumigation for potato disease control, Larkin & Griffin (2007) concluded that biofumigation through release of GSL-hydrolysis compounds such as ITCs are not the only mechanism of action for disease suppression by *Brassica* crops, and in some cases may not be the most important.

9.7.2 Enhanced biodegradation of ITCs

Enhanced microbial biodegradation is a phenomenon in which repeated use of certain pesticides selects for microorganisms able to degrade the compounds and utilize them as a food source. The phenomenon is of particular pertinence to biofumigation because of the known susceptibility of the synthetic compound methyl ITC (Matthiessen *et al.*, 2004) and the recent demonstration of cross-degradation – the ability of the same organisms to rapidly degrade a range of other ITCs from plant sources (Warton *et al.*, 2003). Matthiessen & Kirkegaard (2006) discuss the implications of this in detail and raise several pertinent points. Firstly, as methyl bromide is not susceptible to this phenomenon, any replacement strategy involving either synthetic or plant-based ITC will be susceptible and this should be managed. Secondly, biofumigation may be ineffective as an alternative disease control strategy on soils which have already developed enhanced biodegradation to synthetic methyl ITC due to cross-degradation. Thirdly, and without specific data, they speculate that the likelihood of developing enhanced biodegradation using biofumigation would be significantly less than using high doses of pure compounds due to the array of other compounds.

9.8 Field implementation

Disease suppression using ITC-based biofumigation can be maximized using implementation strategies based on the fundamental aspects of ITC formation, release and activity in soils and the attributes of the target organism outlined in the previous sections. The extent to which the variability in suppression levels previously reported can be ascribed to the failure to adopt such approaches cannot be known, but the obvious involvement of mechanisms other than ITC-related biofumigation emphasize the need to consider biofumigation in an IPM context, as part of an overall disease control strategy. Strategies for success, opportunities for integrated approaches and some specific case studies are used to exemplify this in the following sections.

9.8.1 Strategies to maximize ITC-related biofumigation

Several general guidelines to improve the efficacy of ITC-based disease control can be proposed based on the principles outlined in previous sections as well as the review of the results from published field experiments (Matthiessen & Kirkegaard, 2006).

(a) Establish a correlation between GSLs, ITCs and pest suppression

Pursuing ITC-related biofumigation as a component of disease control may be pointless if there is no clear link between GSL content of the tissue, the ITC released from the tissue and the level of pest suppression. Such relationships have often been investigated *in vitro* using pure compounds, rehydrated or macerated tissue. The numerous impacts of soil on ITC activity suggest some indication of this link in a soil-based assay would be valuable prior to expensive field studies. Experience also suggests some pathogens (e.g. *Pythium*) are less sensitive to ITCs and can proliferate on incorporated organic material and are therefore unlikely targets for success with biofumigation.

(b) Select an appropriate biofumigant for the farming system

Select a biofumigant that produces large quantities of GSLs known to release ITCs most toxic to the target organism (preferably demonstrated in soil) in the time available for its growth. *Brassica* species high in short-chained aliphatic ITCs such as 2-propenyl ITC (e.g. mustards) may be superior to those with longer-chain or aromatic ITCs (rapeseed) due to higher volatility, reduced sorption of those compounds to organic matter in soil, and the resultant maintenance of biocidal activity (Matthiessen & Kirkegaard, 2006). This may override their often lower level of contact toxicity demonstrated in laboratory agar assays (Matthiessen & Shackleton, 2005). The species or variety selected will preferably grow vigorously and remain vegetative or have just commenced flowering at the intended time of incorporation. The best commercial biofumigants can produce >300 mol ha^{-1} of ITC-liberating GSL. Ideally, the biofumigant should not host the targeted disease or other diseases and pests of the commercial crop involved and should have a low risk of becoming a weed.

(c) Grow and incorporate at least 5% W/W of fresh material if possible

Recent studies suggest that increasing the amount of incorporated fresh material up to around 5% W/W improved pathogen suppression, while a diminishing response was observed at higher rates (JA Kirkegaard, unpublished data). This is equivalent to incorporation of 5–6 kg m^{-2} of fresh vegetative biomass into the top 20 cm of soil. As far as ITC-based suppression goes, the diminishing response at higher rates of incorporation may result from the liberated ITC reacting with proteins and amino acids associated with the incorporated plant material itself, rather than with the targeted disease organisms. This may explain the lack of increased effectiveness of biofumigants at higher rates of amendment such as those used for high N seed meals (Brown & Morra, 1997). Suppression which is related to non-GSL mechanisms may improve at higher rates of amendment and in this case the strategy would be to maximize organic matter input.

(d) Incorporation strategy – macerate–incorporate–irrigate

Pulverize the plant tissues to generate as much cellular disruption as possible prior to incorporation to maximize ITC release. Machinery with flailing or crushing actions rather than chopping can generally achieve greater ITC release (Matthiessen *et al.*, 2004). Incorporate the pulverized material rapidly into the soil with thorough mixing to distribute the ITCs. Alternatively, if soil structural preservation is a priority, or application is to perennial vine or tree crops, watering from the top can also move the ITCs into the soil from pulverized material or seed meals. Irrespective of incorporation method, ensure high water availability to facilitate hydrolysis, assist to distribute ITC evenly and to assist in reducing volatile losses from the surface.

(e) Cover the soil to reduce volatile losses

The losses of ITC from the soil surface can be reduced if the material is buried or covered by soil during incorporation, or the soil is tapped or covered following incorporation. In some systems (e.g. strawberries), this can coincide with the placement of the standard plastic mulch on the beds. Opportunities to utilize integrated strategies using solarization or biological dis-infestation can potentially be combined with biofumigation using various covering material (see Chapter 10).

(f) Allow at least 1–2 weeks prior to planting following crops

GSL hydrolysis products including ITCs can have significant phytotoxic effects on plants if sufficient time is not allowed for the compounds to dissipate in the soil. In most cases depending on the conditions, 2 weeks is generally sufficient to avoid negative impacts of persistent allelochemicals. Re-watering during the days immediately after incorporation appears to result in further spikes in ITC release. The required plant-back time can be longer where large amounts of seed meals are applied, in some cases effects on apple replants have still been observed 12 weeks after incorporation (Mazzola *et al.*, 2001).

(g) Target lighter-textured soils with low organic matter
In common with the synthetic ITC-liberating pesticide metham sodium, the characteristics of ITCs mean that disease control is likely to be more effective on lighter-textured soils with lower organic matter. These soils will have lower ITC sorption due to reduced organic matter, less microbial degradation due to lower microbial activity, greater mobility of ITC through the soil matrix in both the liquid and vapor phases due to larger less tortuous pores, and less protection of microorganisms occluded within soil micro-aggregates. Suppression which is related to non-GSL mechanisms may be more important in soils with higher organic matter levels by stimulation of the resident microbial populations.

(h) Adjust strategies to incorporate appropriate IPM approaches
Whether disease suppression using *Brassica* amendments is principally related to ITCs or to other mechanisms, it is unlikely as a single alternative strategy to match the levels of disease control achieved by the synthetic fumigants and pesticides which we are seeking to replace (Lazarovits, 2001; Matthiessen & Kirkegaard, 2006). No examples of this level of control have been reported. As a consequence, opportunities to integrate *Brassica* amendments with other components of an IPM approach should be sought. These include wider rotations, crop hygiene, resistant/tolerant varieties, solarization, bio-control agents and augmented pesticide applications. Many of these approaches have been applied in combination with biofumigation to improve disease control (e.g. solarization Chapter 10). Some examples of this integration are discussed further in the following case studies.

9.8.2 Case studies

In order to highlight some of the practical issues related to developing biofumigation as a component of disease management two case studies of specific applications are considered. A more detailed compilation and review of 18 field studies directed at *Brassica*-based biofumigation for pest and disease control is provided in Matthiessen & Kirkegaard (2006).

9.8.2.1 Brassica green manures in potato production systems

A wide range of potato pests and diseases have been targets for control using Brassicaceous amendments with varying success (Table 9.1). Studies on root-knot nematodes in the Pacific Northwest of USA demonstrated significant suppression by rapeseed (*B. napus*) compared with wheat green manures and provided evidence for the role of GSLs (Mojtahedi *et al.*, 1993). Despite the relatively high levels of suppression achieved compared to non-*Brassica* controls (Table 9.1), infection and damage to tubers remained at levels which required augmenting with the chemical nematicide ethoprop to achieve levels of control similar to that using 1,3-dichloropropene. Rapeseed green manures in the same area were also effective in reducing weed populations from 59% to 90% compared to sudan grass or fallow and increased the yield of subsequent potatoes by 17–25% (Boydston & Hang, 1995). The authors concede that even under the relatively low levels of weed pressure in their experiments, rapeseed green manures achieved commercially acceptable weed control in only 1 of the 2 years and under higher weed pressure may be less effective. Thus reduction in herbicide use, rather than replacement was the target. The yield increase was related to factors other than weed suppression as herbicide-treated weed-free plots following rapeseed also had increased yield. Impacts of biofumigants on various potato fungal and bacterial diseases have been variable in studies worldwide with evidence for both GSL and non-GSL related mechanisms demonstrated (Table 9.1). In general, biofumigation is sought as a means of rapidly reducing disease inoculum to avoid either lengthy rotations out of potatoes or the use of synthetic fumigants. Davis *et al.* (1996) found suppression of *Verticillium dahliae* by rapeseed was intermediate between fallow/pea treatments and the more effective sudangrass/corn treatments and suggested microbial antagonism as the key mechanisms but did not specifically investigate the role of GSLs. Harding & Wicks (2001) found green manures of Indian mustard (*B. juncea*), canola (*B. napus*) and radish (*R. sativus*) all reduced populations of *V. dahliae* to a greater degree than a range of cereals but not more than a clover/ryegrass mixture. They suggested that biofumigants may only be useful where existing soil population levels are close to the disease threshold. Larkin & Griffin (2007) reported a number of laboratory, glasshouse and field experiments investigating the potential for *Brassica* green manures to control a range of soil-borne disease of potatoes in Northeastern USA. In an on-farm trial at a site with a significant powdery scab problem (*Spongospora subterranea*), Indian mustard, rapeseed, canola and ryegrass reduced the disease in the following crop by 15–40%, and canola and rapeseed reduced black scurf (*Rhizoctonia solani*) by 70–80% compared with a standard oat rotation. Disease control was not always associated with high GSL crops (e.g. compare canola and rapeseed) and was observed for ryegrass indicating other non-GSL mechanisms were involved, especially for *Rhizoctonia*. Snapp *et al.* (2007) demonstrated that incorporation of mustard (*B. juncea*) plants was effective in reducing soil-borne fungi, principally *Rhizoctonia* and promoting healthy roots and tubers. Compared to bare fallow, a rye cover crop increased the disease rating of tubers by 37% while an Indian mustard cover crop reduced it by 25% (Table 9.1). The increase in healthy white roots was similar for both cover crops (increase from 63% in fallow to ~90%) despite a much lower amount of incorporated biomass for mustard (2.5 t ha^{-1}) than rye (4.5 t ha^{-1}). Suppression of common scab (*Streptomyces scabiei*) was most effective with a mustard green manure at one site in the USA study using various green manures (Larkin & Griffin, 2007), but was much greater in experiments in South Africa using dried and ground post-harvest residues of *Brassica* vegetables (Gouws & Wehner, 2004), although the role of GSLs was not specifically determined (Table 9.1). The suppression of bacterial wilt (*Ralstonia solanacearum*) in potato crops at 5 sites in the Philippines was also shown to be significantly reduced (40–50%) by a range of *Brassica* amendments incorporated at 5 kg fresh material m^{-2} (JA Kirkegaard, unpublished data) (Table 9.1). In systems such as these in developing countries, utilization of existing residues from vegetable crops (broccoli and radish) rather than purposeful mustard green manures were of interest to avoid loss of income associated with a green manure. Suppression levels of around 32% using sunflower (*Helianthus annus*) as a control in some experiments indicated non-GSL suppression was also operating to suppress this disease at some sites. Related work in northern Australia on much sandier soils demonstrated a much greater role for short-term ITC-related suppression on sandy soils with low organic matter compared to loamy soils

Table 9.1 A summary of reported pest and disease suppression in potato cropping systems by various Brassicaceous amendments. The levels of suppression are calculated as the change in the level of plant infection/damage or soil populations compared with either a non-*Brassica* control where possible, or untreated fallow.

Pest/Disease	Country	Brassica	Suppression (%)	Comments	Reference
Nematode					
Meloidogyne chitwoodi	USA	*B. napus*	78	Chemical augmenting required	Mojtahedi (1993)
Weeds					
Various	USA	*B. napus*	50–96	Chemical augmenting required	Boydston & Hang (1995)
Fungi					
Verticillium dahliae	USA	*B. napus*	0–50	Wilt compared with pea control	Davis *et al*. (1996)
V. dahliae	Australia	*B. juncea*	42	Inoculum levels in soil compared to oat control.	Harding & Wicks (2001)
		B. napus	23		
		R. sativus	26	Only valuable if inoculum close to disease threshold	
		Rye	+66		
Spongospora subterranea	USA (2 sites)	*B. napus* (rape) *B. napus* (canola)	19 and 0 0 and na	Activity higher in high-GSL material such as rape and mustard, although some activity of rye	Larkin & Griffin (2007)
		B. juncea	40 and 0		
		S. alba	0 and 0		
		Rye (Lemtal)	27 and 0		

Rhizoctonia solani	USA (2 sites)	*B. napus* (rape)	73 and 71	Non-GSL mechanisms implicated by effect of rye and similarity in canola (low GSL) and rape (high GSL) effect. Mustard ineffective.	Larkin & Griffin (2007)
		B. napus (canola)	78 and NA		
		B. juncea	0 and 0		
		S. alba	48 and 77		
		Rye (Lemtal)	43 and 100		
R. solani	USA	*B. juncea*	25	Compared with bare, fumigated fallow.	Snapp *et al.* (2007)
		Rye	+37	Incorporation of whole mustard (root + shoot) best.	
Bacteria					
Streptomyces scabies	USA (2 sites)	*B. napus* (rape)	0 and 0	Mustard was most effective of all the brassicas for this disease.	Larkin & Griffin (2007)
		B. napus (canola)	0 and NA		
		B. juncea	0 and 12		
		S. alba	0 and 0		
		Rye (Lemtal)	0 and 0		
Streptomyces scabies	South Africa	*B. oleracea*	90	Post-harvest material; dried ground, incorporated	Gouws & Wehner (2004)
Ralstonia solanacearum	Philippines (5 sites)	*B. juncea* *R. sativus* *B. oleracea*	39 39 51	5 kg m^{-2} fresh shredded material incorporated 4 weeks prior to planting	Kirkegaard *et al.* (unpublished)

where longer-term suppression related to non-GSL mechanisms associated with organic inputs were operating (Matthiessen & Kirkegaard, 2006). The most advanced commercial adoption of biofumigants in potato production is that of mustard green manures (*B. juncea* and *S. alba*) in the wheat/potato rotations of the pacific Northwest in USA. A system involving incorporated mustard green manures to replace conventional metham sodium treatment has provided adequate control of diseases (*Verticillium, Sclerotinia, Helminthosporium, Meloidogyne* and weeds), an improvement in soil structure and soil protection, maintenance of yield and quality, and a cost saving of around USD\$169 ha^{-1} (Gies, 2004; McGuire, 2004). Central to the wide adoption of this on over 16 000 ha (in 2003) has been the identification, commercialization and availability of effective biofumigant varieties and the involvement of growers at all stages of the research to develop and adapt the concept in a practical way to the farming system (Gies, 2004). These various studies worldwide in potato production systems demonstrate several important issues in relation to biofumigation and disease control: (a) the variability in the success of disease control, even for the same pathogens (b) the clear involvement of non-GSL-related mechanisms in disease suppression at several sites (c) the need to combine biofumigation with other strategies to achieve acceptable commercial disease control and (d) the wider farming system benefits which may be associated with biofumigant green manures. Few of the studies utilized all of the best-bet strategies summarized above to maximize ITC-based suppression so some improvements in disease suppression may be possible. The clear involvement of non-ITC mechanisms indicates this is not necessarily inevitable.

9.8.2.2 Control of apple replant disease using Brassica seed meal

A series of recent studies on the mechanisms underlying the suppression of the disease complex responsible for apple replant disease using *Brassica napus* seed meals exemplifies the complexity of the soil processes which can be involved. The studies set out to identify suitable alternatives to chemical soil fumigants to control apple replant disease, a complex involving species in the genera *Cylindrocarpon*, *Phytophthora, Pythium* and *Rhizoctonia,* as well as nematodes (Mazzola, 1998). The goal was to integrate methods that selectively suppress these organisms with procedures that enhance the activity of resident microbial antagonists. A specific aim was to combine applications of rapeseed meal to suppress disease organisms with pre-crops of a wheat variety shown to induce microbial antagonists of the diseases in soil (Mazzola *et al.*, 2001). In studies using high and low GSL rapeseed meal, *Rhizoctonia solani* and *Pratylenchus penetrans* were suppressed by both high and low GSL meal while *Pythium* spp. increased in response to the low GSL meal. The suppression of *Rhizoctonia* by low GSL meal suggested non-GSL mechanisms were operating and dramatic changes were measured in various components of the microbial community. These included a 100-fold increase in *Streptomyces* spp. and a predominance of those which produced nitric oxide (NO) rather than those demonstrating direct antagonism to *R. solani* (Cohen *et al.*, 2005). NO can induce plant systemic resistance and is predominately generated in amended soils by the activity of nitrifying bacteria that oxidize the ammonium released from the incorporated seed meal. A subsequent study provided evidence that both direct antagonism by *Streptomyces* and indirect effects of NO were involved in the suppression of *R. solani* by low GSL rapeseed meal (Cohen & Mazzola, 2006) although the exact mechanism remained uncertain. Preventing *Pythium* proliferation when low-GSL rapeseed meal was applied could be accomplished using a soil drench of mefenoxam, and the combination of the rapeseed meal and the chemical drench could reduce disease symptoms to the same extent as fumigation (Mazzola & Mullinix, 2005). Another approach under investigation involves utilization of mixes comprising mustard (*Brassica juncea*) and rapeseed meal as the principal ITC released from mustard (2-propenyl ITC) has been shown to have greater activity against *Pythium* spp. (Cohen & Mazzola, 2006). This series of studies exemplify the complexity of mechanisms which can be involved in disease control using Brassicaceous amendments. However, it also serves to highlight how careful experimentation with appropriate controls to separate the GSL and non-GSL related mechanisms of disease control operating in disease suppression can lead to integrated approaches which may combine Brassicaceous amendments with other strategies to provide commercially viable control.

9.9 Summary

Further development of biofumigation as a disease management tool will require research within a systematic framework taking account of many factors from the fundamental soil and plant processes, to the practicalities of specific farming systems. The most recent research and development findings summarized in this chapter suggest there may now be impetus for self-sustaining further development and market penetration of the concept. However there needs to be appropriate targeting and realistic expectations of the concept as a component of a disease management system. Biofumigation is not sufficiently powerful or practical in implementation to replace synthetic fumigants and such misdirection would be counterproductive. It has been demonstrated to have efficacy, secondary soil benefits and, notably economic benefits in an appropriate production system that traditionally uses metham sodium, and uptake in at least one regional production system has been high (McGuire, 2004). Successful future implementation could arise from current research in several areas. There appears to be significant scope to improve the effectiveness of ITC-related suppression by purposeful selection or development of varieties with very high concentrations of the relevant GSL, and by using incorporation techniques which maximize the release efficiency and residence time of ITCs in soil. Understanding the fate of both the GSLs and the hydrolysis products in soil is required to minimize environmental consequences, a key aim of non-chemical approaches. Much of the necessary soil and plant data relevant to these processes has been notably absent from many previous studies on disease control using biofumigation. The non-specific biocidal nature of ITCs combined with the significant non-ITC related changes in microbial profiles increasingly reported following biofumigation demands a more thorough understanding of the overall changes in the soil biology generated by biofumigants. New DNA-based soil microbial profiling approaches such as DDGE, RISA, T-RFLP and utilization of specific PCR primers offer new opportunities to understand both the specific and general changes in the soil microbial ecology which can influence disease expression. This understanding is essential particularly where biofumigation is combined with other biological approaches so that synergistic rather than antagonistic processes are encouraged using these integrated approaches. Without this understanding, biofumigation, along with other biological approaches, will be consigned to the list of promising but inconsistent cultural control options which cannot offer the reliable efficacy of the existing chemical options they seek to replace.

9.10 References

Akhtar M, Alam MM, 1991. Integrated control of plant parasitic nematodes on potato with organic amendments, nematicide and mixed cropping with mustard. *Nematologia Mediterranea* **19,** 169–171.

Angus JF, Gardner PA, Kirkegaard JA, Desmarchelier JM, 1994. Biofumigation: ITCs released from *Brassica* roots inhibit growth of the take-all fungus. *Plant and Soil* **162,** 107–112.

Askew MF, 2004. Economic aspects of biofumigation in EU. *Agroindustria* **3,** 229–231.

Bailey KL, Lazarovits G, 2003. Suppressing soil-borne diseases with residue management and organic amendments. *Soil and Tillage Research* **72,** 169–180.

Balesh T, Zapata T, Aune JB, 2005. Evaluation of mustard meal as organic fertiliser on tef (Eragrostis tef (Zucc) trotter) under field and greenhouse conditions. *Nutrient Cycling in Agroecosystems* **73**, 49–57.

Benderoth M, Textor S, Windsor AJ, Mitchell-Olds T, Gershenzon J, Kroymann J, 2006. Positive selection driving diversification in plant secondary metabolism. *Proceedings of the National Academy of Sciences USA* **103**, 9118–9123.

Bending GD, Lincoln SD, 1999. Characterisation of volatile sulphur-containing compounds produced during decomposition of *Brassica juncea* tissues in soil. *Soil Biology and Biochemistry* **31,** 95–703.

Bones AM, Rossiter JT, 1996. The myrosinase–glucosinolate system, its organisation and biochemistry. *Physiologia Plantarum* **97,** 194–208.

Borek, V, Elberson LR, McCaffrey JP, Morra MJ, 1997. Toxicity of rapeseed meal and methyl ITC to larvae of the black vine weevil (Coleoptera: Curculionidae). *Journal of Economic Entomology* **90,** 109–112.

Boydston RA, Hang A, 1995. Rapeseed (*Brassica napus*) green manure crop suppresses weeds in potato (*Solanum tuberosum*). *Weed Technology* **9,** 669–675.

Brown PD, Morra MJ, 1995. Glucosinolate-containing plant tissues as bioherbicides. *Journal of Agricultural and Food Chemistry* **43,** 3070–3074.

Brown PD, Morra MJ, 1997. Control of soil-borne plant pests using glucosinolate-containing plants. *Advances in Agronomy* **61,** 167–231.

Chan KY, Heenan DP, 1996. The influence of crop rotation on soil structure and soil physical properties under conventional tillage. *Soil and Tillage Research* **37,** 113–125.

Chew FS, 1988. Biological effects of glucosinolates. In: Cutler HG, ed. *Biologically Active Natural Products. Potential Uses in Agriculture*. Washington, DC: American Chemical Society, 155–181.

Clossais-Besnard N, Larher F, 1991. Physiological role of glucosinolates in *Brassica napus*. Concentration and distribution pattern of glucosinolates among plant organs during a complete life cycle. *Journal of the Science of Food and Agriculture* **56,** 25–38.

Cohen MF, Mazzola M, 2006. Resident bacteria, nitric oxide emission and particle size modulate the effect of *Brassica napus* seed meal on disease incited by *Rhizoctonia solani* and *Pythium* spp. *Plant and Soil* **286,** 75–86.

Cohen MF, Yamasaki H, Mazzola M, 2005. *Brassica napus* seed meal soil amendment modifies microbial community structure, nitric oxide production and incidence of *Rhizoctonia* root rot. *Soil Biology and Biochemistry* **37**, 1215–1227.

Davis JR, Huisman OC, Westermann DT, Hafez SL, Everson DO, Sorensen LH, Schneider AT, 1996. Effects of green manures on Verticillium wilt of potato. *Phytopathology* **86,** 444–453.

Ellenby, C, 1945. Susceptibility of South American tuber forming species of *Solanum* to the potato root eelworm *Heterodera rostochiensis* Wollenweber. *Empire Journal of Experimental Agriculture* **13,** 158–168.

Fahey JW, Zalcmann AT, Talalay P, 2001. The chemical diversity and distribution of glucosinolates and isothiocyanates among plants. *Phytochemistry* **56,** 5–51.

Fenwick GR, Heaney RK, Mullin WJ, 1983. Glucosinolates and their breakdown products in food and food plants. *Critical Reviews in Food Science and Nutrition* **18,** 123–201.

Gardiner J, Morra MJ, Eberlein CV, Brown PD, Borek V, 1999. Allelochemicals released in soil following incorporation of rapeseed (*Brassica napus*) green manures. *Journal of Agricultural and Food Chemistry* **47,** 3837–3842.

Gies D, 2004. Commercial use of mustards for green manure and biofumigation in the United States. *Agroindustria* **3,** 403–405.

Gimsing AL, Kirkegaard JA, 2006. GSL and ITC concentration in soil following incorporation of *Brassica* biofumigants. *Soil Biology and Biochemistry* **38,** 2255–2264.

Gimsing AL, Kirkegaard JA, 2008. GSLs and Biofumigation: Fate of glucosinolates and their hydrolysis products in soil. *Phytochemistry Reviews*. Available at: http://www.springerlink.com/content/f5715121321 1q6t3/fulltext.pdf

Gimsing AL, Strobel BW, Hansen HCB, 2008. Benzyl and 2-propenyl isothiocyanate sorption and degradation in soil. *Environmental Toxicology and Chemistry* (in press).

Gouws R, Wehner FC, 2004. Biofumigation as alternative control measure for common scab on seed potatoes in South Africa. *Agroindustria* **3,** 309–312.

Harding RB, Wicks TJ, 2001. *In vitro* suppression of soilborne potato pathogens by volatiles released from *Brassica* residues. In: Porter IJ, ed. *Proceedings of the* 2nd *Australasian Soilborne Diseases Symposium*. Lorne, Australia, 148–149.

Hartz TK, Johnstone PR, Mitao EM, Davis RM, 2005. Mustard cover crops are ineffective in suppressing soil-borne disease or improving processing tomato yield. *Hortscience* **40**, 2016–2019.

Holst B, Williamson G, 2004. A critical review of the bioavailability of glucosinolates and related compounds. *Natural Product Reports* **21,** 425–447.

Josefsson E, 1967. Distribution of thioglucosides in different parts of *Brassica* plants. *Phytochemistry* **6,** 1617–1627.

Kirkegaard JA, Matthiessen JN, 1999. Biofumigation research – beyond empiricism. In: Magarey RC, ed. *Proceedings 1*st *Australasian Soilborne Disease Symposium*. Bureau of Sugar Experiment Stations, Brisbane, 155–157.

Kirkegaard JA, Matthiessen JN, 2004. Developing and refining the biofumigation concept. *Agroindustria* **3,** 233–239.

Kirkegaard JA, Sarwar M, 1998. Biofumigation potential of brassicas. I. Variation in glucosinolate profiles of diverse field-grown brassicas. *Plant and Soil* **201,** 71–89.

Kirkegaard JA, Gardner PA, Desmarchelier JM, Angus JF, 1993. Biofumigation – using *Brassica* species to control pests and diseases in horticulture and agriculture. In: Wratten N, Mailer RJ, eds. *Proceedings 9th Australian Research Assembly on Brassicas*. Agricultural Research Institute, Wagga, 77–82.

Kirkegaard JA, Sarwar M, Wong PTW, Mead A, Howe G, Newell M, 2000. Field studies on the fumigation of take-all by *Brassica* break crops. *Australian Journal of Agriculture Research* **51,** 445–456.

Kirkegaard JA, Simpfendorfer S, Holland J, Bambach R, Moore KJ, Rebetzke GJ, 2004. Effect of previous crops on crown rot and yield of durum and bread wheat in northern NSW. *Australian Journal of Agricultural Research* **55,** 321–334.

Larkin RP, Griffin TS, 2007. Control of soilborne potato diseases using Brassica green manures. *Crop Protection* **26**, 1067–1077.

Larkin RP, Honeycut CW, 2006. Effects of different 3-yr cropping systems on soil microbial communities and soilborne disease in potato. *Phytopathology* **96**, 68–79.

Lazarovits G, 2001. Management of soil-borne plant pathogens with organic soil amendments: A disease control strategy salvaged from the past. *Canadian Journal of Plant Pathology* **23,** 1–7.

Lazzeri L, Leoni O, Bernardi R, Malaguti L, Cinti S, 2004. Plants, techniques and products for optimising biofumigation in the full field. *Agroindustria* **3,** 281–287.

Manici LM, Lazzeri L, Palmieri S, 1997. *In vitro* antifungal activity of GSLs and their enzyme derived products towards plant pathogenic fungi. *Journal of Agricultural and Food Chemistry* **45,** 2768–2773.

Manici LM, Lazzeri L, Baruzzi G, Leoni O, Galletti S, and Palmieri S, 2000. Suppressive activity of some GSL enzyme degradation products on *Pythium irregulare* and *Rhizoctonia solani* in sterile soil. *Pest Management Science* **56,** 921–926.

Mari M, Iori R, Leoni O, Marchi A, 1993. *In vitro* activity of glucosinolate-derived ITCs against post-harvest fruit pathogens. *Annals of Applied Biology* **123**, 155–164.

Martin FN, 2003. Development of alternative strategies for management of soilborne pathogens currently controlled with methyl bromide. *Annual Reviews in Phytopathology* **41,** 325–350.

Matthiessen JN, Kirkegaard JA, 2006. Biofumigation and enhanced biodegradation: Opportunity and challenge in soilborne pest and disease management. *Critical Reviews Plant Science* **25,** 235–265.

Matthiessen JN, Shackleton MA, 2005. Biofumigation: Environmental impacts on the biological activity of diverse pure and plant-derived isothiocyanates. *Pest Management Science* **61,** 1043–1051.

Matthiessen JN, Warton B, Shackleton MA, 2004. The importance of plant maceration and water addition in achieving high *Brassica*-derived isothiocyanate levels in soil. *Agroindustria* **3,** 277–280.

Mazzola M, 1998. Elucidation of the microbial complex having a causal role in the development of apple replant disease in Washington. *Phytopathology* **88**, 930–938.

Mazzola M, Mullinix K, 2005. Comparative field efficacy of management strategies containing *Brassica napus* meal or green manure for the control of apple replant disease. *Plant Disease* **89**, 1207–1213.

Mazzola M, Granatstein DM, Elfving DC, Mullinix K, 2001. The suppression of specific apple root pathogens by *Brassica napus* seed meal amendment regardless of glucosinolate content. *Phytopathology* **91,** 673–679.

McCully ME, Miller C, Sprague SJ, Huang CX, Kirkegaard JA, 2008. The distribution of GSLs and sulphur-rich cells in roots of field-grown canola (*Brassica napus*). *New Phytologist* **180**, 193–205.

McLeod RW, Steele CC, 1999. Effects of *Brassica* leaf green manures and crops on activity and reproduction of *Meloidogyne javanica. Nematology* **1,** 613–624.

McLeod RW, Kirkegaard JA, Steele CC, 2001. Invasion, development, growth and egg-laying by *Meloidogyne javanica* in Brassicaceae crops. *Nematology* **3,** 463–472.

McGuire AM, 2004. Mustard green manure crops replace fumigant and improve infiltration in potato cropping system. *Agroindustria* **3,** 331–333.

Mithen RF, 2001. Glucosinolates and their degradation products. *Advances in Botanical Research* **35,** 213–262.

Mojtahedi H, Santo GS, Wilson JH, Hang AN, 1993. Managing *Meloidogyne chitwoodi* on potato with rapeseed as green manure. *Plant Disease* **77,** 42–46.

Morra M, 2004. Controlling soil-borne plant pests using glucosinolate-containing tissues. *Agroindustria* **3,** 251–255.

Morra MJ, Kirkegaard JA, 2002. Isothiocyanate release from soil-incorporated *Brassica* tissues. *Soil Biology and Biochemistry* **34,** 1683–1690.

Müller J, 1999. The economic importance of *Heterodera schachtii* in Europe. *Helminthologia* **36,** 205–213.

Njoroge SMC, Riley MB, Keinath AL, 2008. Effect of incorporation of *Brassica* spp residues on population densities of soilborne microorganisms and on damping-off and Fusarium wilt of watermelon. *Plant Disease* **92**, 287–294.

Palmieri S, 2004. Chemistry and technology of glucosinolates: Isolation, formulation and use. *Agroindustria* **3,** 245–249.

Patalano G, 2004. New practical perspectives for vegetable biocidal molecules in Italian agriculture: Bluformula brand for commercialisation of biocidal green manure and meal formulations. *Agroindustria* **3,** 409–412.

Potter MJ, Davies K, Rathjen AJ, 1998. Suppressive impact of glucosinolates in *Brassica* vegetative tissues on root lesion nematode *Pratylenchus neglectus. Journal of Chemical Ecology* **24,** 67–80.

Potter MJ, Davies KA, Kirkegaard JA, Rathjen AJ, 1999 Reduced susceptibility of *Brassica napus* to *Pratylenchus neglectus* in plants with elevated root levels of 2-phenylethyl glucosinolate. *Journal of Nematology* **31**, 291–298.

Price AJ, Charron CS, Saxton AM, Sams CE, 2005 Allyl ITC and carbon dioxide produced during degradation of *Brassica juncea* tissue in different soil conditions. *Hortscience* **40,** 1734–1739.

Rahman L, Somers T, 2005. Suppression of root knot nematode (*Meloidogyne javanica*) after incorporation of Indian mustard cv. Nemfix as green manure and seed meal in vineyards. *Australasian Plant Pathology* **34,** 77–83.

Rask L, Andreasson E, Ekbom B, Eriksson S, Pontoppidan B, Meijer J, 2000. Myrosinase: Gene family evolution and herbivore defense in Brassicaceae. *Plant Molecular Biology* **42,** 93–113.

Rollin P, Palmieri S, 2004. Sulfur-containing metabolites in Brassicales. *Agroindustria* **3**, 241–244.

Rosa EAS, Heaney RK, Fenwick GR, Portas CAM, 1997. GSLs in crop plants. *Horticultural Reviews* **19,** 99–215.

Rumberger A, Marschner P, 2003. 2-Phenylethylisothiocyanate concentration and microbial community composition in the rhizosphere of canola. *Soil Biology and Biochemistry* **35,** 445–452.

Rumberger A, Marschner P, 2004. 2-Phenylethylisothiocyanate concentration and bacterial community composition in the rhizosphere of field-grown canola. *Functional Plant Biology* **31,** 623–631.

Sang JP, Minchinton P, Johnstone P, Truscott RJW, 1984. Glucosinolate profiles in the seed, root and leaf tissue of cabbage, mustard, rapeseed, radish and swede. *Canadian Journal of Plant Science* **64,** 77–93.

Sarwar M, Kirkegaard JA, 1998. Biofumigation potential of brassicas. II. Effect of environment and ontogeny on glucosinolate production and implications for screening. *Plant and Soil* **201,** 91–101.

Sarwar M, Kirkegaard JA, Wong PTW, Desmarchelier JM, 1998. Biofumigation potential of brassicas. III. *In vitro* toxicity of isothiocyantes to soil-borne fungal pathogens. *Plant and Soil* **201,** 103–112.

Schlathoelter M, 2004. Biofumigation with nematode resistant crops. *Agroindustria* **3,** 407.

Smith BJ, Kirkegaard JA, 2002. *In-vitro* inhibition of soil microorganisms by 2-phenylethyl isothiocyanate. *Plant Pathology* **51,** 585–593.

Smith BJ, Kirkegaard JA, Howe GN, 2004. Impacts of *Brassica* break-crops on soil biology and yield of following wheat crops. *Australian Journal of Agricultural Science* **56,** 1–11.

Smolinska U, 2000. Survival of *Slerotium cepivorum* and *Fusarium oxysporum* chlamydospores in soil amended with cruciferous residues. *Journal of Phytopathology* **148**, 343–349.

Smolinska U, Horbowicz M, 1999. Fungicidal activity of volatiles from selected cruciferous plants against resting propagules of soil-borne fungal pathogens *Journal of Phytopathology* **147**, 119–124.

Smolinska U, Morra MJ, Knudsen GR, Brown PD, 1997. Toxicity of glucosinolate degradation products from *Brassica napus* seed meal toward *Aphanomyces euteiches* f. sp. *pisi*. *Phytopathology* **87**, 77–82.

Snapp SS, Date KU, Kirk W, O'Neil K, Kremen A, Bird G, 2007. Root, shoot tissues of *Brassica juncea* and *Cereal secale* promote potato health. *Plant and Soil* **294**, 55–72.

Stephens PM, Davoren CW, Wicks T, 1999. Effect of methyl bromide, metham sodium and the biofumigants Indian mustard and canola on the incidence of soilborne fungal pathogens and growth of grapevine nursery stock. *Australasian Plant Pathology* **28,** 187–196.

Stirling GR, Stirling AM, 2003. The potential of *Brassica* green manure crops for controlling root-knot nematode (*Meloidogyne javanica*) on horticultural crops in a subtropical environment. *Australian Journal of Experimental Agriculture* **43,** 623–630.

Subbarao KV, Kabir Z, Martin FN, Koike ST, 2007. Management of soilborne diseases in strawberry using vegetable rotations. *Plant Disease* **91**, 964–972.

Thorup-Kristensen K, Magid J, Jensen LS, 2003. Catch crops and green manures as biological tools in nitrogen management in temperate zones. *Advances in Agronomy* **51,** 227–302.

van Dam NM, Tytgat OG, Kirkegaard JA, 2008. Root and shoot glucosinolates – a comparison of their diversity, function and interactions in natural and managed ecosystems. *Phytochemistry Reviews*. Available at: http://www.springerlink.com/content/j811790504k5n603/fulltext.pdf

Walker JC, Morell S, Foster H, 1937. Toxicity of mustard oils and related sulphur compounds to certain fungi. *American Journal of Botany* **24,** 536–541.

Warton B, Matthiessen JN, Shackleton MA, 2003. Cross-enhancement: Enhanced biodegradation of ITCs in soils previously treated with metham sodium. *Soil Biology and Biochemistry* **35,** 1123–1127.

Yulianti T, Sivasithamparam K, Turner DW, 2006. Response of different forms of propagules of *Rhizoctonia solani* AG2-1 (ZG5) exposed to the volatiles produced in soil amended with green manures. *Annals of Applied Biology* **148**, 105–111.

Chapter 10
Control of plant disease through soil solarization

Abraham Gamliel[1] *and Jaacov Katan*[2]

[1]Agricultural Research Organization Volcani Center, Bet Dagan 50250, Israel
[2]The Hebrew University, Rehovot 76100, Israel

10.1 Introduction

Soil-borne pests, including pathogens, weeds and arthropods, cause heavy losses to all major crops, especially when they are grown frequently in the same soil. Many methods for the management of these soil pests have been developed, with differing levels of success. These include chemical, physical, biological and cultural methods. In addition to effective pest control, pest-management programs need to consider the environmental, economic, technological and legal issues.

Soil disinfestation is one of the most effective means of controlling soil-borne pests and improving plant health. Soil disinfestation is a drastic means applied to soil before planting in order to reduce or eliminate populations of soil pests (Katan, 1984). The basic idea is to eradicate the pests at the desired soil (or substrate) depth, before planting. This is usually achieved by applying a drastic physical or chemical tool which has the capacity to penetrate the soil and reach the inoculum at each site within it. Usually, these eradicants exhibit a wide range of control over many organisms, and are detrimental to the crop as well. They should therefore be applied such that they will have enough time to dissipate before planting. Until the 1970s, there were two major approaches for soil disinfestation, both of which had been developed by the end of the nineteenth century, in the early days of modern plant pathology: physical soil disinfestation, mainly by heating the soil with steam, and chemical soil disinfestation, using fumigants. However, fumigants have always been the major tool for soil disinfestation, methyl bromide (MB) being foremost among them since the 1950s and until recently. A third approach, soil solarization (also called solar heating of the soil, SH) was introduced in 1976 (Katan *et al.*, 1976; Katan & DeVay, 1991).

The main principle of SH is to harness solar energy in order to raise the temperature of a moistened soil. Mulching (covering, tarping) the soil with transparent polyethylene or any other transparent plastic sheet is, at present, the most common means of carrying out this task. Future technologies may provide simpler, more effective and less costly tools to capture solar energy for plant protection. The use of sprayable plastic polymers (see further on), instead of laying plastic films, may revolutionize SH. Attempts to use solar energy for controlling biotic agents in soil and in plant material have been known for decades, even centuries (Katan & DeVay, 1991). For example, Hagan (1933) heated

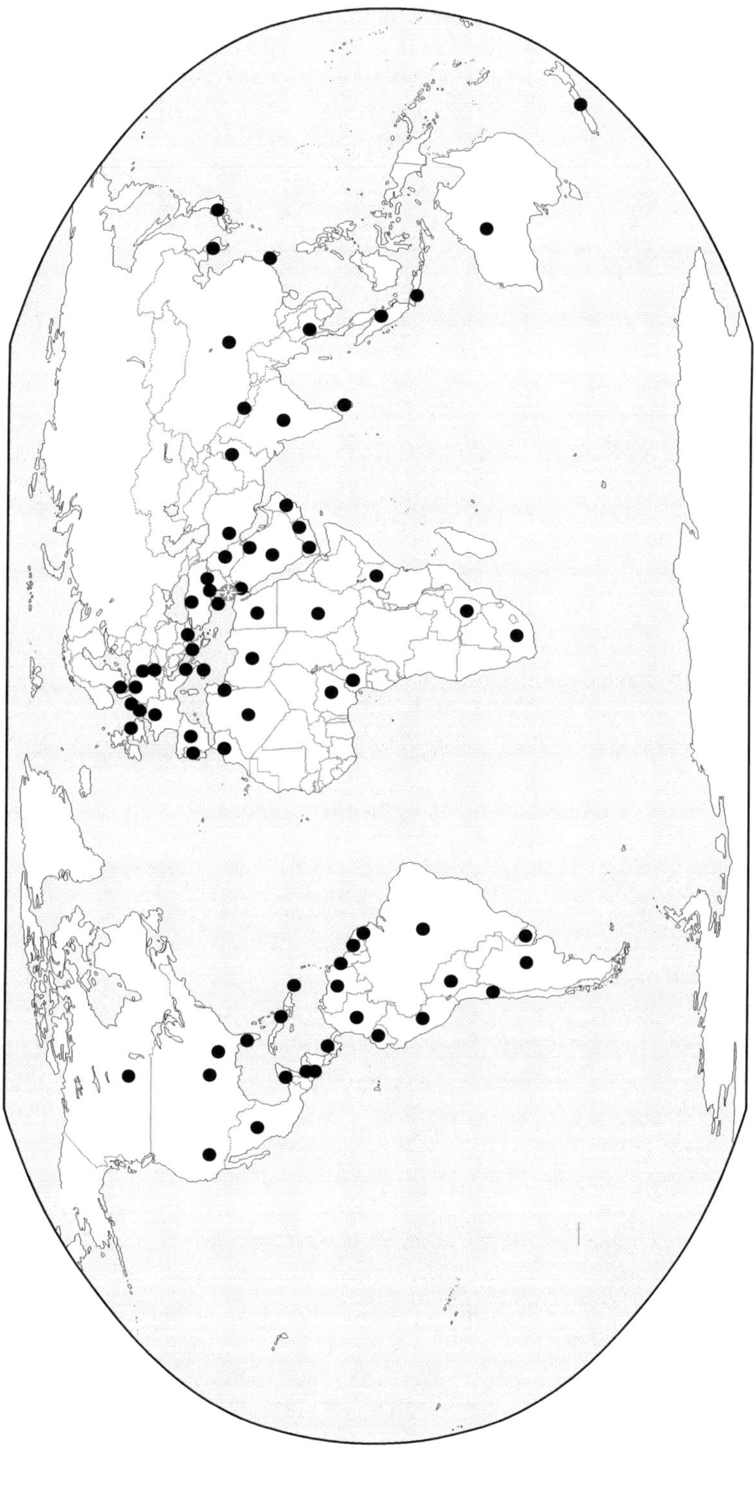

Figure 10.1 World distribution of sites where solarization has been investigated.

soils in Hawaii by mulching soils with cellophane, apparently controlling *Heterodera radicicola*. Adams (1971) was probably the first to report the potential of plastic material for heating soil and controlling pathogens. In 1986, 10 years after the first publication, nearly 180 publications or reports from 22 countries had already been published on the topic, indicating a deep interest in new methods of control (Katan *et al.*, 1987). In the last three decades, since the establishment of SH in its present form, hundreds of studies on this topic have been published and this approach has been adopted or investigated in over 60 countries, both developed and developing (Katan, 1981; Stapleton & DeVay, 1986; Katan & DeVay, 1991; Katan, 1996; McGovern & McSorley, 1997; Stapleton *et al.*, 1998) (Figure 10.1). Owing to their sheer number, we refer to only some of these studies in this chapter. Many educational and extension materials for introducing SH in new areas have also been developed. The studies on SH deal with several major issues, such as exploring the potential effectiveness of this technique in various regions, seasons and climatic conditions, mechanisms of disease control and yield increase, and technological and application issues, among others; some of these issues are discussed herein.

10.2 Principles of soil solarization

Soil solarization is a climate-dependent method which needs to be adapted to the specific region and season in which it is applied. However, there are basic principles which are common to most uses and are summarized as follows (Katan, *et al.*, 1976; Stapleton & DeVay, 1986; Katan, 1996):

(a) Soil mulching should be carried out during periods of high temperature and intense solar irradiation, and low or no precipitation.
(b) The soil should be kept moist to increase the thermal sensitivity of resting structures and to improve heat conduction.
(c) The thinnest polyethylene tarp that can be used (25- to 50-μm thickness) is recommended, since it is both cheaper and somewhat more effective than thicker ones. Because the upper soil layer is heated more intensely than the lower ones, the mulching period should be sufficiently long, usually 4 weeks or more, to achieve pest control at all desired depths. The longer the mulching period, the greater the depth of effective activity, and the higher the pathogen-killing rates (Katan *et al.,* 1976).
(d) Solarization heats the soil through repeated daily cycles. At increasing soil depths, maximal temperatures decrease, are reached later in the day, and are maintained for longer periods of time (Figure 10.2). In solarized plots in which effective disease and weed control were obtained, typical maximal temperatures were within the range of 45–50°C and 38–45°C at depths of 10 and 20 cm, respectively, although higher temperatures have been recorded in certain regions. The temperatures in the solarized soil are 5–15°C higher than those in comparable non-solarized ones.
(e) The best time for SH, when climatic conditions are the most favorable, can be determined experimentally by mulching the soil and measuring the resulting temperatures. Meteorological data from previous years and predictive models (see further on) further facilitate this task. Monitoring changes in the time span naturally occurring pathogen populations, or intentionally introduced pathogen inocula, is an additional approach to determining the effectiveness of SH, and this method can be regarded as

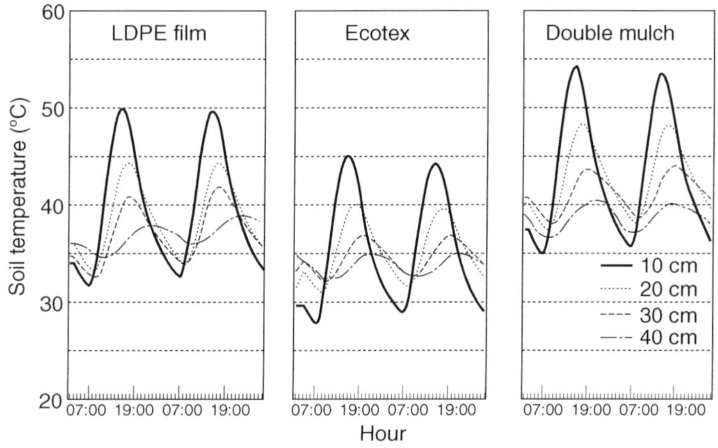

Figure 10.2 Soil temperatures at various depths during soil solarization at the summer time in Israel using various mulches. LDPE – low-density polyethylene sheets, Ecotex – black sprayable polymer; double mulch – clear LDPE, over Ecotex mulch.

a bioassay. French-Monar *et al.* (2007) demonstrated the importance of using more than one technique for assessing pathogen populations in such studies. Assessing weed control is another simple approach to evaluating the effectiveness of solarization. Systems which reliably simulate SH under controlled conditions (Klein *et al.*, 2007) can be very useful for SH research.

(f) The ultimate test for the effectiveness of SH, or of any control method, is to assess its effectiveness in controlling diseases (and other pests, e.g. weeds) and increasing yield and quality under field conditions and local agricultural practices. Such tests must be done, repeated and carefully analyzed before drawing conclusions regarding solarization potential or recommending it for use by farmers.

(g) As with any new method of control, the introduction of SH to a new region has to be done gradually by professionals in extension with continuous monitoring.

(h) As with any disinfestation method, recontamination of the soil after the termination of solarization, for example, by infected propagation material, or infested soil or water, should be absolutely avoided.

(i) The main goal of any pest-management method is to ensure economic reduction in pest populations of the relevant crop. This depends, among other things, on the level of pest control, yield increase, the price of a yield unit and the cost of the management.

10.3 Pathogen and weed control

All agricultural crops are plagued by numerous soil-borne pathogens, including fungi, bacteria, nematodes and parasitic plants. Many pathogens can cause severe damage to crops at all growth and production stages. Thus, SH under appropriate climatic conditions is a potentially effective tool for mitigating this production limitation. Lists describing the pathogens which are controlled (or not controlled) by SH have been reported

(Davis, 1991; Davis *et al.*, 2008). The list of pathogenic fungi which are controlled by solarization includes many key pathogens, such as *Verticillium, Rhizoctonia* and *Fusarium.* The first focus of solarization was aimed at controlling diseases caused by pathogenic fungi, such as Verticillium wilt in eggplants and tomatoes (Katan *et al.*, 1976), Fusarium wilt in cotton (Katan *et al.*, 1983) and Rhizoctonia in potatoes (Elad *et al.*, 1980). The wide spectrum of pathogen control was evident in the early days of solarization when combined control of both *V. dahliae* and *Pratylenchus thornei* was achieved in potatoes, along with a 33% increase in yield and successful weed control (Grinstein *et al.*, 1979). Similar results regarding the control of Verticillium wilt were observed by Davis & Sorensen (1986) in Idaho.

The heat-tolerant pathogens *Monosporascus cannonballus* and *Macrophomina phaseolina* are not controlled by solarization. *Fusarium oxysporum* f. sp. *dianthi* is also considered one of the wilt pathogens that is not easily controlled by solarization (Tjamos *et al.*, 1999).

The control of phytopathogenic bacteria by solarization has only been reported in a relatively few cases. Soil-borne bacteria, including *Agrobacterium* and *Streptomyces*, are among the bacteria controlled by solarization. *Streptomyces scabies* is an important pathogen of potatoes and peanuts, which has also been reported to respond to solarization (Davis & Sorensen, 1986; Grinstein *et al.*, 1995). Solarization for 8 weeks in tomato plastic houses drastically reduced symptoms caused by *Clavibacter michiganensis* ssp. *michiganensis* (Antoniou *et al.*, 1995). Solarization reduced populations of Gram-positive bacteria by 64–99% (Stapleton & Garza-Lopez, 1988). It has been shown in a detailed study in Oregon that *Agrobacterium* spp. population densities declined within solarized plots and incidence of crown gall on cherry root stock planted in solarized soil was reduced significantly (Pinkerton *et al.*, 2000). Negative effects, due to control of beneficial rhizobia, have also been reported (Abdel-Rahim *et al.*, 1988).

Many phytopathogenic nematodes are controlled by solarization (Stapleton & Heald, 1991). Reports of effective control of nematodes for a short time, such as in annual crops, are well established. The efficacy of solarization for long-term suppression of nematodes, such as in perennial crops, is inconclusive. Certain nematodes, such as the root knot *Meloidogyne* sp., are not always effectively controlled by solarization. However, ectoparasite nematodes such as *Pratylenchus* and *Ditylenchus* are well controlled (Grinstein *et al.*, 1979; Siti *et al.*, 1982). Many phytopathogenic nematodes cannot survive at temperatures above 40°C, and solarization therefore kills them in the upper soil layers, but not necessarily at depths below 40 cm. One of the most important, and as-yet unexplored, factors in nematode suppression is their upward migration from the deep soil layers to the root zone following plant establishment during crop growth. This effect might be connected with failure of nematode control by SH and other disinfestation methods.

Solarization has been found to be effective in reducing the viability of various weeds. The spectrum of controlled weeds includes species of winter and summer annual weeds (Elmore, 1991; Cohen & Rubin, 2007). A parasitic weed of the genus *Orobanchae* was controlled by solarization, (Jacobsohn *et al.*, 1980; Abdel Rahim *et al.*, 1988). Similar results were obtained in the laboratory with the parasitic weed *Striga* (Elmore, 1991). In contrast, differential responses are achieved with solarization in perennial weeds. Weeds from the genus *Cyperus* are inconsistently controlled by solarization (Elmore, 1991).

The perennial weeds *Cynodon dactylon, Sorghum hlepense* and *Convolvulus arvensis* are also less sensitive to solarization (Elmore, 1991). Moreover, some annual weeds, such as *Melilotus sulcatus* Desf., are not controlled by solarization (Cohen & Rubin, 2007).

It should be emphasized that since SH is climate-dependent, it is not surprising that different results pertaining to the effectiveness of control of a certain pathogen are reported from different climatic regions.

10.4 Mechanisms of control and plant-growth improvement

10.4.1 Mechanisms of pathogen and disease control

The first studies on SH were already showing that disease and pathogen control could be obtained, even under situations that are considered marginal from a climatic point of view. This indicated that pest control by solarization is more than merely physical killing of pests at elevated temperatures, and that additional processes, for example, biocontrol, are involved. The long-term effect of solarization reported in various studies (e.g. Katan *et al.*, 1983; Tjamos & Paplomatas, 1988; Satour *et al.*, 1989; Stevens *et al.*, 2003) supported the notion of a shift in the soil equilibrium that is suppressive to the pathogen, which could be connected to the physical, chemical and biological processes occurring in the soil during and after solarization. Many studies have shown that biological control processes are induced or stimulated in the solarized soil, thus contributing to pathogen control.

The reduction in disease incidence occurring in plants grown in solarized soils, as with any soil treatment, results from the effects exerted on each of the three living components involved in the disease – the pathogen, the surrounding soil microorganisms and the host plant, as well as on the physical and chemical environment which, in turn, affects the activities and interrelationships of these biotic components (Cook & Baker, 1983). Although these processes occur primarily during solarization, they may continue to various extents and in different ways after removal of the polyethylene sheets and planting. SH may affect inoculum density, inoculum potential or both. It is not possible to cover all of the studies related to biocontrol processes induced by solarization. Below are just some typical examples.

The 'weakening' effect is an important mechanism involved in solarization and possibly other disinfestation methods. Thus, propagules of the pathogen may survive sublethal heating (and therefore, inoculum density will show no initial major reduction), but the partially damaged (weakened) propagules become more vulnerable to the biocontrol processes and microbial activities taking place in the soil. Consequently, inoculum density of the pathogen will show later a more rapid decline. This has been shown with various pathogens, including *F. oxysporum* f. sp. *niveum* (Freeman & Katan, 1988), *Sclerotium rolfsii* (Lifshitz *et al.*, 1983), *Armillaria mellea* (Munneke *et al.*, 1976), *Sclerotinia minor* (Philips, 1990), *V. dahliae* (Tjamos & Fravel, 1995), *F. oxysporum* f. sp. *ciceri* (Arora *et al.*, 1996) and others. Solarization and organic amendments had a stronger effect on disease severity by *M. phaseolina* than on inoculum density, possibly indicating a weakening effect (Ndiaye *et al.*, 2007). Flow-cytometric, physiological and micoscopic studies on the viability of sublethally heated conidia of *Fusarium* showed that, although apparently not affected when examined shortly after heating, their population was nevertheless reduced later, suggesting delayed mortality (Assaraf *et al.*, 2002). These studies suggest that assessing the level of pathogen populations shortly after exposure to a control agent does not necessarily reflect the full control potential, or the potential of the weakening phenomenon induced by sublethal treatment.

Various studies have shown stimulation of populations of beneficial microorganisms (e.g. biocontrol agents and plant-growth promoters) in solarized soils. These include microorganisms such as *Trichoderma* spp., fluorescent pseudomonads, *Bacillus* spp., *Talaromyces* and others (Elad *et al.*, 1980; Stapleton & DeVay, 1982; Greenberger *et al.*, 1987; Gamliel & Katan, 1991; Tjamos *et al.*, 1991; Stevens *et al.*, 2003). Moreover, frequently (but not in all cases), solarized soils become more suppressive to pathogens, which is apparently connected with the aforementioned long-term effect (Kassaby, 1985; Greenberger *et al.*, 1987; Freeman *et al.*, 1990; Gamliel & Katan, 1993). An additional mechanism might be partial or complete nullification of fungistasis in the absence of the host, thus exposing the vulnerable germinating propagules to the antagonistic action of soil biota (Greenberger *et al.*, 1987). An analogous situation was described by Cohen *et al.* (2004) with the invasive plant *Acacia saligna*. They found that solarization was very effective in reducing the population of this plant, despite the fact that its seeds are highly tolerant to elevated temperatures. A possible explanation is that solarization nullified the dormancy of the seeds and consequently, the heat-sensitive germinating seeds were killed upon exposure to the high temperatures induced by solarization, essentially by 'suicidal germination'.

10.4.2 Induced resistance as a mechanism of disease control

Most of the studies on mechanisms of disease control by solarization concentrate on either the direct effect on the pathogen, namely, thermal killing, or the indirect effect via stimulation of microbial antagonistic activity in the soil as detailed above. However, another indirect effect, via induced resistance in the plant, should also be considered (Katan, 1981). In recent years, many studies have shown that certain biocontrol agents induce resistance and consequently disease reduction. There are studies which demonstrate physiological, including hormonal, changes in plants growing in solarized soils (Grunzweig *et al.*, 1993, 2000). In these soils, plant growth is frequently stimulated and mineral nutrient levels are increased, as detailed further on. These effects have also the potential to affect plant resistance. SH has been shown to reduce some foliar diseases despite the fact that only the roots are in contact with the solarized soil (Hassan & Younis, 1984; Daelemans, 1989; Stevens *et al.*, 1992, 1996; Lopez-Herrera *et al.*, 1994; Levy *et al.*, 2005). The works of Stevens *et al.* (1996) and Levy *et al.* (2005) indicate that reduction of foliar diseases in solarized soil is connected with induced resistance. Since only the roots are exposed to the solarized soil, signals may move upward in the plant, causing physiological changes (Grunzweig *et al.*, 1993). Solarization frequently stimulates rhizobacteria, such as fluorescent pseudomonads and *Bacillus* (Stapleton & DeVay, 1984; Gamliel & Katan, 1991; Stevens *et al.*, 2003), which are potential inducers of resistance (see Chapter 4). These can also indirectly contribute to induced resistance in the solarized soil.

It can be concluded that SH can affect disease incidence via a variety of mechanisms, beyond its direct effect on the pathogen.

10.4.3 Increased growth response (IGR) of plants in solarized soil

Improved plant growth and consequent increases in yield are expected to occur when soil-borne pests are controlled by any disinfestation method, as has been shown in many studies. However, less expected is the phenomenon of plant-growth improvement by disinfestation of soils *with no known major pathogens*. This was detected in the early days of soil disinfestation, more than a hundred years ago, and has been observed with all soil-disinfestation methods, including solarization (Chen & Katan, 1980; Cook & Baker, 1983; Stapleton *et al.*, 1985; Chen *et al.*, 1991; Gamliel & Katan, 1991; and other studies). Different mechanisms that are not related to pathogen control have been suggested to explain this IGR in disinfested soils: increased micro- and macro-elements in the soil solution (Chen & Katan, 1980; Stapleton *et al.*, 1985; Patricio *et al.*, 2006), elimination of minor pathogens or parasites, destruction of phytotoxic substances in the soil, release of growth-regulator-like substances, and stimulation of mycorrhizae or other beneficial microorganisms (Chen *et al.*, 1991). In another study, the level of soluble organic substances, that is humic substances, was higher in the solarized soil (Chen *et al.*, 2000). These soluble substances increased the growth of corn plants as well as populations of fluorescent pseudomonads, indicating an additional mechanism for improved plant growth. In another study (Grunzweig *et al.*, 1993), plant improvement was recorded in shoots of tomato seedlings 15 days after transplanting, and only 2 weeks later in roots, and significantly higher levels of chlorophyll and protein contents were detected in plants from solarized soil as compared to those from control soil. In addition, the degradation of these compounds, the decrease in net photosynthesis at near-saturation light intensity and of photochemical yield with ageing, were delayed in plants growing in solarized soil, as compared to controls. Delayed leaf senescence appeared to be a plant response contributing to plant improvement in this case. The association between plant-growth improvement in the solarized soil and microbial activities, such as the stimulation of beneficial rhizobacteria and suppression of minor pathogens, has also been reported (Gamliel & Katan, 1991).

The increases in plant growth in these studies range from a few to several hundred percent, depending on the soil, the plant, and the parameter used. The effect on yield components has not always been analyzed.

The definition of IGR depends largely on that of a healthy plant. It is now widely accepted that plant health involves much more than simply disease control. As Browning (1983) states: 'Plant health is far more than the opposite of plant disease as used in plant pathology.'

The IGR has very important economic implications which should be taken into account when considering the use of solarization. The major difficulty is that we cannot predict whether and to what extent a soil will respond with an IGR. Developing predictive methods for this purpose and developing measures to further enhance the IGR will improve the economic benefits of solarization (and other disinfestation methods).

10.5 Integrated management

Certain control methods, such as fumigation and the use of resistant cultivars, are very effective and some of them, like the latter, are not harmful to the environment and are easy to apply. Nevertheless, each method has its limitations, and additionally, cannot be used in all situations. For example, fumigants may have environmental and toxicological problems, whereas the appearance of new physiological races may hamper the use of resistant cultivars. The integrated pest management (IPM) approach simultaneously addresses pest control, environmental, economic, legal and public issues in an attempt to achieve effective, economical, environmentally and publicly acceptable pest control by using a diversity of methods which are adapted to the specific cropping system (Kendrick, 1988; Gupta, 1996; Katan, 1996; Davis *et al.*, 2008). An appropriate IPM program can achieve better control with minimal use of pesticides and environmental hazards, a broader spectrum of control and even a long-term effect. The diversity of approaches reduces the risks involved when a cropping system depends on a single control method, as in fact happened when most of the intensive crops became dependent on methyl bromide (MB) (see further on). An appropriate IPM program has the potential to provide a wide spectrum of control similar to that of MB, but this requires appropriate and reliable diagnostic tools in order to choose and adjust the appropriate control measure for each situation. Combining and alternating methods of control are at the heart of IPM. The following issues should be considered when developing IPM programs (Katan, 1996; Davis *et al.*, 2008):

(a) All pests of the crop, with an emphasis on the major ones, need to be taken into account.
(b) The IPM program should be in harmony with the cropping system.
(c) In combining methods of control, priority should be given to non-chemical methods.
(d) Emphasis should be placed on reducing pesticide usage, not necessarily with the aim of eliminating it entirely.
(e) Environmental, economic, legal and social considerations need to be taken into account.
(f) When possible, decision-making tools should be used (see below).

Gupta (1996) indicated that 'the integrated disease-management system involves the simultaneous manipulation of a number of available strategies for reducing plant disease, with the aim of causing the least possible damage to the environment. Being part of the agroecosystem, one has to manage the whole system rather than just the individual malady.' Therefore, IPM is a holistic approach.

The combining of solarization with other methods of control should be considered from two different sides: (a) as a way of improving solarization (as detailed below) and (b) as a way of improving the other method. For example, from the latter point of view, a suitable combination of solarization (or any non-chemical method) with *a reduced dosage* of pesticide makes the pesticide less harmful (and more acceptable), without reducing control effectiveness, and maybe even increasing it. McGovern & McSorley (1997), emphasized that combination of solarization with other pest management practices may be necessary to ensure acceptable reduction of difficult-to-control pests, especially in suboptimal climates.

The benefit of combining SH with other methods should be studied with a view on the long term. Even if its benefit is realized in only some of the growing seasons, it will be justified since the combination will be regarded as insurance for those situations in which the climatic conditions are less favorable for solarization.

Many studies have combined control methods with solarization, frequently with good results. Such combinations may result in either synergistic or additive effects. The previously described weakening phenomenon is such a mechanism, potentially leading to synergism, or to improved control.

Combining methods of control is more than just mixing two methods. The combination has to be optimal. For example, combining short SH (for 10 days) with a reduced dosage of MB or metham sodium resulted in higher effectiveness in controlling pathogens than each method alone. However, if solarization was applied first and MB afterwards, this combination was much more effective than when the opposite sequence was applied (Eshel *et al.,* 2000).

Certain elements of integrated management are powerful tools for improving solarization. Therefore, combining solarization with other methods of control enables us to address the limitations of the former, that is climate dependency, uncertainty due to climatic variations, occupation of the mulched soil with plastic for 4–6 weeks and the inability of solarization to control thermotolerant pathogens, for example, *Macrophomina, Monosporascus* and possibly others. Examples will be given below.

Solarization can also be combined with cultural methods. For example, combining solarization with a certain crop sequence improved the control of Fusarium wilt of cotton (Katan *et al.,* 1983).

10.5.1 Combining solarization with fumigants

Combining solarization with a fumigant has many potential advantages: improved control due to heating of the chemical and consequently allowing a reduction in fumigant dosage, and cost reduction because the same plastic can be used for both purposes: solarization and fumigation. Combining solarization with chemicals at reduced dosages, or other measures, for example biocontrol agents, can reduce the limitations of solarization. The control efficacy may be increased due to additive effects, or to a synergistic effect caused by the hotter environment, which increases vapor pressure and chemical activity of the added pesticide. Another possible mechanism for improved activity of the pesticide is weakening of the pathogenic resting structure by heat. Solarization combined with fumigants could shorten the required duration of solarization, thus making the method more acceptable for farmers. A combination of solarization and metham (Ben-Yephet *et al.,* 1988) killed more propagules of *F. oxysporum* f. sp. *vasinfectum*, and faster than solarization alone. Thus, the period during which solarization is effective might be longer than previously thought, when a suitable improvement, in the form of combined treatment, is made. There are many other examples. Furthermore, sublethal fumigation in combination with solarization is especially useful for areas that are marginal for the application of solarization. Studies on combining solarization with other chemical or non-chemical methods include those combining solarization with MB for controlling yield decline in *Gypsophila* (Gamliel *et al.,* 1993). An MB–chloropicrin mixture combined with solarization was effective in controlling *Pseudomonas solanacearum* in tomato (Chellemi *et al.,* 1994). Additional examples are combining solarization with dazomet for the control of pink root of onion (Porter *et al.,* 1989), or metham sodium for the control of delimited shell spot of peanut (Frank *et al.,* 1986) and soil-borne diseases of strawberry (Hartz *et al.,* 1993). Combining fumigants with SH under virtually impermeable films improved the control of fungi, nematodes and yellow and purple nutsedge (*Cyperus*) (Chellemi *et al.*, 1997; Gamliel *et al.*, 2000b).

10.5.2 Combining solarization with organic amendments (biofumigation)

Combining organic amendments with SH is another integrated approach to improving the control of soil-borne pests. Heating of soils covered with plastic film and amended with appropriate organic material actuates a chain reaction of chemical and microbial degradation, leading to the generation of toxic compounds in the vapor and liquid soil phases. The plastic mulch traps the volatile compounds and creates an atmosphere in the soil with a high concentration of volatile compounds. Thus, organic amendments which generate volatile compounds are especially desirable. Solarization of soil amended with chicken manure, chicken litter or plant residues was reported to be effective in controlling several soil-borne pathogens and with some pathogens, this combination was more effective than one treatment alone (Ramirez-Villapudua & Munnecke 1988; Gamliel & Stapleton 1993; Gamliel *et al.*, 2000a; Stevens *et al.*, 2003). The temperature of solarized soil amended with compost usually increases by 2–3°C over that in solarized, non-amended soil (Gamliel & Stapleton, 1993). This additional rise in temperature may be an important factor in improving the control of pathogenic organisms such as *M. incognita* as well as heat-tolerant organisms. SH reduced inoculum density, viability and incidence of disease caused by *S. cepivorum* in garlic and increased yield (Ulacio-Osorio *et al.*, 2006).

Combining chemical and non-chemical methods of soil disinfestation (including solarization) improved the spectrum of pest control in tomato and reduced fumigant rate (Chellemi & Mirusso, 2006). A very detailed long-term study on integrating SH with chemical, biological and cultural control for the management of soil-borne diseases of vegetables was carried out by Stevens *et al.* (2003). Combining SH with reduced dosages of fungicide, a biocontrol agent (*Trichoderma virens*) or chicken litter much improved control of diseases such as root knot nematodes and southern blight caused by *S. rolfsii* in tomato and led to increases in yield. Better long-term effectiveness for more than 2 years was also achieved in that study using the integrated approach. It is difficult to control *M. phaseolina* by solarization since the pathogen is thermotolerant. The use of organic amendments followed by solarization resulted in better control of charcoal rot and a higher yield increase in cowpeas in the Sahel desert (Ndiaye *et al.*, 2007). On the other hand, solarization may increase the phytotoxicity of herbicides by suppressing their degradation in the soil (Avidov *et al.*, 1985). This has been demonstrated with the combination of solarization and the herbicide dacthal used with collards (Stevens *et al.*, 1990).

The combination of solarization with organic amendments gives effective results, which under optimal conditions equal those obtained with the use of fumigants. Benlioglu *et al.* (2005) found optimal combination of solarization with chicken manure for the control of soil-borne diseases and weeds of strawberry in the Western Anatolia region of Turkey. The effects of SH for 30 days and treatment with goat manure at 0, 20 and 40 $t\,ha^{-1}$ were studied in terms of their impact on weed management and muskmelon yield (Lira-Saldivar *et al.*, 2004). Solarization reduced the emergence and growth of weeds. Goat manure also had an antagonistic effect on weed density, but this effect was not clear in solarized plots.

10.5.3 Combining solarization with biocontrol agents

Combining biological control agents, as mentioned above (Stevens *et al.*, 2003), or other beneficial organisms with solarization is an especially attractive approach. Its effectiveness has been studied with *T. harzianum* combined with solarization for the control of Fusarium crown and root rot of tomato (Sivan & Chet, 1993) and of *Rhizoctonia* (Elad *et al.*, 1980; Chet *et al.*, 1982), and with *Gliocladium virens* combined with solarization for controlling *S. rolfsii* (Ristaino *et al.*, 1991). Solarization controlled *Armillaria* borne in coarse plant material and improved control was obtained when it was combined with application of *T. harzianum* (Otieno *et al.*, 2003).

Recently, solarization has been reported as a promising alternative for field-grown cut flowers in the US (McSorley *et al.*, 2006), when integrated with other measures such as biorational fungicides and biocontrol agents; for strawberries in Spain when combined with *Trichoderma* (Porras *et al.*, 2007), and for peppers in the US when combined with cover crops (Wang *et al.*, 2006). Combining SH with the biocontrol agent *Streptomyces griseovirides* was highly effective in controlling soil-borne pathogens in tomatoes (Minuto *et al.*, 2006).

10.6 Modelling of soil solarization and decision-making tools

10.6.1 Modelling

A main difficulty with SH is its dependence on climate. Thus, there is a need to be able to predict the effectiveness of solarization in various climatic regions and seasons. This can be achieved experimentally, by following soil temperatures and pathogen mortality in the solarized soil, and by modeling.

A variety of models for predicting temperatures of solarized soils under various climatic conditions have been developed and validated. The first models were developed for arid conditions (Mahrer, 1979, 1991; Mahrer & Katan, 1981). Simplified models (Cenis, 1989) and a model which is also suitable for more humid areas (Wu *et al.*, 1996) were also developed. However, information on soil temperature in the solarized soil, although necessary and very helpful, is not sufficient for predicting the effectiveness of SH for pathogen control, since it does not take into consideration pathogen mortality that is not due to heat, as detailed in the section on mechanisms of pest control. Modelling pathogen control by solarization is usually based on studies of thermal inactivation of pathogens, utilizing data collected under constant elevated temperatures. Based on chemical reaction kinetics at constant temperatures, an exponential inactivation relationship is typically expected (Toledo, 1999). Traditionally, thermal inactivation of microorganisms is considered a first-order reaction, characterized by a logarithmic change in the organism's population with time. A logarithmic relationship was indeed found between time and temperature (at constant temperatures) for the thermal inactivation of four soil-borne plant pathogens (Pullman *et al.*, 1981). Studies on thermal inactivation under fluctuating temperatures, such as those naturally prevailing in the field, are much more complicated to perform, because the partial effects of varying temperatures on pathogens are difficult to weigh, and they require numerical integration to account for their complexity (Shlevin *et al.*, 2003). There are other approaches to simulating and modelling pathogen control by solarization, for example by plotting the level of mortality versus accumulated hours above a certain temperature, that is the degree-hours (DH) (Chellemi *et al.*, 1994). Shlevin *et al.* (2003) developed a model describing the process of pathogen control with time under structural (dry) solarization. In an other study on modelling regular (wet) SH, the common DH approach for predicting the rate of heat inactivation was further improved by giving different weights to the different temperatures, thus refining the correlation between temperature data and pathogen survival from $R^2 = 0.324$ to $R^2 = 0.86$ (Shlevin *et al.*, 2005).

10.6.2 Decision-making tools

We should aim to control pests only when it is effective, and economically and environmentally justified, as required by the IPM approach. This can be achieved through the use of decision-making tools (Katan, 1996; Davis *et al.*, 2008). Such tools are used for foliar diseases, but much less for soil-borne pathogens. They have the potential to improve, control and reduce harmful effects on the environment. They can be especially useful for solarization.

The necessary decision-making tools for the management of soil-borne pathogens by solarization are:

(a) Monitoring pathogen populations in the soil, for which sensitive assays are needed.
(b) Studying relationships between inoculum density and disease incidence, as well as between disease incidence and yield.
(c) Predictive models for soil heating, which have already been developed, combined with models on pathogen control under a solarization regime, as described above, are potentially powerful tools for this purpose.
(d) Economic considerations, especially regarding yield reduction by the pathogen (Yaron *et al.*, 1991).
(e) Consideration of the level of soil suppressiveness, wherever possible.

10.7 Improvements by intensifying soil heating

Solarization is the simplest method to apply for soil disinfestation. Nevertheless, since the day of its introduction, solarization has undergone continual improvement and technological innovation in order to adapt the method to a wide variety of conditions and cropping systems. Improvements are also needed to overcome difficulties, limitations, and existing negative attributes which have already been encountered or could potentially occur. Advanced solarization seeks to achieve additional goals, such as improving the level of control and creating a more long-term effect, lasting throughout a season or over successive seasons. Some of the other reasons for improvements are cost reduction, and increasing the reliability and reproducibility of the method, which is climate-dependent; however, others include shortening the period during which the soil is occupied with mulch, and making solarization possible for longer periods during the year and more acceptable.

Improvement of SH can be achieved by either using improved plastic or modifying the application technology. For example, by solarizing shallow layers of growth substrates, the temperatures can be increased to very high levels, thus leading to control of even the thermo-tolerant pathogen *Monosporoascus cannonballus* (Pivonia *et al.*, 2002).

Over the years, many efforts have been made to move from the clear polyethylene plastic sheets to other polymers and other formulations in order to maximize heat absorbance under the tarp. The following describes some of the current advances in film formulation and technology.

10.7.1 Double tarps

The use of a single layer of plastic film is the common application method for SH. Transparent films allow transition of the solar irradiation, which heats the soil layer adjacent to the film. Being thin and with no insulation layer, the heat flux also escapes to the atmosphere through the film. Thus the efficacy of soil heating is reduced.

10.7.1.1 Two transparent films

One way to increase soil heating is the use of double-layer mulch, which heats the soil to higher levels than a single one. This was shown by Raymundo & Alcazar (1986), who achieved an increase of 12.5°C at a depth of 10 cm using a double-layer film compared to a single-layer one (60°C versus 47°C, respectively). Ben Yephet *et al.* (1987) achieved a similar increase in soil temperature by using double-layer films. They observed a 98% reduction in the viability of *F. oxysporum* f. sp. *vasinfectum* after 30 days under the double mulch compared with a 58% reduction under single mulch, at a depth of 30 cm. Double-layer films form a static air space under and between the plastic layers. This construction apparently acts as an insulator for heat loss from the soil to the atmosphere, especially during the night. The use of a double-layer film also offers opportunities for applying solarization in areas and climatic conditions which are not favorable for solarization when a single layer is used. In central Italy, in an area which is climatically marginal for solarization, double-film solarization was very effective in reducing the viability of *Pythium, Fusarium* and *Rhizoctonia* in a forest nursery (Annesi & Motta, 1994). This approach was also followed successfully in Australia with nursery potting mix (Duff & Connelly, 1993). Its one drawback, however, is cost: a double-layer film is more expensive than one layer of tarp. SH in a closed greenhouse is another version of the double-layer film which improved disease control (Garibaldi & Gullino, 1991).

10.7.1.2 Transparent-over-black double film

A different approach to increasing solar heating with a double layer is the use of a sprayable black polymer as the lower mulch, laid over the soil surface. Stevens *et al.* (1999) reported a 5°C increase in soil temperature when applying solarization in strips in a cloudy climate. Similarly, Arbel *et al.* (2003) achieved an increase in soil temperature by mulching transparent polyethylene sheet over a layer of sprayable black mulch (Figure 10.2). In field experiments, they observed that the mortality of resting structures of *F. oxysporum, S. rolfsii* and *R. solani* was higher than in the plots which were solarized by a single plastic layer. Consequently, in the solarized plots with double plastic

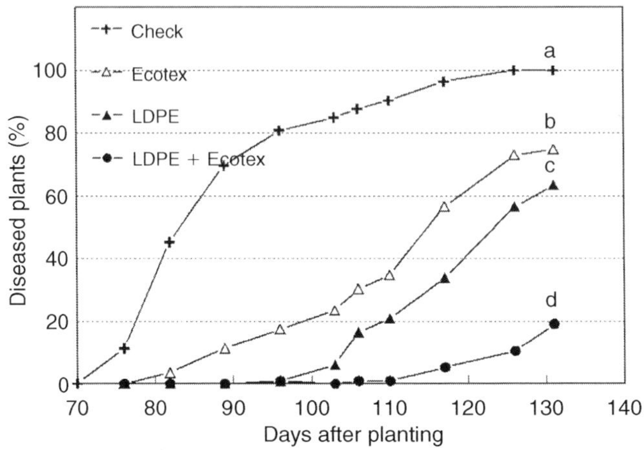

Figure 10.3 Control of crown and root rot of tomatoes by solarization using various mulches. LDPE: low-density polyethylene sheets; Ectorex: black sprayable polymer; double mulch: transparent LDPE, over Ecotex mulch. Different letters denote significant differences among treatments ($P < 0.05$).

mulch, effective control of Fusarium crown and root rot of tomatoes (Figure 10.3) and vine collapse of melons (caused by *M. cannonballus)* was achieved, while solarization with regular films was not effective (Arbel *et al.*, 2003). The use of double mulch which consists of a transparent film over a black polymer coating is based on the same principle as solar collectors for water heating in sunny countries. However, the cost of the double mulching and the technology that needs to be developed for simultaneous tarping limit the implementation of such an approach. Alternatively, this approach can be successfully applied in strip solarization and small farm plots.

10.7.2 Improved films

The use of new films that intensify solar heating can offer an additional option for improved control of soil-borne pathogens in various limiting crop-management systems.

Efforts were made to produce plastic films which prevent heat loss through the film with infrared (IR) blocking material. Chase *et al.* (1999) observed improved heating using IR films under rainy and cloudy conditions in Florida, but other studies reported minor or no differences in soil heating using IR films. The use of IR films was also studied by Stevens *et al.* (1999).

Another successfully tested film was a polyethylene film which was formulated with the addition of anti-drip (AD) components (Arbel *et al.*, 2003). This formulation prevents condensation of water droplets on the film surface, leading to a 30% increase in irradiation transmittance over regular film. Soil temperatures under AD film were 2–7°C higher than under regular film (Figure 10.4). Solarization with AD film in field experiments resulted in effective control of sudden wilt of melons, while solarization with common transparent film had no effect on disease level (A. Gamliel, unpublished data). Virtually impermeable films were more effective in raising soil temperatures and killing *Fusarium* than regular polyethylene (Chellemi *et al.*, 1997).

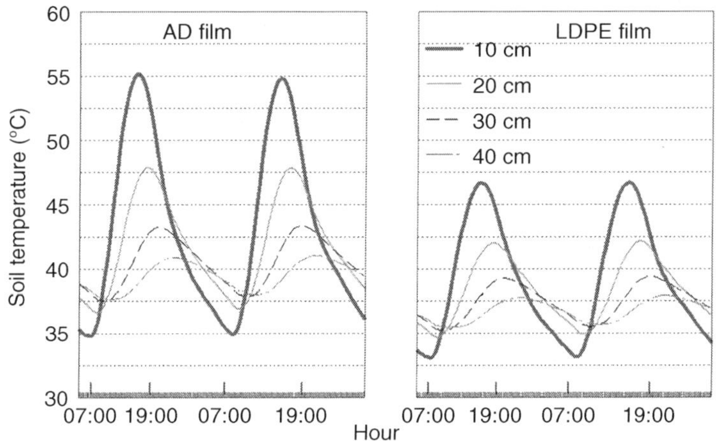

Figure 10.4 Soil temperatures at various depth during soil solarization at the summer time in Israel using two mulches. LDPE: low-density polyethylene sheets; AD: low-density polyethylene sheets with addition of antidrip material to the polymer.

10.8 Implementation and application

The most common plastic film used for solarization is polyethylene. The thickness of the film has no considerable effect on soil heating. The sheet is applied at thicknesses as low as 20 μm with appropriate plastic material. Therefore, the farmer can consider using the thinnest plastic (which is also the cheapest) available that is still strong enough for his application technology. Used polyethylene sheets are also effective for SH (Avissar *et al.*, 1986), at practically no cost.

Continuous mulch is desirable for effective SH. It reduces the 'border effect' (Grinstein *et al.*, 1995) and also improves control efficacy and reduces soil reinfestation compared to strip solarization. However, strip solarization applied to 20-cm high, 0.9-m wide beds, increased soil temperatures and eliminated the border effect (Chellemi *et al.*, 1997). This is another example of improved technology of solarization. The optimal area for covering a solarized plot is, of course, the whole field. Such application causes yield improvement and provides a long-term effect, thus spreading the cost of disinfestation over several crops. The economic advantages resulting from complete coverage of the plot should be assessed against the expenses involved.

Relatively small plots can be covered manually. The edges of the sheet should be firmly embedded in the soil, while ensuring film tightness. A continuous plastic covering for relatively small plots can be achieved manually by anchoring the edge of two adjacent sheets together in one furrow (Grinstein & Hetzroni, 1991). Mechanized plastic mulch is essential for use in large plots. This is applied essentially as soil fumigation is performed. The tarp layer machine unrolls plastic strips (2–4 m width), each of which is anchored to the soil at one side and connected the other sheet by means of glue or welding. Sheet gluing is most common in soil-fumigation rigs. The glue is sprayed in a 5- to 8-cm strip along the edge of the previously laid sheet near the border of the anchored area. During the glue spraying, a new film is unrolled and one side is pressed over the top of the sprayed glue, while its other edge is buried in the soil. Since solarization requires a minimum period of 30 days, durable glue is required. In recent years, glues have been developed that are appropriate for use in continuous mulching.

An alternative system to weld polyethylene sheets together was developed in Israel (Grinstein & Hetzroni, 1991). Sheet fusing is accomplished with hot air streams, emitted from a combustion chamber and directed onto the plastic sheets. The machine is tractor-mounted with a capacity similar to that achieved by the gluing method. The formulation of the film-additive mixture was found to be of great importance in determining the strength of the welded seam. Pure polyethylene film is easily welded, but some additives (e.g. some UV-absorbent ones) were found to interfere with the fusing process, leading to peeling after only a few days.

An important consideration in large field applications is damage to the mulched film. Perforation and ruptures in the film increase heating loss from the mulch and decrease the effective size of the treated area. Once the film is laid, tearing can be caused by birds, animals and wind. Tears often extend over a large area, thus reducing the efficacy of the solarization treatment. This is a drawback of continuous solarization over soil fumigation.

10.8.1 Sprayable films

Sprayable polymers offer a feasible and cost-effective alternative to plastic tarps for SH. The plastic-based polymers are sprayed on the soil surface in the desired quantity and form a membrane film, which can maintain its integrity in soil and elevate soil temperatures. Nevertheless, the formed membrane is porous and allows overhead irrigation. Initial research with these polymers was conducted by Stapleton & Gamliel (1993), who achieved effective soil heating and a reduction in the viability of *Pythium* propagules. The application, however, was not cost-effective, since a high amount of polymer was required to achieve a continuous and uniform coating. In Israel, a sprayable polymer product, 'Ecotex,' was developed together with the technology to apply it economically on soils for various purposes (Skutelsky *et al.*, 2000). Soil coating using this technology with a black polymer formulation resulted in a membrane film that could raise soil temperatures to close to solarization levels (Figure 10.2). Soil heating with sprayable mulch is faster than that with plastic film, but the soil also cools down to lower temperatures at night. Overall, soil temperatures under sprayable mulch are lower than those obtained under plastic film. The thickness of the sprayed coat is critical to obtaining effective heating (Skutelsky *et al.*, 2000). SH using sprayable mulches was effective in controlling Verticillium wilt and potato scab in potato (Gamliel *et al.*, 2001), at a level matching that achieved by solarization using plastic films. However, there is room for further improvement in the use of sprayable polymers.

10.9 Special uses of solarization

The use of solar heating for the control of soil-borne pests has expanded over the years beyond the soil and the pests dwelling in it. The killing properties of heat and humidity have been used for unusual purposes, some of which are listed below.

A tent for SH was developed by Stapleton (2000) as a method of eradicating phytoparasitic pests in closed horticultural applications (Stapleton, 2000; Stapleton *et al.*, 2002). Disinfestation is accomplished via heating, the latter induced by covering with a greenhouse structure; a double-layer tent can be created to increase heating efficacy. The soil is either mulched or left bare, and this type of application can be used to disinfest the soil inside the structure or in container media. The temperature inside the tent rises to above 70°C, a level of heat which, even if applied for only a short time, can completely eradicate both phytoparasitic and free-living nematodes. This application of solarization could potentially come into common use, particularly in developing countries, for disinfesting soil seedbeds, containerized planting media, and cold-frames in small and simple structures. These are ideal niches for solarization, since the individual areas to be treated are small, soil temperature can be greatly increased, the cost of application is low, the period of solarization is short, and the value of the plants produced is high. Note that the production of disease-free planting stock is critical for producing healthy crops.

Unlike soil treatment, solarization of structures aims to control the inoculum which is left within the greenhouse or any other structure and its structural components (the structure itself, irrigation lines, wires, etc.), which might be contaminated with inoculum (Gamliel *et al.*, 1996). This is achieved by closing the greenhouse during the summer time, thereby elevating the temperatures (dry heating) to 60, and even close to 70°C (Figure 10.5). Structural solarization is carried out under dry conditions and is considered solar sanitation (Shlevin *et al.*, 2003).

Other special solarization techniques involve: SH in existing, perennial crops such as pistachio (Ashworth *et al.*, 1982) and olive (Tjamos *et al.*, 1991), disinfestation of wooden tomato stakes of *Didymella lycopersici* (Besri & Diop, 1985), solarization of soils in 40-cm high piles, which was effective in controlling root knot nematodes in olive nurseries in Spain (Nico *et al.*, 2003), and using solar collectors for soil disinfestation (Ghini, 1993). SH is the only method for soil disinfestation that can be used in organic farming (CDFA, 2004). Indeed SH is used by farmers practicing organic methods.

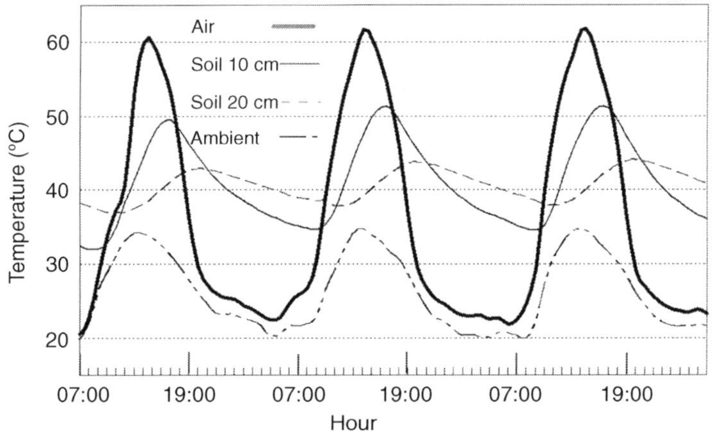

Figure 10.5 Soil and air temperatures during solarization of closed greenhouse during the summer time in Israel (for structural solarization), compared to ambient air temperature.

10.10 Solarization and the MB crisis

Soil fumigation with chemicals may have negative effects on the environment, thus leading to their phase-out, as in the case of MB. As a consequence, farmers are left in desperate need of innovative and environmentally accepted alternative approaches. Indeed, SH has replaced MB in certain hot regions of the world and it continues to be further adopted as an alternative to MB, particularly in combination with another technique(s). In Costa Rica, for example, an estimated 20% of the melon cropping area (about 2000 ha) is now being solarized, and this has proven particularly successful when combined with reduced doses of metham sodium (MBTOC, 2007). Solarization is also widely used in Greece and Turkey. Improving solarization performance by combining it with fumigation could solve many problems, since such combinations are suitable for areas deemed marginal for the application of solarization alone.

In recent years, several studies have examined the effects of long-term, large-scale use of SH and organic amendments on weed populations, nematodes, yields and soil fertility with peppers (*Capsicum annuum*) and cucumbers (*Cucumis sativus*) (Roe *et al.*, 2004; Ozores-Hampton *et al.*, 2005). Benlioglu *et al.* (2005) studied solarization in Turkey and compared it to chemical treatments to control *Rhizoctonia* spp., *Phytophthora cactorum* and *V. dahliae* in strawberry production. Solarization provided 163% higher yield than the control and over 50% more than a high rate of the fumigants tested. SH was shown to be cost-effective, compatible with other pest-management tactics and a valid alternative to pre-plant fumigation with MB under the tested conditions in Florida (Chellemi *et al.*, 1997).

The MB phase-out showed that dependence on a single method or chemical can lead to an agricultural crisis. Thus, integrated approaches are the best solution for pest control. Solarization can be a major component in such integrated programs.

10.11 Concluding remarks

Today, the biggest challenge in crop protection sciences is to effectively control pests, while avoiding environmental hazards and degradation of natural resources. SH is an additional tool for achieving this task, when it is used in appropriate situations. Although the positive effects of SH outweigh the negative ones (e.g. plant-growth retardation due to harmful effects on beneficial microorganisms, such as *Rhizobium* or mycorrhizae), the emphasis in research should be placed on the negative effects and on developing means to detect and avoid them. After more than 130 years of soil disinfestation, the arsenal of chemical disinfestants is still very limited, and the arsenal of non-chemical agents for soil disinfestation even more so. Therefore, the integration of pest management methods, rather than relying on one powerful control agent, is not only desirable but also the only feasible solution for coping with our need for methods of controlling soil-borne pathogens in an atmosphere of environmental awareness, concerns, and pressure. SH is needed not only as an alternative to MB, but also for many other purposes. There are many challenges awaiting the further development of SH: improvements in implementation technology and control effectiveness, thereby shortening the mulching period and extending the period for solarization; a better understanding of control mechanisms which can lead to more effective disturbance of the pathogens' life cycles; further development of the models and decision-making tools, and last but not least, use of the proper extension tools for introducing SH into new regions. Other examples for improving SH have been provided in this chapter. SH is only one more way-station on our long journey toward better and safer sustainable agriculture; additional milestones are awaiting us (Katan & DeVay, 1991).

An interesting approach, suggested by Grinstein & Ausher (1991), is the use of SH as a 'cleaning tool' for infested soils, in the framework of crop rotation. It is based on the aforementioned finding that SH frequently has a long-term effect. The idea is to apply solarization in a field every 3–5 years, before the field becomes heavily infested and prior to planting a cash crop, to provide the most benefit from SH. The soil is again solarized prior to planting the next cash crop. This procedure also reduces the cost of SH per crop and maintains low infestation, provided that a proper crop rotation is practiced.

Our ultimate goal in developing a new method for pest management is to place it in the hands of the farmer. This involves a professional approach and combined efforts by researchers, extension personnel and the involved industry. Chellemi (2005) emphasized that a 'successful integration of soil disinfestation programs into commercial horticultural production systems involves many aspects that extend far beyond standard efficacy studies conducted in laboratories or small research plots. An understanding of the entire pest complex and the associated ecology is essential to ensuring the long-term stability of a disinfestation program.'

10.12 References

Abdel-Rahim MF, Satour MM, Mickail KY, El Eraki SA, Grinstein A, Chen Y, Katan J, 1988. Effectiveness of soil solarization in furrow-irrigated Egyptian soils. *Plant Disease* **72,** 143–146.

Adams PB, 1971 Effect of soil temeratures and soil amendments on *Thielaviopsis* root rot of sesame. *Phytopatholgy* **61,** 93–97.

Annesi T, Motta E, 1994. Soil solarization in an Italian forest nursery. *European. Journal of Forest Pathology* **24,** 203–209.

Antoniou P, Tjamos EC, Panagopoulos CG, 1995. Use of soil solarization for controlling bacterial canker of tomato in plastic houses in Greece. *Plant Pathology* **44,** 438–447.

Arbel A, Siti M, Barak M, Katan J, Gamliel A, 2003. Innovative plastic films enhance solarization efficacy and pest control. *Proceedings of the 10*th *Annual International Research Conference on Methyl Bromide Alternatives and Emission Reduction.* San Diego, CA, USA, 91-1–91-3.

Arora DK, Randey AK, Srivastava AK, 1996. Effects of heat stress on loss of *F. graminearum* and pathogenicity from chlamydospores of *Fusarium oxysporum* f. sp. *ciceri. Soil Biology and Biochemistry* **28,** 399–407.

Ashworth LJ, Gaona SA, 1982. Evaluation of clear polyethylene mulch for controlling Verticillium wilt in established pistachio nut groves. *Phytopathology* **72,** 243–246.

Assaraf MP, Ginzburg C, Katan J, 2002. Weakening and delayed mortality of *Fusarium oxysporum* by heat treatment: Flow cytometry and growth studies. *Phytopathology* **92,** 956–963.

Avidov E, Aharonson N, Katan J, Rubin B, Yarden O, 1985. Persistence of terbutryn and atrazine in soil as affected by soil disinfestation and fungicides. *Weed Science* **33,** 457–461.

Avissar R, Naot O, Mahrer Y, Katan J, 1986. Field aging of transparent polyethylene mulches. II. Influence on the effectiveness of soil heating. *Soil Science Society America Journal* **50,** 205–209.

Benlioglu S, Boz O, Yildiz A, Kaskavalci G, Benlioglu K, 2005. Alternative soil solarisation treatments for the control of soilborne diseases and weeds of strawberry in the western Anatolia of Turkey. *Journal of Phytopathology* **153,** 423–430.

Ben-Yephet Y, Stapleton JJ, Wakeman RJ, DeVay JE, 1987. Comparative effects of soil solarization with single and double layers of polyethylene on survival of *Fusarium oxysporum* f. sp. *vasinfectum. Phytoparasitica* **15,** 181–185.

Ben-Yephet Y, Melero-Vera JM, DeVay JE, 1988. Interaction of soil solarization and metham-sodium in the destruction of *Verticillium dahliae* and *Fusarium oxysporum* f. sp. *vasinfectum*. *Crop Protection* **7,** 327–331.

Besri M, Diop M, 1985. Control of *Didymella lycopersici* in tomato by storing the supports in plastic tunnels: New application of the solar heating or solarization. *Review Horticulture* **58,** 99–102.

Browning A, 1983. Goal for plant health in the age of plants: A national plant heath system. In: Kammedahl T, Williams PH, eds. *Challenging Problems in Plant Health*. St. Paul, MN: APS Press, 45–57.

CDFA, 2004. Approved treatment and handling procedures to ensure against nematode pest infestation of nursery stock. Nursery Inspection Procedures Manual, NIPM Item 7. In: *California Department of Food and Agriculture (CDFA), Plant Health and Pest Prevention Services, Pest Exclusion Branch, Sacramento*. [http://www.cdfa.ca.gov/phpps/pe/nipm.htm].

Cenis JL, 1989. Temperature evaluation in solarized soils by Fourier analysis. *Phytopathology* **79,** 506–510.

Chase CA, Sinclair TR, Chellemi DO, Gilreath JP, Locascio ST, Olson SM, 1999. Heat-retentive films for increasing soil temperature in a humid, cloudy climate. *HortScience* **34,** 1085–1089.

Chellemi DO, 2005. Integrating soil disinfestation programs into crop production systems. *Acta Horticulturae* **698,** 105–114.

Chellemi DO, Mirusso J, 2006. Optimizing soil disinfestation procedures for fresh market tomato and pepper production. *Plant Disease* **90,** 668–674.

Chellemi DO, Olson SM, Mitchell DJ, 1994. Effects of soil solarization and fumigation on survival of soilborne pathogens of tomato in northern Florida. *Plant Disease* **78,** 1167–1172.

Chellemi DO, Olsen SM, Mitchel DJ, Secker I, McSorly R, 1997. Adaptation of soil solarization to the integrated management of soilborne pests of tomato under humid conditions. *Phytopathology* **87,** 250–258.

Chen Y, Katan J, 1980. Effects of solar heating of soils by transparent polyethylene mulching on their chemical properties. *Soil Science* **130,** 271–277.

Chen Y, Gamliel A, Stapleton JJ, Aviad T, 1991. Chemical, physical, and microbial changes related to plant growth in disinfested soils. In: Katan J, DeVay JE, eds. *Soil Solarization*. Boca Raton, FL: CRC Press, 103–129.

Chen Y, Katan J, Gamliel A, Aviad T, Schnitzer M, 2000. Involvement of soluble organic matter in increased plant growth in solarized soils. *Biology and Fertility of Soils* **31,** 28–34.

Chet I, Elad Y, Kalfon A, Hadar Y, Katan J, 1982. Integrated control of soilborne pathogens in iris. *Phytoparasitica* **10,** 229–2366.

Cohen O, Rubin B, 2007. Soil solarization and weed control. In: Upadhyaya MK, Blackshaw RE, eds. *Non-chemical Weed Management*. Wallingford, UK: CAB International, 177–200.

Cohen O, Bar P, Katan J, Riov J, Gamliel A, 2004. Soil solarization as a method for reducing viability of invasive woody seed bank – the case of *Acacia saligna*. *Proceedings of the 10th MEDOCS Conference,* 25 April–1 May 2004, Rhodes, Greece, 1–7.

Cook RJ, Baker KF, 1983. *The Nature and Practice of Biological Control of Plant Pathogens*. St. Paul, MN: APS Press.

Daelemans A, 1989. Soil solarization in west Cameron; effect on weed control, some chemical properties and pathogens of the soil. *Acta Horticulturae* **255,** 169–175.

Davis JR, 1991. Soil solarization: Pathogen control and increase in crop yield and quality, short and long term effects and integrated control. In: Katan J, DeVay JE, eds. *Soil Solarization*. Boca Raton, FL: CRC Press, 39–50.

Davis JR, Sorensen LH, 1986. Influence of soil solarization of moderate temperatures on potato genotypes with differing resistance to *Verticillium dahliae*. *Phytopathology* **76,** 1021–1026.

Davis JR, Katan J, Gamliel A, 2008. The potential of using soil solarization as a means of controlling diseases caused by soilborne pathogens and weeds in potatoes. In: Petiss RD, ed. *Current Concepts in Potato Disease Management*. Kerala, India: Research Signpost (in press).

Duff JD, Connelly MI, 1993. Effect of solarization using single and double layers of clear plastic mulch in *Pythium*, *Phythophthora* and *Sclerotium* species in a potting mix. *Australasian Plant Pathology* **22,** 28–35.

Elad Y, Katan J, Chet I, 1980. Physical, biological and chemical control integrated for soilborne diseases in potatoes. *Phytopathology* **70,** 418–422.

Elmore C, 1991. Weed control by soil solarization. In: Katan J, DeVay JE, eds. *Soil Solarization*. Boca Raton, FL: CRC Press, 61–72.

Eshel D, Gamliel A, Grinstein A, Di Primo P, Katan J, 2000. Combined soil treatments and sequence of application in improving the control of soilborne pathogens. *Phytopathology* **90,** 751–757.

Frank ZR, Ben-Yephet Y, Katan J, 1986. Synergistic effect of metham and solarization in controlling delimited shell spots of peanut pods. *Crop Protection* **5,** 199–202.

Freeman S, Katan J, 1988. Weakening effect on propagules of *Fusarium* by sublethal heating. *Phytopathology* **78,** 1656–1661.

Freeman S, Sztenjnberg A, Shabi E, Katan J, 1990. Long-term effect of soil solarization for the control of *Rosellinia necatrix* in apple. *Crop Protection* **9,** 312–316.

French-Monar RD, Jones JB, Ozores-Hampton M, Roberts PD, 2007. Survival of inoculum of *Phytophthora capsici* in soil through time under different soil treatments. *Plant Disease* **91,** 593–598.

Gamliel A, Katan J, 1991. Involvement of fluorescent pseudomonads and other microorganisms increased growth response of plants in solarized soils. *Phytopathology* **81,** 494–502.

Gamliel A, Katan J, 1993. Suppression of major and minor pathogens by fluorescent pseudomonads in solarized and nonsolarized soils. *Phytopathology* **83,** 68–75.

Gamliel A, Stapleton JJ, 1993. Effect of chicken compost or ammonium phosphate and lettuce growth. *Plant Disease* **77,** 886–891.

Gamliel A, Hadar E, Katan J, 1993. Improvement of growth and yield of *Gypsophila paniculata* by solarizaiton or fumigation of soil or container medium in continuous cropping systems. *Plant Disease* **77**, 933–938.

Gamliel A, Katan T, Younis H, Katan J, 1996. Fusarium wilt and crown rot of basil: Involvement of soilborne and airborne inoculum. *Phytopathology* **86,** 56–62.

Gamliel A, Austerweil M, Kritzman G, 2000a. Non chemical approach to soilborne pest management – organic amendments. *Crop Protection* **19,** 847–853.

Gamliel A, Grinstein A, Zilberg V, Beniches M, Ucko O, Katan J, 2000b. Control of soilborne diseases by combining soil solarization and fumigants. *Acta Horticulturae* **532,** 157–164.

Gamliel A, Skutelski Y, Peretz-Alon Y, Becker E, 2001. Soil solarization using sprayable plastic polymers to control soilborne pathogens in field crops. *Proceedings of the* 8th *Annual International Research Conference on Methyl Bromide Alternatives and Emission Reduction.* San Diego, CA, 10-1–10-3.

Garibaldi A, Gullino ML, 1991. Soil solarization in southern European countries with emphasis in soilborne disease control of protected crops. In: Katan J, DeVay JE, eds. *Soil Solarization.* Boca Raton, FL: CRC Press, 227–235.

Ghini R, 1993. A solar collector for soil disinfestation. *Netherlands Journal of Plant Pathology* **99,** 45–50.

Greenberger A, Yogev A, Katan J, 1987. Induced suppressiveness in solarized soils. *Phytopathology* **77,** 1663–1667.

Grinstein A, Ausher R, 1991. Soil solarization in Israel. In: Katan J, DeVay JE, eds. *Soil Solarization.* Boca Raton, FL: CRC Press, 193–204.

Grinstein A, Hetzroni A, 1991. The technology of soil solarization. In: Katan J, DeVay JE, eds. *Soil Solarization.* Boca Raton, FL: CRC Press, 159–170.

Grinstein A, Orion D, Greenburger A, Katan J, 1979. Solar heating of the soil for the control of *Verticillium dahliae* and *Pralylencus thornei* in potatoes. In: Schippers B, Gams W, eds. *Soilborne Plant Pathogens.* London: Academic Press, 431–438.

Grinstein A, Kritzman G, Hetzroni A, Gamliel A, Mor M, Katan J, 1995. The border effect of soil solarization. *Crop Protection* **14**, 315–320.

Grunzwieg JM, Rabinowitch HD, Katan J, 1993. Physiological and developmental aspects of increased plant growth in solarized soils. *Annals of Applied Biology* **122,** 579–591.

Grunzweig JM, Rabinowitch HD, Katan J, Wonder M, Ben Tal Y, 2000. Involvement of endogenous gibberellins in the regulation of increased tomato shoots growth in solarized soil. *Plant Growth Regulation* **30,** 233–239.

Gupta UK 1996. Integrated management of soil-borne diseases. In: Utkhede RS, Gupta VK, eds. *Management of Soil-Borne Diseases.* New Delhi, India: Kalyani Publishers, 40–64.

Hagan HR, 1933. Hawaiian pineapple field soil temperatures in relation to the nematode *Heterodera radicicola* (Greef). *Soil Science* **36,** 83–95.

Hartz TK, DeVay JE, Elmore CL, 1993. Solarization is an effective soil disinfestation technique for strawberry production. *HortScience* **28,** 104–106.

Hassasn MS, Younis MA 1984. Cucumber cultivation with soil solarization and plastic mulching. *Arab Journal of Plant Protection* **2,** 65–69 (Arabic with English summary).

Jacobsohn R, Greenberger A, Katan J, Levi M, Alon H, 1980. Control of Egyptian broomrape (*Orobanche aegyptiana)* and other weeds by means of solar heating of the soil by polyethylene mulching. *Weed Science* **28,** 312–316.

Kassaby FY, 1985. Solar heating soil for control of damping off diseases. *Soil Biology Biochemistry* **17,** 429–434.

Katan J, 1981. Solar heating (solarization) of soil for control of soilborne pests. *Annual Review of Phytopathology* **19,** 211–236.

Katan J, 1984. The role of soil disinfestation in achieving high production in horticultural crops. *Proceedings of the British Crop Protection Conference,* 1189–1196.

Katan J, 1996. Soil solarization: Integrated control aspects. In: Hall R, ed. *Strategies for Managing Soilborne Plant Pathogens*. St Paul, MN: APS Press, 250–278.

Katan J, DeVay JE, 1991. *Soil Solarization.* Boca Raton, FL: CRC Press.

Katan J, Greenberger A, Alon H, Grinstein A, 1976. Solar heating by polyethylene mulching for the control of diseases caused by soilborne pathogens. *Phytopathology* **66,** 683–689.

Katan J, Fishler G, Grinstein A, 1983. Short-and long-term effects of soil solarization and crop sequence on Fusarium wilt and yield of cotton in Israel. *Phytopathology* **73,** 1215–1219.

Katan J, Grinstein A, Greenberger A, Yarden O, DeVay JE, 1987. The first decade of soil solarization (solar heating): A chronological bibliography. *Phytoparasitica* **15,** 229–255.

Kendrick JB, 1988. A view point of integrated pest management. *Plant Disease* **72,** 647–648.

Klein E, Austerweil M, Katan J, Gamliel A, 2007. Controlled laboratory system to study soil solarization and organic amendment effects on plant pathogens. *Phytopathology* (in press).

Levy N, Elad Y, Katan J, 2005. Induced resistance in plants treated with solarized soil or *Trichoderma. Phytoparasitica* **33,** 288.

Lifshitz R, Tabachnik M, Katan J, Chet I, 1983. The effect of sublethal heating on sclerotia of *Sclerotium rolfsii. Canadian Journal of Microbiology* **29,** 1607–1610.

Lira-Saldivar RH, Salas MA, Cruz J, Coronado A, Hernandez FD, Guerrero E, Gallegos G, 2004. Solarization and goat manure on weed management and melon yield. *Phyton* **205,** 211.

Lopez-Herrera CJ, Verdu-Valiente B, Melero-Vara JM, 1994. Eradication of primary inoculum of *Botrytis cinerea* by soil solarization. *Plant Disease* **78,** 594–597.

Mahrer Y, 1979. Prediction of soil temperature of a soil mulched with transparent polyethylene. *Journal of Applied Meteorology* **18,** 1263–1267.

Mahrer Y, 1991. Physical principles of solar heating of soils by plastic mulching in field and in glasshouses are simulation models. In: Katan J, DeVay JE, eds. *Soil Solarization.* Boca Raton, FL: CRC Press, 75–86.

Mahrer Y, Katan J, 1981. Spatial soil temperature regime under transparent polyethylene mulch: Numerical and experimental studies. *Soil Science* **131,** 82–87.

MBTOC, 2007. *2006 Report of the Methyl Bromide Technical Options Committee (MBTOC).* Nairobi, Kenya: UNON, Publishing section services.

McGovern RJ, McSorley R, 1997. Physical methods of soil sterilization for disease management including soil solarization. In: Rechcigel NA, Rechcigel JE, eds. *Environmentally Safe Approaches to Disease Control.* Boca Raton: CRC Lewis Publishers, 283–313.

McSorley R, Wang KH, Kokallis-Burelle N, 2006. Solarization as an alternative to methyl bromide in Florida floriculture. In: *Annual International Research Conference on Methyl Bromide Alternatives and Emissions Reductions,* 6–9 November 2006, Orlando, Florida, USA.

Minuto A, Spadaro D, Garibaldi A, Gullino ML, 2006. Control of soilborne pathogens of tomato using a commercial formulation of *Streptomyces griseoviridis* and solarization. *Crop Protection* **25,** 468–475.

Munnecke DE, Wilbur W, Darley EF, 1976. Effect of heating or drying on *Armillaria mellae* or *Trichoderma viride* and the relation to survival of *A. mellae* in soil. *Phytopathology* **66,** 1363–1368.

Ndiaye M, Termorshuizen AJ, Van Bruggen AHC, 2007. Combined effects of solarization and organic amendment on charcoal rot caused by *Macrophomina phaseolina* in the Sahel. *Phytoparasitica* **35,** 392–400.

Nico AI, Jimenez-Diaz RM, Castelo L, 2003. Solarization of soils in piles for the control of *Meliodogyne incognita* in olive nurseries in southern Spain. *Plant Pathology* **52,** 770–778.

Otieno W, Termorshuizen A, Jeger M, Othieno CO, 2003. Efficacy of soil solarization, *Trichoderma harzianum,* and coffee pulp amendment against *Armillaria* sp. *Crop Protection* **22,** 325–331.

Ozores-Hampton M, Stansly PA, McSorley R, Obreza TA, 2005. Effects of long-term organic amendments and soil solarization on pepper and watermelon growth, yield, and soil fertility. *HortScience* **40,** 80–84.

Patrıcio FRA, Sinigagli C, Barros BC, Freitas CC, Tessarioli Neto G, Cantarella H, Ghini R, 2006. Solarization and fungicides for the control of drop, bottom rot and weeds in lettuce. *Crop Protection* **25,** 31–38.

Philips AJL, 1990. The effect of soil solarization on sclerotial populations of *Sclerotinia sclerotiorum. Plant Pathology* **39,** 38–43.

Pinkerton JN, Ivors KL, Miller ML, Moore LW, 2000. Effect of soil solarization and cover crops on populations of selected soilborne plant pathogens in Western Oregon. *Plant Disease* **84,** 952–960.

Pivonia S, Cohen R, Levita R, Katan J, 2002. Improved solarization of containerized medium for the control of *Monosporascus* collapse in melon. *Crop Protection* **21,** 907–912.

Porras M, Barrau C, Romero F, 2007. Effects of soil solarization and Trichoderma on strawberry roduction. *Crop Protection* **26,** 782–787.

Porter IJ, Merriman PR, Keane PJ, 1989. Integrated control of pink root *(Pyrenochaeta terrestris)* of onion by dazomet and soil solarization. *Australian Journal of Agricultural Research* **40,** 864–869.

Pullman GS, DeVay JE, Garber RH, Weinhold AR, 1981. Soil solarization effects of Verticillium wilt of cotton and soilborne populations of *Verticillium dahliae*, *Pythium* spp., *Rhizoctonia solani*, and *Thielaviopsis basicola*. *Phytopathology* **71,** 954–959.

Ramirez-Villapudua J, Munnecke DE, 1988. Effect of solar heating and soil amendments of cruciferous residues on *Fusarium oxysporum* f. sp. *conglutinans* and other microorganisms. *Phytopathology* **78,** 289–295.

Raymundo SA, Alcazar J, 1986. Increasing efficiency of soil solarization in controlling root knot nematode by using two layers of plastic mulch. *Nematology* **18,** 626.

Ristaino JB, Perry KB, Lumsden RD, 1991. Effect of solarization and *Gliocladium virens* on scleroia of *Sclerotium rolfsii*, soil microbiota, and the incidence of southern blight of tomato. *Phytopathology* **81,** 1117–1124.

Roe N, Ozores-Hampton M, Stansly PA, 2004. Solarization effects on weed populations in warm climates. *Acta Horticulturae* **638,** 197–200.

Satour MM, Abdel-Rahim MF, Grinstein A, Rabinowitch HD, Katan J, 1989. Soil solarization in onion fields in Egypt and Israel: Short and long term effects. *Acta Horticulturae* **255,** 151–159.

Shlevin E, Saguy S, Mahrer M, Katan J, 2003. Modeling the survival of two soilborne pathogens under dry structural solarization. *Phytopathology* **93,** 1247–1257.

Shlevin E, Mahrer I, Saguy S, Katan J, Siti M, Arbel A, Gamliel A, 2005. Modeling thermal inactivation of soilborne pathogens under thermal structural (dry) and soil (wet) solarization. *Acta Horticulturae* **698,** 167–173.

Siti E, Cohen E, Katan J, Mordechai M, 1982. Control of *Ditylenchus dipasci* in garlic by bulb and soil treatments. *Phytoparasitica* **10,** 93–100.

Sivan A, Chet I, 1993. Integrated control of Fusarium crown and root rot of tomato with *Trichoderma harizianum* in combination with methyl bromide or soil solarization. *Crop Protection* **12,** 380–386.

Skutelsky Y, Gamliel A, Kritzman G, Peretz-Alon I, Becker E, Katan J, 2000. Soil solarization using sprayable plastic polymers to control soilborne pathogens in field crops. *Phytoparasitica* **28,** 269–270.

Stapleton JJ, 2000. Soil solarization in various agricultural production systems. *Crop Protection* **19,** 837–841.

Stapleton JJ, De Vay JE, 1982. Effect of soil solarization on population of selected soilborne microorganisms and growth of deciduous fruit seedlings. *Phytopathology* **72,** 323–326.

Stapleton JJ, De Vay JE, 1984. Thermal components of soil solarization as related to changes in soil and root microflora and increased plant growth response. *Phytopathology* **74,** 255–259.

Stapleton JJ, DeVay JE, 1986. Soil solarization: A nonchemical approach for the management of plant pathogens. *Crop Protection* **5,** 190–198.

Stapleton JJ, Gamliel A, 1993. Feasibility of soil fumigation by sealing soil amended with fertilizers and crop residues containing biotoxic volatiles. *Plant Protection Quarterly* **3,** 10–13.

Stapleton JJ, Garza-Lopez JG, 1988. Mulching of soil with transparent (solarization) and black polyethylene films to increase growth of annual and perennial crops in Southwest Mexico. *Tropical Agriculture* **65,** 29–33.

Stapleton JJ, Heald C, 1991. Management of phytopathogenic nematodes by soil solarization. In: Katan J, De Vay JE, eds. *Soil Solarization*. Boca Raton, FL: CRC Press, 51–59.

Stapleton JJ, Quick J, De Vay JE, 1985. Soil solarization: Effects on soil properties, crop fertilization and plant growth. *Soil Biology Biochemistry* **17,** 369–373.

Stapleton JJ, DeVay JE, Elmore CL, 1998. Soil solarization and integrated management of soilborne pests. *Proceedings of the 2nd International Conference on Soil Solarization*. FAO No. 147, Rome, 657 pp.

Stapleton JJ, Prather TS, Mallek SB, Ruiz TS, Elmore CL, 2002. High temperature solarization for production of weed-free container soils and potting mixes. *HortTechnology* **12,** 697–700.

Stevens C, Khan VA, Okoronkuro T, Tang AY, Wilson MA, Lu J, Brown JE, 1990. Soil solarization and dachtal: Influence on weeds, growth, and root microflora of collards. *HortScience* **25,** 1260–1262.

Stevens C, Kahn VA, Collins D, Rodrigquez-Kabana R, Ploper LD, Adeyeye O, Brown J, Backman P, 1992. Use of soil solarization to reduce the severity of early blight, southern blight and root knot in tomatoes. *Phytopatholgy* **82,** 500.

Stevens C, Kahn VA, Ploper LD, Backman P, Rodrigquez-Kabana R, Collins D, Brown J, Wilson MA, Igwegbe ECK, 1996. Reducing tomato early blight by combining soil solarization and biological control strategies. *Proceedings of the 26th National Agricultural Plastic Congress*, 14–18 June 1996, Atlantic City, NJ, USA, 72–76.

Stevens C, Khan VA, Wilson MA, Brown JE, Collins DJ, 1999. Use of Thermo film-IR single layer and double layer soil solarization to improve solar heating in a cloudy climate. *Plasticulture* **118,** 20–34.

Stevens C, Khan VA, Rodriguez-Kabana R, Proper LD, Backman PA, Collins DJ, Brown JE, Wilson MA, Iqebe ECK, 2003. Integration of soil solarization with chemical, biological and cultural control for the management of soilborne diseases of vegetables. *Plant and Soil* **253,** 493–506.

Tjamos EC, Fravel DR, 1995. Detrimental effects of sublethal heating and *Talaromyces flavus* on microsclerotia of *Verticillium dahliae*. *Phytopathology* **85,** 388–392.

Tjamos EC, Paplomatas EJ, 1988. Long-term effect of soil solarization in controlling Verticillium wilt of globe artichokes in Greece. *Plant Pathology* **37,** 507–515.

Tjamos EC, Biris DA, Paplomatas EJ, 1991. Recovery of olive trees with Verticillium wilt after individual application of soil solarization in established olive orchards. *Plant Disease* **75,** 557–562.

Tjamos EC, Grinstein A, Gamliel A, 1999. Disinfestation of soil and growth media. In: Albajes R, Lodovica Gullino M, Van Lentern JC, Elad Y, eds. *Integrated Pest and Disease Management in Greenhouse Crops*. Dordrecht, The Netherlands: Kluwer Academic Publishers, 130–149.

Toledo RT, 1999. *Fundamental of Food Process Engineering*. Berlin: Springer.

Ulacio-Osorio D, Zavaleta-Meya E, Martinez-Garza A, Petroza-Sandoval A, 2006. Strategies for management of *Sclerotium cepivorum* Berk. in garlic. *Journal of Plant Pathology* **88,** 253–261.

Wang KH, McSorley R, Kokallis-Burelle N, 2006. Effect of solarization and cover cropping on nematodes, weeds and pepper yield. In: *Annual International Research Conference on Methyl Bromide Alternatives and Emissions,* 6–9 November 2006, Orlando, Florida, p. 21.

Wu Y, Perry KB, Ristaino J, 1996. Estimating temperature of mulched and bare soil from meteorological data. *Agricultural and Forest Meteorology* **81,** 299–323.

Yaron D, Regev A, Spector R, 1991. Economic evaluation of soil solarization as disinfestation. In: Katan J, DeVay JE, eds. *Soil Solarization*. Boca Raton, FL: CRC Press, 171–190.

Chapter 11
Plant disease control by nutrient management: sulphur

Silvia Haneklaus, Elke Bloem and Ewald Schnug

Institute for Crop and Soil Science, Julius kühn Institut,
Federal Research Centre for Cultivated Plants, D-38116 Braunschweig, Germany

11.1 Introduction

In 2006, more than 30 000 tonnes of herbicides, insecticides and fungicides were used in Germany; the total number of approved pesticides was 665 in the same year (Anon, 2006). Although the risk of health impairments by combination toxicology has been rated as insignificant (Carpy *et al.*, 2000), contamination of foodstuffs with pesticide residues is a particular threat for the health of children (Goldman & Koduru, 2000). A growing public concern about adverse health effects of pesticides and environmental issues has led to an increasing demand for products from organic farming. However, not only in organic farming, but also in conventional agriculture, there is an increasing interest in alternative methods for disease control. These imply soil tillage operations, plant cultivation practices, plant breeding and plant nutrition, of which the latter will be addressed in this chapter.

11.1.1 Nutrient-induced resistance against diseases

It was Justus von Liebig (1873), who initially identified the nutritional status of a plant as being one of the key factors regulating its susceptibility against diseases. Interactions between mineral elements and plant diseases are well known for essential macro and micro plant nutrients, and aluminium and silicon (Datnoff *et al.*, 2007). Although the significance of individual nutrients for maintaining or promoting plant health saw some interest in the 1960s and 1970s (Bergmann, 1983), research in the field of nutrient-induced resistance mechanisms has been limited by its complexity and its practical significance remained unrecognised due to the availability of effective pesticides. Substantial progress in identifying the underlying triggers and processes of nutrient-induced resistances against plant diseases can be expected since elaborate analytical facilities, all coming under the term 'omics', for example, ionomics and metabolomics, are available nowadays.

The mineral nutrient supply is supposedly the primary and pivotal barrier against infection, which also influences the course of pathogenesis. In general, the greatest benefit to the plant in terms of health can be expected when full nutrient sufficiency is provided; however, the response to a particular nutrient may be different when going from deficiency to sufficiency than from sufficiency to excess (Huber & Haneklaus, 2007). Not only the supply of an individual nutrient, but balanced, crop-specific nutrient ratios, which adequately supply the plant during its development, and in varying environmental conditions, are crucial for improving plant health. Through an understanding of disease interactions with each specific nutrient, the effects on the plant, pathogen and environment can be effectively modified to improve disease control, enhance production efficiency and increase crop quality (Huber & Haneklaus, 2007). In Figure 11.1, a schematic illustration of interacting compartments involved in plant disease is shown.

Under unfavourable conditions, time-dependent interactions between plant, pathogen and environment will result in infection of the plant by fungal pathogens (Figure 11.1). The nutritional status with essential plant nutrients and their balance contribute significantly to the physiological predisposition of a plant against fungal infections. Experimentation in the field of sulphur (S) nutrition and plant disease provided added indications that the S status of a plant is particularly critical during the initial phase of pathogenesis, and if several stress factors coincide, while its significance declines with spread of the disease in the host (Salac *et al.*, 2006; Bloem *et al.*, 2007; Haneklaus *et al.*, 2007a). In this context it should be noted that evaluating the nutritional status of a plant correctly is not as easy as commonly assumed. Nevertheless, such assessment is indispensable for determining the influence of the nutritional status on susceptibility against fungal pathogens. There is not even one exclusive critical nutrient value for each crop, as it depends on the growth conditions, the developmental stage of the plant at sampling, the collected plant part, the determined S species, the targeted yield and mathematical approach for calculating it

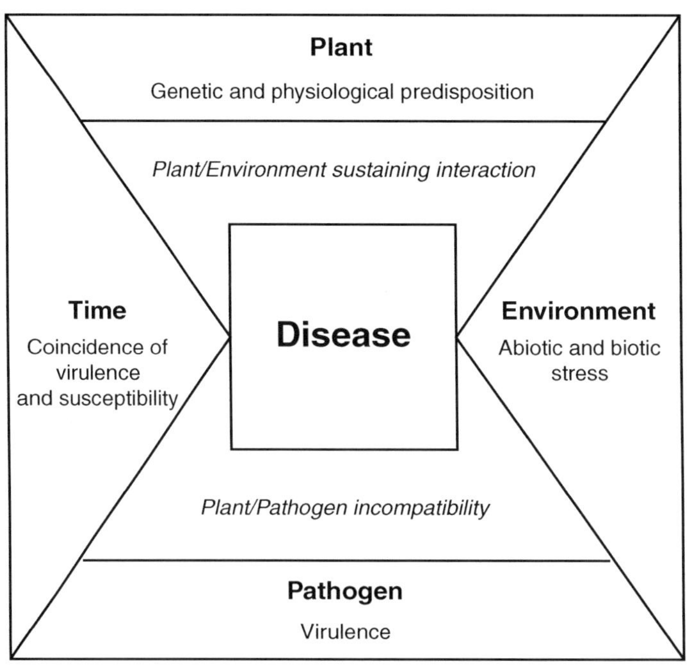

Figure 11.1 Schematic illustration of interacting components involved in plant disease.

synthesised *de novo* and are part of active or induced resistance responses to pathogen attack. A detailed description of resistance mechanisms in plants can be found elsewhere (e.g. Toyoda *et al.*, 2002; Mayer, 2004).

The S-containing metabolites glutathione and glucosinolates can be classed as phytoanticipins, while elemental S depositions, pathogenesis-related (PR) proteins such as plant defensins of class PR-12, thionins of class PR-13 and lipid transfer proteins of class PR-14 (van Loon *et al.*, 1994), and last but not the least low-molecular-weight antibiotics such as camalexin, are part of the active resistance armoury. Van Loon *et al.* (1994) based his classification of PR proteins on amino acid sequences, serological relationships, and/or enzymatic and biological activity. So far, it is unclear which group, phytoanticipins or active resistance responses, the release of H_2S of plants can be assigned to. A relationship between S supply/S status of the plant and H_2S emissions in non-infected plants, which yields fungicidal levels of H_2S, would point to preformed resistance. An interaction between S supply or S status and non-protein cysteine, and H_2S emissions of infected plants as a direct, rapid and efficient response of the plant to the pathogen, would point to a mechanism of active or induced resistance.

There are two main prerequisites for an S-containing compound to be involved in SIR: firstly, a significant relationship between plant S status and metabolite content, and secondly, direct or indirect fungitoxicity of the component.

11.2.2 Relationship between S nutritional status and S metabolite content

Macroscopic S deficiency can occur on all soil types and is generally exacerbated by the following: high yields; soils with light texture, high permeability and low organic matter content; sites poorly connected to capillary ascending, S-rich groundwater; high rate of leaching; reduced root growth and rooting intensity; soil compaction or low temperatures (Haneklaus *et al.*, 2006a). Schnug & Haneklaus (2005) and Haneklaus *et al.* (2006b) provide a detailed description of the symptomatology of S deficiency in agricultural crops. The level of S uptake depends on plant available sulphate in the soil and S fertilisation. The efficacy of S fertilisation to increase the S nutritional status depends on the initial S content (Schnug & Haneklaus, 1994). In the range of severe and moderate S deficiency (Haneklaus *et al.*, 2006a), Paulsen (1999), using an S dose of 40 kg ha^{-1} applied to oilseed rape, obtained an increase in the total S content of younger leaves at the start of stem elongation, of 58–83 μg g^{-1} S per kg S applied. Application of 25 kg ha^{-1} S to winter wheat and winter barley increased the total S content in shoots at stem extension by 52 and 68 μg g^{-1} S per kg S, respectively (Paulsen, 1999).

A significant relationship between S fertilisation and GSH, and glucosinolate content, was found in greenhouse and field experimentation (Haneklaus *et al.*, 2006a). Haneklaus *et al.* (2006a) compared and summarised relevant data from the literature for these metabolites. Under field conditions, the glutathione and glucosinolate content of oilseed rape might increase by up to 64 and 150 nmol g^{-1} dry weight per kg ha^{-1} S, respectively (Haneklaus *et al.*, 2006a). S fertilisation increased the free cysteine content in younger leaves of oilseed rape by up to 8.7 nmol g^{-1} dry weight per kg ha^{-1} S applied (Salac, 2005).

First experiments with field-grown oilseed rape indicate that the release of H_2S is related to S supply (Table 11.1). The highest peak in the release of H_2S was found at start

Table 11.1 H_2S emissions of oilseed rape cultivars differing in susceptibility against *Pyrenopeziza brassicae* in relation to S fertilisation and growth stage (adapted from Bloem *et al.*, 2007, with permission from Wiley-Blackwell).

Soil-applied S ($kg\,ha^{-1}$)[a]	Cultivar[b]	Growth stage[c] (BBCH)	H_2S emission ($pg\,g^{-1}$ d.w. min^{-1})
0	Genotype 15157	55	−3.4
0	*Lion*	55	6.2
0	Genotype 15157	65	−4.6
0	*Lion*	65	0.0
0	Genotype 15157	70	0.8
0	*Lion*	70	4.6
150	Genotype 15157	55	162.9
150	*Lion*	55	105.5
150	Genotype 15157	65	26.9
150	*Lion*	65	14.5
150	Genotype 15157	70	46.3
150	*Lion*	70	4.0

[a]S was applied weekly at rates of 10 $kg\,ha^{-1}$ S during growth before winter and with start of the vegetation period.
[b]*Lion* rated as resistant, genotype 15157 rated as highly susceptible to *Pyrenopeziza brassicae*.
[c]BBCH scale (Strauss *et al.*, 1994).

of flowering (BBCH 55) when plants had received S regularly since sowing (Table 11.1). At full flowering (BBCH 65) and end of flowering (BBCH 70) plants emitted distinctly less H_2S. In comparison, plants that received no S showed very low H_2S emissions. Negative values indicate that H_2S emissions from the soil were greater than from the plants; no H_2S was measured in the ambient air. While the cultivar Lion, which is resistant to the pathogen *Pyrenopeziza brassicae,* showed higher H_2S emissions when no S was applied, the susceptible genotype 15157 emitted distinctly higher rates with S fertilisation (Table 11.1).

No data are available on the influence of S supply on phytoalexin synthesis, so any impact can only be inferred from its influence on their precursors.

11.2.3 Fungitoxicity of S metabolites

PR-proteins and low-molecular-weight antibiotics have a proven fungitoxic effect (Smith, 1982; Vidhyasekaran, 1997). The significance of phytoalexins for disease resistance depends upon the speed and magnitude with which they are synthesised rather than the selective toxicity or selectivity of elicitation (Kuć, 1994).

For S^0, free cysteine and methionine, indirect evidence for their antifungal effect is given by the fact that concentrations were higher in resistant than non-resistant plant tissues (Cooper *et al.*, 1996; Vidhyasekaran, 2000). A distinct increase in cysteine in host tissue and emissions of H_2S was found in the initial phase of pathogenesis (Kruse *et al.*, 2005; Bloem *et al.*, 2007). It might be speculated that enrichment with S^0 affects the pathogen most in the initial phase of development, after adhesion of conidia and germ tube formation, but before penetration by appressoria, since Kassemeyer (2003) found the highest efficiency of foliar-applied S^0 against powdery mildew of grapes at this stage of germling development.

H_2S is described in the literature as being highly fungitoxic (Beauchamp *et al.*, 1984; Carlile *et al.*, 2004), but no critical values are available for H_2S concentrations and duration of fumigation in relation to individual pathogens, which would permit an evaluation of H_2S emissions of plants with view to their role in SIR. First experiments with H_2S fumigation of fungal cultures, however, reveal that its effect is obviously not only related to concentration and duration of fumigation, but also to the taxonomic group of the pathogen (Figure 11.2).

Growth of *S. sclerotiorum* was significantly reduced after 24 and 48 hours at all H_2S concentrations; only at 10 $\mu l\ l^{-1}$ H_2S were differences not significant statistically after 24 hours (Figure 11.2). With increasing H_2S concentration colony growth of *Rhizoctonia solani* was significantly inhibited by the time the fumigation was terminated (Figure 11.2). This effect was also observed three and four days later and was consistent for all durations of fumigation (Yang *et al.*, 2006). An inverse effect of the duration of fumigation was found in that it significantly promoted colony growth (Figure 11.2). The impact was greatest immediately after fumigation, with a mean colony diameter that was 80% higher when the mycelium was fumigated for the longest period of 16 hours (Yang *et al.*, 2006). The effects proved to be consistent for all H_2S concentrations (Yang *et al.*, 2006). Basidiomyceta are known to degrade H_2S and dimethyl sulphide (DMS) by oxidation to sulphate and dimethylsulphoxide (Phae & Shoda, 1991), so that it might be assumed that in this taxonomic group of fungal pathogens, H_2S is not involved in SIR.

Comparing these data with that from H_2S emissions from plants (Table 11.1), it is evident that even an excessive S supply might not increase emissions to a level that they become fungitoxic. However, it is possible that increased H_2S emissions are concentrated locally around the site of infection and thus not yet measurable. Riemenschneider (2006), for instance, has shown that in leaves of *A. thaliana,* H_2S concentrations of 4–10 μM could be detected in the mesophyll.

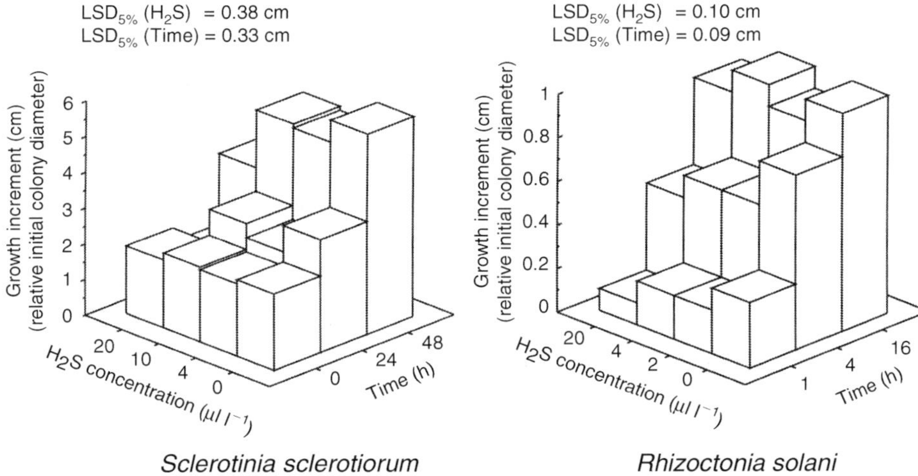

Figure 11.2 Influence of fumigation with rated H_2S concentrations on growth of mycelia of *Sclerotinia sclerotiorum* and *Rhizoctonia solani* (adapted from Yang *et al.* (2006) and Haneklaus *et al.* (2007c), with permisstion from *Phyton* journal).

11.2.4 Changes in S metabolism during pathogenesis

For linking mineral nutrition with disease, different approaches have been used: determination of the influence of fertilisation on disease incidence and/or severity; comparative analysis of mineral concentrations in healthy or resistant tissues and infected or susceptible tissues; analysis of conditions controlling the availability of the nutrient and disease (Huber & Haneklaus, 2007). In general, it can be expected that the greatest benefit to the plant is provided when nutrient supply is in the sufficiency range, but the response to a pathogen may well be different when going from deficiency to sufficiency, or sufficiency to excess (Huber & Haneklaus, 2007).

Raj & Srivastava (1977) showed that total S content in infected tissues of *Brassica juncea* plants was inversely correlated with the pathogenicity of different isolates. While the S content was 50% and 26% lower in plants infected with highly and medium virulent isolates, respectively, a two-fold increase in S level was determined in plants infected with weakly virulent isolates (Raj & Srivastava, 1977). The authors suggested metabolism of host S by pathogens of high and medium virulence, which is, however, limited so that higher plant S concentrations may exert toxic effects on the pathogen. In contrast, weakly virulent isolates lacked the ability to utilise plant S. Earlier experiments by Yarwood & Jacobson (1955) showed similar results.

Glutathione is supposedly a systemic messenger in the hypersensitive response as GSH is rapidly accumulated after fungal attack (Edwards *et al.*, 1991; Foyer & Rennenberg, 2000; Gullner & Kömives, 2001). Participation in anti-oxidative defence reactions is the primary role of GSH in plant tissues, either by direct reactions with reactive oxygen species or through the ascorbate–GSH cycle (Winterbourn & Metodiewa, 1999; Foyer & Rennenberg, 2000). In general, a rapid accumulation of GSH occurs after infection (Vanacker *et al.*, 2000). A positive relationship between GSH content and protection against fungal diseases exists (Gullner & Kömives, 2001), but speed of GSH accumulation might be equally important (Vanacker *et al.*, 2000). Field experiments conducted by Salac *et al.* (2003) revealed that infections of oilseed rape by the black leg fungus (*Leptosphaeria maculans*) led to increased synthesis of GSH, whereby initiation of GSH synthesis appeared to be dependent on some threshold infection. Furthermore, GSH concentration and accumulation rate might not only depend on the availability of its precursor cysteine, but a sufficient sulphate pool in the soil.

Long-term field experiments in Northern Germany and Scotland have shown that an infection by *P. brassicae* may yield both significant increases and decreases of the cysteine and GSH contents in oilseed rape leaf discs (Salac *et al.*, 2005). The authors assumed a connection to the different inoculum pressures in both countries; when inoculum pressure was extraordinarily high, a decline in cysteine and GSH occurred. It is not known whether such a decrease is a transient phenomenon and related to the consumption of these metabolites during defence processes, or else a shift to catabolic processes when infection severity is extremely high (Salac *et al.*, 2005). Kuzniak & Sklodowska (2005) observed such a decrease of the GSH content after successful establishment of the pathogen, together with a decline in ascorbate and associated enzyme activities.

The field trials of Burandt *et al.* (2001) and Bloem *et al.* (2004) suggested a relationship between the activity of H_2S releasing enzymes, S status of the crop and infection with fungal diseases. Fungal infection yielded a higher potential for H_2S release, since

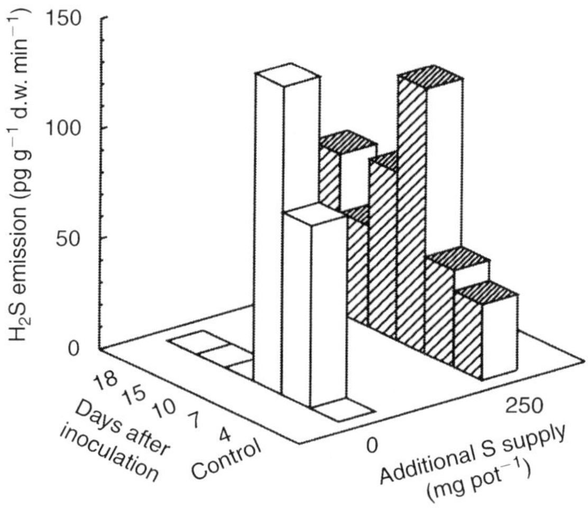

Figure 11.3 Release of H_2S after infection of grapes with *Uncinula necator* (adapted from Bloem *et al.*, 2007 with permission from Wiley-Blackwell).

L-cysteine-desulphhydrase (LCD) activity increased following fungal infection. In a greenhouse experiment with grapes that had been sufficiently supplied with S, Bloem *et al.* (2007) determined H_2S peaks of 140 and 120 pg g^{-1} min^{-1} seven days after inoculation without and with additional S supply (Figure 11.3). Although the peak value was higher when no additional S was supplied, the increase in H_2S emissions in the previous three days was about 20% higher when S was added. At later stages, H_2S emissions were below the minimum amount of total H_2S release when no S was added (255 pg), while afterwards, H_2S emissions remained at a level of about 50 pg g^{-1} min^{-1} (Figure 11.3).

A steep and fast increase during the initial phase of pathogenesis was not only determined for H_2S, but also cysteine, GSH and phytoalexins (see above). Although the sequence, magnitude and efficacy of individual S metabolites involved in the activation and strengthening of plant defences by S fertilisation are not yet known, these could be released in a chain reaction triggered by the pathogen and mediated by the S status of the plant (Figure 11.4).

It seems possible that infection triggers the activation of all effective resistance mechanisms of the host. The accumulation of salicylic acid, which initiates and maintains systemic acquired resistance (Parker, 2000), is linked to plant S metabolism as synthesis of salicylic acid requires coenzyme A (CoASH) in the β-oxidation pathway (Ryals *et al.*, 1996); cysteine is one of the precursors of CoASH synthesis (Luckner, 1990). In the course of accumulation of cysteine it is possible that the release of H_2S is a by-product to keep the level below the phytotoxicity threshold, or a targeted stimulation of LCD activity may result in an accordingly higher H_2S emission that requires elevated cysteine concentrations. A storage experiment with broccoli revealed, however, that enzyme activity was not limiting the release of H_2S, while higher substrate availability coincided with higher emission rates (Derbali & Makhlouf, 1998). Additionally, the free cysteine pool is the precursor for all of the relevant S-containing metabolites putatively involved in SIR.

It was shown previously that an increasing S supply to plants was associated with a higher concentration of cysteine, glutathione, glucosinolates and H_2S, so that these plants

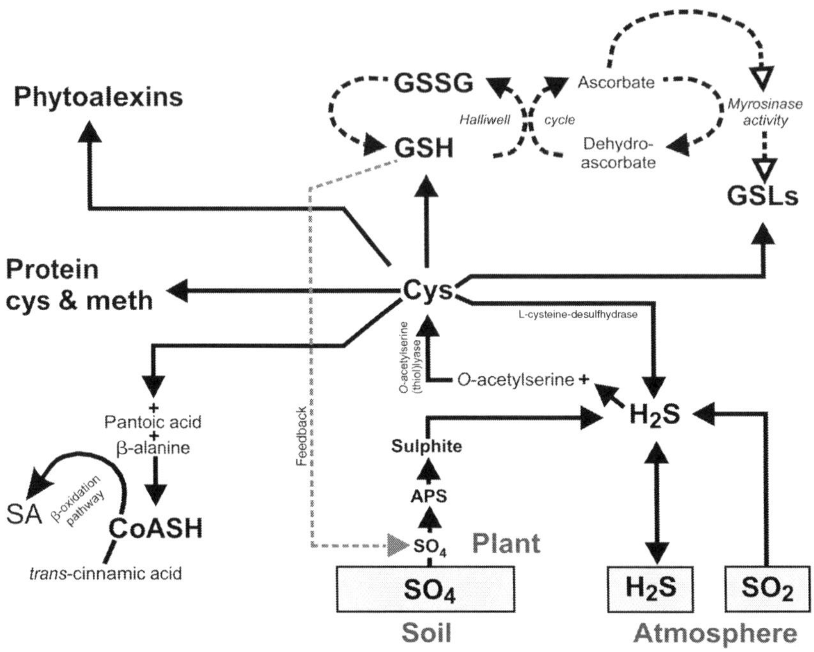

Figure 11.4 Sulphur metabolites and pathways putatively involved in SIR and released in a chain reaction in *Brassica* species (adapted from Haneklaus *et al.*, 2007a). Reprinted by permission from Mineral Nutrition Plant Disease (2007). LE Datnoff, WH Elmer and DM Huber, eds. American Phytopathological Society, St Paul, MN.

with a higher content of phytoanticipins might not only have *a priori* a better protection against pathogens, but also be able to activate resistance mechanisms more rapidly and intensely. In addition, an S supply in the soil that is sufficiently high to instantly satisfy an elevated S demand after fungal attack and which might well exceed the nutrient demand, may play a pivotal role in SIR (Figure 11.4).

11.2.5 Prospective significance of SIR in agriculture

Particularly in S-deficient regions of the world, SIR already plays an unregarded role in enhancing plant health though S fertilisation usually focuses on yield and quality of agricultural crops and its potential regularly remains undiscovered because of the use of effective pesticides.

11.2.5.1 S nutritional status of agricultural crops

With clean air acts coming into force at the start of the 1980s, atmospheric S depositions were reduced drastically and rapidly in Western Europe and declined further in the 1990s after the political transition of Eastern European countries (Haneklaus *et al.*, 2005). On production fields, S deficiency can be retraced to the beginning of the 1980s (Schnug & Pissarek, 1982). Since then, severe S deficiency has become the main nutrient disorder in agricultural crops. Here, the highest efficacy of SIR can be expected.

Important threshold markers for the S nutritional status are: the symptomatological value, which reflects the S concentration below which deficiency symptoms become visible; the critical nutrient value, which stands for the S concentration above which the plant is sufficiently supplied with S for achieving the maximum potential yield or yield reduced by 5%, 10% and 20%; and the toxicological value, which indicates the S concentration above which toxicity symptoms can be observed (Haneklaus *et al.*, 2006a). But there is no one exclusive critical nutrient value for each crop, since it depends on the growth conditions, the developmental stage of the plant at sampling, the collected plant part, the determined S species, the targeted yield and mathematical approach for calculating it. Thus, numerous widely differing thresholds for the S nutritional status of the same crop exist. Based on a profound and robust statistical procedure, critical S concentrations have been calculated for cereals, oilseed rape and sugar beet (Haneklaus *et al.*, 1998, 2006a, 2006b). These threshold values are in good agreement with median values that have been calculated from compiled and categorised available, individual S nutritional data of studies with varying experimental conditions (Haneklaus *et al.*, 2006a). This database should be used for evaluating the S nutritional status and adjusting S fertiliser rates where reliable site-specific values for individual crops are not available.

11.2.5.2 Potency and spectrum of SIR against fungal pathogens

Sulphur fertilisation reduced the disease index for various host/pathogen relationships by 5–50% and 17–35% in greenhouse and field experiments, respectively (Haneklaus *et al.*, 2006c). This finding reflects the as-yet untapped potential of SIR, and only a breakthrough in promoting SIR steadily in production fields will facilitate an integration into current plant protection schemes.

So far, SIR has been demonstrated against biotrophic and necrotrophic pathogens belonging to the Ascomycetes and Basidiomcyetes, as well as Oomycetes. Results obtained for members of the Deuteromycotina have been not conclusive.

In greenhouse experiments, S nutrition significantly influenced the disease index for different host/pathogen combinations (Wang *et al.*, 2003). The disease index decreased 8–20 days after fungal inoculation by 5% for infections of oilseed rape with stem rot (*Sclerotinia sclerotiorum*), by 21% for infection of corn with southern blight (*Bipolaris maydis*) and by 44% for infection of winter wheat with sharp eyespot (*Rhizoctonia cerealis*) (Wang *et al.*, 2003). For infections of cotton with *Fusarium oxysporum* and *Verticillium dahliae,* results proved to be inconsistent (Wang *et al.*, 2003). In other experiments, the S nutritional status of oilseed rape was related to resistance against black leg (*Leptosphaeria maculans*), grey mould blight (*Botrytis cinerea*), and *Phytophthora brassicae* (Dubuis *et al.*, 2005). In field trials with oilseed rape and potatoes, S fertilisation significantly reduced the disease index for light leaf spot (*Pyrenopeziza brassicae*) and black scurf (*Rhizoctonia solani*) (Schnug *et al.*, 1995; Klikocka *et al.*, 2005). Infection rate and disease severity of potato tubers with *R. solani* were reduced, for example, by 41% and 29%, respectively (Klikocka *et al.*, 2005). In the case of oilseed rape, SIR ensured that the yield potential could be more or less fully maintained (Schnug *et al.*, 1995).

11.3 Perspectives in research

A major shortcoming of SIR is that the spectrum and potency are only known for some host/pathogen relationships. Further systematic studies from lab to field level are required to quantify interactions between plant, pathogen and environment.

11.3.1 Experimental design, analytical procedures and interpretation modules

Cross-talk between nutrients is a well-known phenomenon in plant nutrition and in terms of SIR, interactions between S metabolism and other nutrients should be considered in experimentation. A promising approach seems to be cross-talk between manganese (Mn) and S metabolic pathways. Mn is a redox-active nutrient and the oxidation of Mn^{2+} is a characteristic of virulent pathogens (Thompson & Huber, 2007). A sufficient content of reduced Mn may be essential for the activation of resistance mechanisms in plants and it might be hypothesised further that the ascorbate/GSH cycle plays a key role in reducing Mn^{4+} to Mn^{2+}.

Further research on the interactions between S nutritional status and the local/systemic release of H_2S in relation to infection, together with fumigation studies on fungal cultures will finally unravel the role of H_2S in SIR.

For an early, quantitative diagnosis before appearance of fungal mycelia, enzyme-linked immunosorbent assays (ELISA) are required. Visual rating of disease incidence and severity is regularly applied to follow up development of the disease in field experiments. While a close relationship was found between ELISA test and visual scoring (Lind, 1992; Newton & Reglinski, 1993), other authors found visual scoring to be more sensitive than the ELISA test (Delapena & Murray, 1994), or vice versa (Balesdent *et al.*, 1995; ForoughiWehr *et al.*, 1996). For the identification of metabolites and pathways underlying SIR, it is crucial to analyse the infected plant tissue during the initial phase when resistance mechanisms are activated, which is, however, difficult to achieve under field conditions.

Common field experimentation, together with routine statistical treatment of results, might not be appropriate to uncover general interactions between host and pathogen (Salac *et al.*, 2003). Geostatistics has been successfully applied for the evaluation of the spatial distribution and spatial simulation of insects (Ribes-Dasi *et al.*, 2001), the prognosis for the spread of plant viruses (Nelson *et al.*, 1994) and the spatial distribution of plant diseases at plot and field scale (Chellemi *et al.*, 1988; Larkin *et al.*, 1995; Wu *et al.*, 2001; Morgan *et al.*, 2002). With regard to SIR, Salac *et al.* (2003) showed in a field experiment that elevated levels of GSH coincided with a high risk of strong fungal infections of oilseed rape with black leg (*Leptosphaeria maculans*), whereby initiation of GSH synthesis appeared to be dependent on some threshold infection. In contrast, interpretation of analytical results by employing standard statistical procedures revealed no significant effect.

11.4 References

Anon, 1988. Pilzkrankheiten des Rapses. *Aktuelles aus Acker- und Pflanzenbau*. Kiel: Landwirtschaftskammer Schleswig-Holstein, 58–63.

Anon, 2003. Verordnung über das Inverkehrbringen von Düngemitteln, Bodenhilfsstoffen, Kultursubstraten und Pflanzenhilfsmitteln. Düngemittelverordnung – DüMV vom 26. November 2003.

Anon, 2006. *Statistisches Jahrbuch über Ernährung, Landwirtschaft und Forsten der Bundesrepublik Deutschland 2006*. Münster-Hiltrup: Landwirtschafts-Verlag.

Balesdent MH, Desthieux I, Gall C, Robin P, Rouxel T, 1995. Quantification of *Leptosphaeria maculans* growth in cotyledons of *Brassica napus* using ELISA. *Journal of Phytopathology* **143**, 65–73.

Ball L, Accotto GP, Bechtold U, Creissen G, Funck D, Jimenez A, Kular B, Leyland N, Mejia-Carranza J, Reynolds H, Karpinski S, Mullineaux PM, 2004. Evidence for a direct link between glutathione biosynthesis and stress defense gene expression in Arabidopsis. *Plant Cell* **16**, 2448–2462.

Beauchamp RO, Bus JS, Popp JA, Boreiko CJ, Andjelkovich DA, 1984. A critical review of the literature on hydrogen sulfide toxicity. *Critical Reviews of Toxicology* **13**, 25–97.

Beffa T, 1993. Inhibitory-action of elemental sulfur (S-degrees) on fungal spores. *Canadian Journal of Microbiology* **39**, 731–735.

Bergmann H, 1983. *Ernährungsstörungen bei Kulturpflanzen*. Stuttgart: Gustav Fischer Verlag.

Beyer M, Klix MB, Klink H, Verreet J-A, 2006. Quantifying the effects of previous crop, tillage, cultivar and triazole fungicides on the deoxynivalenol content of wheat grain – a review. *Journal of Plant Diseases and Protection* **113**, 241–246.

Bloem E, Haneklaus S, Salac I, Wickenhäuser P, Schnug E, 2007. Facts and fiction about sulphur metabolism in relation to plant–pathogen interactions. *Journal of Plant Biology* **9**, 596–607.

Bloem E, Riemenschneider A, Volker J, Papenbrock J, Schmidt A, Salac I, Haneklaus S, Schnug E, 2004. Sulfur supply and infection with *Pyrenopeziza brassicae* influence L-cysteine desulfhydrase activity in *Brassica napus* L. *Journal of Experimental Botany* **55**, 2305–2312.

Boerner H, 1997. *Pflanzenkrankheiten und Pflanzenschutz*. Stuttgart: Ulmer Verlag.

Bourbos VA, Skoudridakis MT, Barbopoulou E, Venetis K, 2000. Ecological control of grape powdery mildew (*Uncinula necator*). Http://www.landwirtschaft-mlr.baden-wuerttemberg.de/la/lvwo/kongress/SULFUR.html.

Burandt P, Papenbrock J, Schmidt A, Bloem E, Haneklaus S, Schnug E, 2001. Genotypical differences in total sulfur contents and cysteine desulfhydrase activities in *Brassica napus*. *Phyton* **41**, 75–86.

Carlile MJ, Watkinson SC, Gooday GW, 2004. *The Fungi*. Amsterdam: Elsevier Academic Press.

Carpy SA, Kobel W, Doe J, 2000. Health risk of low-dose pesticide mixtures: A review of the 1985–1998 literature on combination toxicology and health risk assessment. *Journal Toxicology and Environmental Health, Part B, Critical Reviews* **3**, 1–25.

Chellemi DO, Rohrbach KG, Yost RS, Sonoda RM, 1988. Analysis of the spatial pattern of plant pathogens and diseased plants using geostatistics. *Phytopathology* **78**, 221–226.

Coleno A, 1987. Role fungicide du soufre. *Proceedings of the International Symposium on Elemental Sulphur in Agriculture*, Nice, Vol. 1, 31–37.

Cook RJ, 1987. Sulphur as a cereal fungicide. *Proceedings of the International Symposium on Elemental Sulphur in Agriculture*, Nice, Vol. 2, 417–424.

Cooper RM, Williams JS, 2004. Elemental sulphur as an induced antifungal substance in plant defence. *Journal of Experimental Botany* **55**, 1947–1953.

Cooper RM, Resende MLV, Flood J, Rowan MG, Beale MH, Potter U, 1996. Detection and cellular localization of elemental sulphur in disease-resistant genotypes of *Theobroma cacao*. *Nature* **379**, 159–162.

Datnoff LE, Elmer WH, Huber DM, 2007. *Mineral Nutrition and Plant Diseases*. St. Paul, MN: APS Press.

Delapena RC, Murray TD, 1994. Identifying wheat genotypes resistant to eyespot disease with a beta-glucuronidase-transformed strain of *Pseudocercosporella herpotrichoides*. *Phytopathology* **84**, 972–977.

Derbali E, Makhlouf J, 1998. Biosynthesis of sulphur volatile compounds in broccoli seedlings under anaerobic conditions. *Postharvest Biology and Technology* **13**, 191–204.

Dubuis P-H, Marazzi C, Stadler E, Mauch F, 2005. Sulphur deficiency causes a reduction in antimicrobial potential and leads to increased disease susceptibility of oilseed rape. *Journal of Phytopathology* **153**, 27–36.

Edwards R, Blount JW, Dixon RA, 1991. Glutathione and elicitation of the phytoalexin response in legume cultures. *Planta* **184**, 403–409.

Foyer CH, Rennenberg H, 2000. Regulation of glutathione synthesis and its role in abiotic and biotic stress defence. In: Brunold C, Rennenberg H, De Kok LJ, Stulen I, Davidian J-C, eds. *Sulfur Nutrition and Sulfur Assimilation in Higher Plants*. Berne, CH: Paul Haupt Publishers, 127–153.

ForoughiWehr B, Zuchner S, Rabenstein F, 1996. Enzyme-linked immunosorbent assay for detection *Rhynchosporium secalis* (Oud) J J. Davis in winter barley. *Journal of Plant Disease and Protection* **103**, 267–271.

Goldmann LR, Koduru S, 2000. Chemicals in the environmental and developmental toxicity to children: A public health and policy perspective. *Environmental Health Perspectives* **108**, 443–448.

Gullner G, Kömives T, 2001. The role of glutathione and glutathione-related enzymes in plant–pathogen interactions. In: Grill D, Tausz M, De Kok LJ, eds. *Significance of Glutathione in Plant Adaptation to the Environment*. Dordrecht: Kluwer Academic Publishers, 202–239.

Haneklaus S, Knudsen L, Schnug E, 1998. Minimum factors in the mineral nutrition of field grown sugar beets in northern Germany and eastern Denmark. *Aspects of Applied Biology* **52**, 57–64.

Haneklaus S, Walker KC, Schnug E, 2005. A chronicle of sulfur research in agriculture. In: Saito K, DeKok LJ, Stulen I, Hawkesford MJ, Schnug E, Sirko A, Rennenberg H, eds. *Sulfur Transport and Assimilation in Plants in the Post Genomic Era.* Leiden: Backhys Publishers, 249–256.

Haneklaus S, Bloem E, Schnug E, De Kok L, Stulen I, 2006a. Sulphur. In: Barker AV, Pilbeam DJ, eds. *Handbook of Plant Nutrition.* Boca Raton, FL: CRC Press, 183–238.

Haneklaus S, Bloem E, Schnug E, 2006b. Sulphur interactions in crop ecosystems. In: Hawkesford MJ, DeKok LJ, eds. *Sulfur in Plants – An Ecological Perspective.* Dordrecht: Springer, 17–58.

Haneklaus S, Bloem E, Schnug E, 2006c. Disease control by sulphur induced resistance. *Aspects of Applied Biology* **79**, 221–224.

Haneklaus S, Bloem E, Schnug E, 2007a. Sulfur and plant disease. In: Datnoff L, Elmer W, Huber D, eds. *Mineral Nutrition and Plant Diseases.* St. Paul, MN: APS Press, 101–118.

Haneklaus S, Bloem E, Funder U, Schnug E, 2007b. Effect of foliar-applied elemental sulphur on Fusarium infections in barley. *FAL Agricultural Research* **57(3),** 213–218.

Haneklaus S, Bloem E, De Kok LJ, Yang Z, Wang S, Schnug E, 2007c. The potential of sulfur induced resistance against plant diseases of oilseed rape. *Proceedings of the 12th International Rapeseed Congress 'Sustainable Development in Cruciferous Oilseed Crops Production.* Wuhan, China, March 26–30, 2007. Monmouth Junction: Science Press, USA, 43–46.

Heitefuss R, 1975. *Pflanzenschutz.* Stuttgart: Thieme Verlag.

Hell R, Kruse C, 2006. Sulfur in biotic interactions of plants. In: Hawkesford MJ, DeKok LJ, eds. *Sulfur in Plants – An Ecological Perspective.* Dordrecht: Springer-Verlag, 197–224.

Hoy MA, 1987. Sulfur as a control agent for pest mites in agriculture. *Proceedings of the International Symposium on Elemental Sulphur in Agriculture*, Nice, Vol. 1, 51–61.

Huber DM, Haneklaus S, 2007. Managing nutrition to control plant disease. *FAL Agricultural Research* (in press).

Jost R, Altschmied L, Bloem E, Bogs J, Gershenzon J, Hähnel U, Hänsch R, Hartmann T, Kopriva S, Mendel RR, Papenbrock J, Reichelt M, Rennenberg H, Schnug E, Schmidt A, Textor S, Tokuhisa J, Wachter A, Rausch T, Hell R, 2005. Expression profiling of metabolic genes in response to methyl jasmonate reveals regulation of genes of primary and secondary sulfur-related pathways in *Arabidopsis thaliana*. *Photosynthesis Research* **86**, 491–508.

Kassemeyer H-H, 2003. Untersuchungen zur Entwicklung des Myzels und der Konidien von *Uncinula necator* und deren Einfluß auf die Epidemiologie. http://www.infodienst.bwl.de/mlr/Forschung/2000/Pflanzenschutz.

Klikocka H, Haneklaus S, Bloem E, Schnug E, 2005. Influence of sulfur fertilization on infections of potato tubers (*Solanum tuberosum*) with *Rhizoctonia solani* and *Streptomyces scabies*. *Journal of Plant Nutrition* **28**, 819–833.

Kruse C, Kopp B, Hartmann M, Jost R, Hell R, 2005. Sulfate-based performance of *Arabidopsis thaliana* in response to pathogen infection. In: Saito K, DeKok LJ, Stulen I, Hawkesford MJ, Schnug E, Sirko A, Rennenberg H, eds. *Sulfur Transport and Assimilation in Plants in the Post Genomic Era.* Leiden: Backhys Publishers, 217–220.

Kuć J, 1994. Relevance of phytoalexins – a critical review. *Acta Horticulturae* **381**, 526–539.

Kuzniak E, Sklodowska M, 2005. Fungal pathogen-induced changes in the anti-oxidant systems of leaf peroxisomes from infected tomato plants. *Planta* **222**, 192–200.

Larkin RP, Gumpertz ML, Ristaino JB, 1995. Geostatistical analysis of *Phytophthora* epidemic development in commercial bell pepper field. *Phytopathology* **85**, 191–203.

Lind V, 1992. Measuring the severity of eyespot disease induced by *Pseudocercosporella herpotrichoides* (Fron) Deighton in wheat cultivars at different growth-stages. *Plant Breeding* **108**, 202–209.

Luckner M, 1990. *Secondary Metabolism in Microorganisms, Plants, and Animals*. Jena: VEB Gustav Fischer Verlag.

Mayer AM, 2004. Resistance to herbivores and fungal pathogens: Variations on a common theme? A review comparing the effect of secondary metabolites, induced and constitutive, on herbivores and fungal pathogens. *Israel Journal of Plant Sciences* **52**, 279–292.

Mitchell AF, Walters DR, 2004. Potassium phosphate induces systemic protection in barley to powdery mildew infection. *Pest Management Science* **60**, 126–134.

Morgan GD, Stevenson WR, MacGuidwin AE, Kelling KA, Binning LK, Zhu J, 2002. Plant pathogen population dynamics in potato fields. *Journal of Nematology* **34**, 189–193.

Mortvedt JJ, Berger KC, Darling HM, 1963. Effects of manganese and copper on the growth of *Streptomyces scabies* and the incidence of potato scab. *American Potato Journal* **40,** 96–102.

Nelson MR, Felix-Gastelum R, Orum TV, Stowell LJ, Myers DE, 1994. Geographic information systems and geostatistics in the design and validation of regional plant virus management programs. *Phytopathology* **84**, 898–905.

Newton AC, Reglinski T, 1993. An enzyme-linked-immunosorbent-assay for quantifying mildew biomass. *Journal of Plant Disease and Protection* **100**, 176–179.

Parker JE, 2000. Signalling in plant disease resistance. In: Dickinson M, Beynon J, eds. *Molecular Plant Pathology. Annual Plant Reviews* **4**, 144–174.

Paulsen HM, 1999. Produktionstechnische und oekologische Bewertung der landwirtschaftlichen Verwertung von Schwefel aus industriellen Prozessen. *FAL Agricultural Research* **197** (special issue).

Pezet R, Pont V, 1977. Elemental sulfur – accumulation in different species of fungi. *Science* **196**, 428–429.

Phae CG, Shoda M, 1991. A new fungus which degrades hydrogen sulfide, methanethiol, dimethyl sulfide and dimethyl disulfide. *Biotechnology Letters* **13**, 375–380.

Rai JN, Srivastava SK, 1977. Changes in the total sulphur content of *Brassica juncea* plants infected with isolates of *Macrophomina phaseolina* and its possible relationship with the virulence of isolates. *Indian Phytopathology* **30**, 486–488.

Reuveni M, 2001. Activity of trifloxystrobin against powdery and downy mildew diseases of grapevines. *Canadian Journal of Plant Pathology* **23**, 52–59.

Ribes-Dasi M, Albajes R, Sarasua MJ, Avilla J, 2001. The use of geostatistics to study the spatial distribution of *Cydia pomonella* and *Pandemis heparana* in Lleida (Spain). *Integrated Fruit Production* **24**, 185–188.

Riemenschneider A, 2006. Isolation and characterization of cysteine-degrading and H_2S-releasing proteins in higher plants. PhD Thesis, University of Hanover, Germany.

Ryals JA, Neuenschwander UH, Willits MG, Molina A, Steiner H-Y, Hunt MD, 1996. Systemic acquired resistance. *The Plant Cell* **8**, 1809–1819.

Salac I, 2005. Influence of the sulphur and nitrogen supply on S metabolites involved in sulphur induced resistance (SIR) of *Brassica napus* L. PhD Thesis, *FAL Agricultural Research* **277** (special issue), 134 pp.

Salac I, Haneklaus S, Gassner A, Bloem E, Schnug E, 2003. Applicability of geostatistics for identifying spatial patterns indicating the risk of fungal infection and factors involved in sulphur induced resistance (SIR) of oilseed rape. *Phyton* **44**, 215–229.

Salac I, Haneklaus S, Bloem E, Booth EJ, Sutherland KG, Walker KC, Schnug E, 2005. Sulfur nutrition and its significance for crop resistance – a case study from Scotland. *FAL- Agricultural Research* **283** (special issue), 111–119.

Salac I, Haneklaus S, Bloem E, Booth E, Sutherland K, Walker K, Schnug E, 2006. Influence of sulfur fertilization on sulfur metabolites, disease incidence and severity of fungal pathogens in oilseed rape in Scotland. *FAL Agricultural Research* **56**, 1–4.

Sasaki-Sekimoto Y, Taki N, Obayashi T, Aono M, Matsumoto F, Sakurai N, Suzuki H, Hirai MY, Noji M, Saito K, Masuda T, Takamiya K, Shibata D, Ohta H, 2005. Coordinated activation of metabolic pathways for antioxidants and defence compounds by jasmonates and their roles in stress tolerance in Arabidopsis. *Plant Journal* **44**, 653–668.

Schnug E, 1997. Significance of sulphur for the nutritional and technological quality of domesticated plants. In: Cram WJ, De Kok LJ, Stulen I, Brunold C, Rennenberg H, eds. *Sulfur Metabolism in Higher Plants – Molecular, Ecophysiological and Nutritional Aspects*. Leiden: Backhuys Publishers, 109–130.

Schnug E, Haneklaus S, 1994. Sulphur deficiency in *Brassica napus* – biochemistry, symptomatology, morphogenesis. *FAL – Agricultural Research* **144** (special issue).

Schnug E, Haneklaus S, 2005. Sulphur deficiency symptoms in oilseed rape (*Brassica napus* L.) – the aesthetics of starvation. *Phyton* **45**, 79–95.

Schnug E, Pissarek H-P, 1982. Kalium und Schwefel, Minimumfaktoren des schleswig-holsteinischen Rapsanbaus. *Kali-Briefe (Buentehof)* **16**, 77–84.

Schnug E, Booth E, Haneklaus S, Walker KC, 1995. Sulphur supply and stress resistance in oilseed rape. *Proceedings of the 9th International Rapeseed Congress,* Cambridge, Vol. 1, 229–231.

Smith DA, 1982. Toxicity of phytoalexins. In: Bailey JA, Mansfield JW, eds. *Phytoalexins*. Glasgow: Blackie, 218–252.

Strauss R, Bleiholder H, van den Boom T, Buhr L, Hack H, Hess M, Klose R, Meier U, Weber E, 1994. Einheitliche Codierung der phaenologischen Entwicklungsstadien mono- und dikotyler Pflanzen. Ciba-Geigy AG, Basel, 56 pp.

Thompson IA, Huber DM, 2007. The role of manganese. In: Datnoff LE, Elmer WE, Huber DM, eds. *Mineral Nutrition and Plant Disease*. St. Paul, MN: APS Press.

Toyoda K, Collins NC, Takahashi A, Shirasu K, 2002. Resistance and susceptibility of plants to fungal pathogens. *Transgenic Research* **11**, 567–582.

Vanacker H, Carver TLW, Foyer C, 2000. Early H_2O_2 accumlation in mesophyll cells leads to induction of glutathione during the hypersensitive response in the barley–powdery mildew interaction. *Plant Physiology* **123**, 1289–1300.

van Loon LC, Pierpoint WS, Boller T, Conejero V, 1994. Recommendations for naming plant pathogenesis-related proteins. *Plant Molecular Biology Reporter* **12**, 245–264.

Vidhyasekaran P, 1997. *Fungal Pathogenesis in Plants and Crops*. Boca Raton, FL: CRC Press.

Vidhyasekaran P, 2000. *Physiology of Disease Resistance in Plants*. Vol. II. Boca Raton, FL: CRC Press.

Vlitos AJ, Hooker WJ, 1951. The influence of sulphur on populations of *Streptomyces scabies* and other streptomycetes in peat soil. *American Journal of Botany* **38,** 678–683.

Walters DR, Walsh D, Newton A, Lyon G, 2005. Induced resistance for plant disease control: Maximizing the efficacy of resistance elicitors. *Phytopathology* **95**, 1368–1373.

Wang J, Zhang J, Ma Y, Wang Li, Yang L, Shi S, Liu L, Schnug E, 2003. Crop resistance to diseases as influenced by sulphur application rates. *Proceedings of the 12th World Fertilizer Congress*, August 3–9, 2001, Beijing, China, 1285–1296.

WHO, 2003. *WHO Guidelines on Good Agricultural and Collection Practices (GACP) for Medicinal Plants*. Geneva: WHO.

Winterbourn CC, Metodiewa D, 1999. Reactivity of biologically important thiol compounds with superoxide and hydrogen peroxide. *Free Radical Biology and Medicine* **27**, 322–328.

Wu BM, Van Bruggen AHC, Subbarao KV, Pennings GGH, 2001. Spatial analysis of lettuce downy mildew using geostatistics and geographic information systems. *Phytopathology* **91**, 134–142.

Yang Z, Haneklaus S, De Kok LJ, Schnug E, Singh BR, 2006. Effect of H_2S and DMS on growth and enzymatic activities of *Rhizoctonia solani* and its implications for sulfur-induced resistance (SIR) of agricultural crops. *Phyton* **46**, 55–70.

Yarwood CE, Jacobson L, 1955. Accumulation of chemicals in diseased areas of leaves. *Phytopathology* **45**, 43–48.

Chapter 12
Control of plant disease by disguising the leaf surface

Dale Walters

Scottish Agricultural College, Edinburgh, UK

12.1 Introduction

Parasitism is essentially a nutritional relationship and if pathogens are to gain access to the nutrients available in the apoplast and within plant cells, they must first get through the outer layers of the plant. On leaf surfaces, the first layer the pathogen will encounter is the cuticle, the insoluble polymeric compounds of which constitute the main physical obstacle to penetration by fungal pathogens (Royle, 1975). Plant surfaces may also be a determinant in the recognition and attachment of fungal spores. For example, evidence provided by Wright *et al.* (2002) suggests that chemical interactions on the barley leaf surface are required for adhesion of powdery mildew conidia, with consequences for subsequent development of the powdery mildew germling. Interestingly, for rust fungi, germinated uredospores 'sense' the leaf surface and several steps are involved in germ tube tropism on such surfaces: adhesion, directional growth, appressorium formation over stomatal openings, directional emergence of infection pegs and adherence of haustorial mother cells (Figure 12.1) (Wynn & Staples, 1981). This germ tube tropism is greatly influenced by surface features of the leaf and any alteration of the leaf surface can alter topography and influence the tropism of uredospore germ tubes. Thus, removal of epicuticular waxes may induce tropism-related mistakes and result in reduced infection frequencies (Wynn, 1981). More recent work (Collins *et al.*, 2001) provided evidence for the involvement of both topographical and chemical signals associated with the wheat stomatal complex, in the induction of appressoria by the stem rust fungus *Puccinia graminis* f. sp. *tritici*.

The importance of leaf surface features in the early development and establishment of foliar pathogens suggests that interference with leaf topography will disrupt pathogen development and lead to reductions in infection. Research on a variety of agents which coat the leaf surface has shown that interference with leaf topography can indeed lead to reductions in pathogen infection. The following sections of this chapter deal with the various approaches that have been used to coat or disguise leaf surfaces in order to control foliar pathogens.

12.2 Controlling disease using film-forming polymers

Film-forming polymers are widely used in agriculture and horticulture as antitranspirants and as spray adjuvants. Their main use as adjuvants is as filming agents to reduce weathering

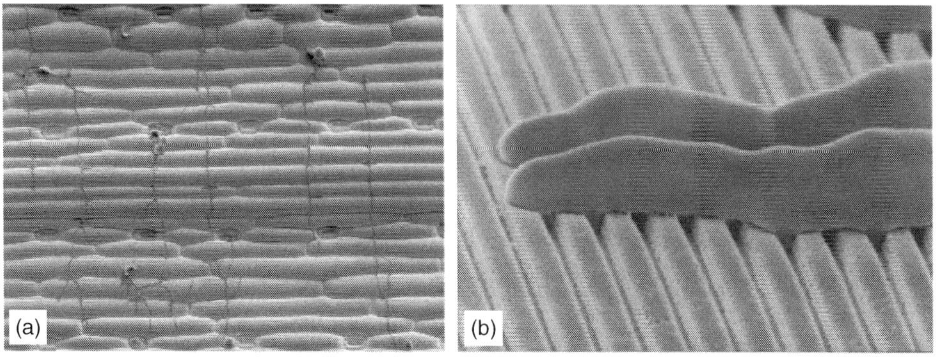

Figure 12.1 (a) Scanning electron micrograph of germ tubes growing from uredospores (u) of *Puccinia graminis* on an inert replica of a wheat leaf. Growth of the germ tubes (seen as dark, narrow lines) is orientated at right angles to the pattern of ridges and grooves of the leaf surface cells. Lines of stomata (s) occur at intervals across the leaf surface replica. (b) Scanning electron micrograph of germ tube tips of *Puccinia graminis* growing perpendicular to precisely spaced ridges and grooves of a polystyrene replica of a microfabricated silicon wafer. The germ tubes are about 4 μm in diameter; the lower region of their tips is flattened against the replica, presumably enabling them to sense the topography. Courtesy of Professor Nick Read, University of Edinburgh.

and extend pesticide efficacy, and as stickers or spreaders to improve distribution and adherence of agrochemicals (Backman, 1978). Film-forming polymers used as antitranspirants include waxes, silicons and a variety of plastic polymers. When used in this role, they form a film over the stomata, increasing resistance to water vapour loss (Gale & Hagan, 1966) and are used particularly on seedlings and transplants to decrease water stress and wilting and to improve water use efficiency in arid conditions (Quarles, 1991).

The first report of disease control provided by an antitranspirant was made nearly 50 years ago. During field trials of the effect of an antitranspirant on the water balance of sugar beet, Gale & Poljakoff-Mayber (1962) found that the incidence of powdery mildew was reduced on treated plots. A number of studies on film-forming polymers have found several to be effective in controlling foliar pathogens of a variety of plants, including cereals (Ziv & Frederiksen, 1983, 1987; Walters, 1992), vegetables (Osswald *et al.*, 1984; Han, 1990) and fruit (Han, 1990). Elad *et al.* (1989) examined the effects of four film-forming polymers (Vapor Gard® [1-p-menthene], Wilt Pruf® [β-pinene], Safe Pack® [a wax emulsion] and Colfix® [40% polyvinyl]) and a number of commercial fungicides, including fenarimol, on control of powdery mildew on cucumber. In pot experiments they found that Vapor Gard and Wilt Pruf reduced powdery mildew on cucumber by up to 82% and 55%, respectively, while under cover, Safe Pack and the fungicide fenarimol reduced infection by up to 67% and 96%, respectively (Elad *et al.*, 1989). Cucumber powdery mildew was controlled most effectively by a mixture of Safe Pack and fenarimol. According to Elad *et al.* (1989), film-forming polymers can only form an effective barrier on the leaf surface if applied at concentrations greater than 3%. Since powdery mildew infection of cucumber was reduced by concentrations of polymers between 0.5% and 3%, they suggested that the polymers were exerting a direct effect on the pathogen. In fact, their studies showed that film-forming polymers reduced germination of powdery mildew conidia when used at concentrations as low as 0.5% (Elad *et al.*, 1989). Similar results were obtained with the grey mould pathogen *Botrytis cinerea*. Here, the film-forming polymers Wilt Pruf and Colfix reduced germination of conidia by 33% and 74%, respectively, when used at a concentration of 0.2% (Elad *et al.*, 1990). Working on lily leaf blight caused by the pathogen *B. elliptica*, Hsieh & Huang (1999) demonstrated that six film-forming, cationic polyelectrolytes controlled disease under both glasshouse and field conditions. Some of the most active compounds were found to reduce conidial germination and inhibit germ tube growth and interestingly, to suppress esterase production by germ tubes (Hsieh & Huang, 1999). Esterases are thought to be important for infection by this pathogen (Doss *et al.*, 1988) and Hsieh & Huang (1999) observed a significant correlation between esterase production by germ tubes and lesion number on inoculated tissue. The authors suggested that the fungicidal activity of the polyelectrolytes was due, in part, to suppression of esterase production by the fungus.

In a study of the effects of a number of biocompatible products in controlling cucurbit powdery mildew caused by the pathogen, *Sphaerotheca fusca*, McGrath & Shishkoff (1999) found that JMS Stylet-Oil was particularly effective. JMS Stylet-Oil suppressed powdery mildew on winter squash and muskmelon and increased yield compared with non-treated plants under field conditions, although it did not provide the degree of full-season control obtained with a conventional fungicide programme (McGrath & Shishkoff, 1999). Interestingly, this product was also effective in controlling powdery mildew on grape clusters and leaves (Wilcox & Riegel, 1996, 1997) and powdery mildew on bell pepper (Smith, 1996). Although JMS Stylet-Oil is a medium viscosity mineral oil, whether it forms a film on plant surfaces and indeed, its precise mode of action, are not yet known.

In a comparison of three different film-forming polymers, Ethokem® (ethoxylated tallow amine), Bond® (synthetic latex) and Vapor Gard, Sutherland & Walters (2002) found that all the three compounds provided significant control of powdery mildew infection on barley under controlled conditions (Figure 12.2). In this work, Bond and Vapor Gard

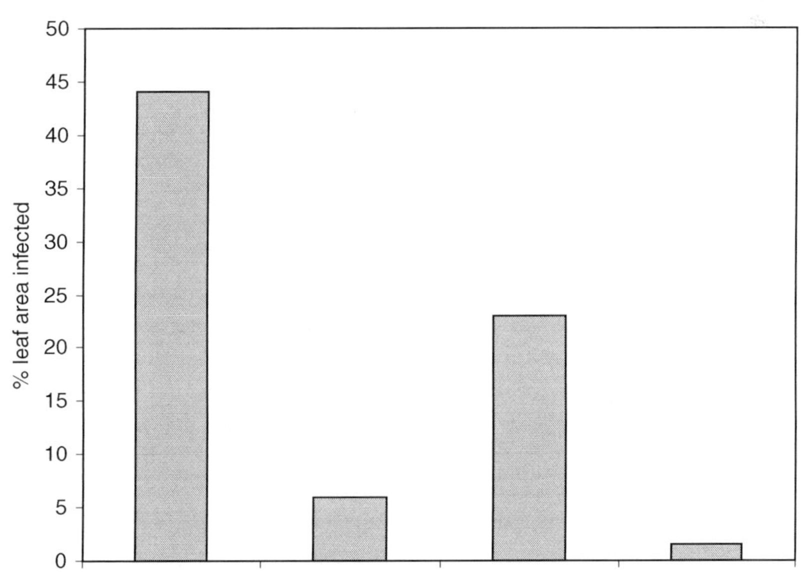

Figure 12.2 Effects of Ethokem, Bond and Vapor Gard on powdery mildew infection of barley. All treatments are significantly different from the untreated control at $P < 0.01$. Data taken from Sutherland & Walters (2002), with kind permission of Springer Science and Business Media.

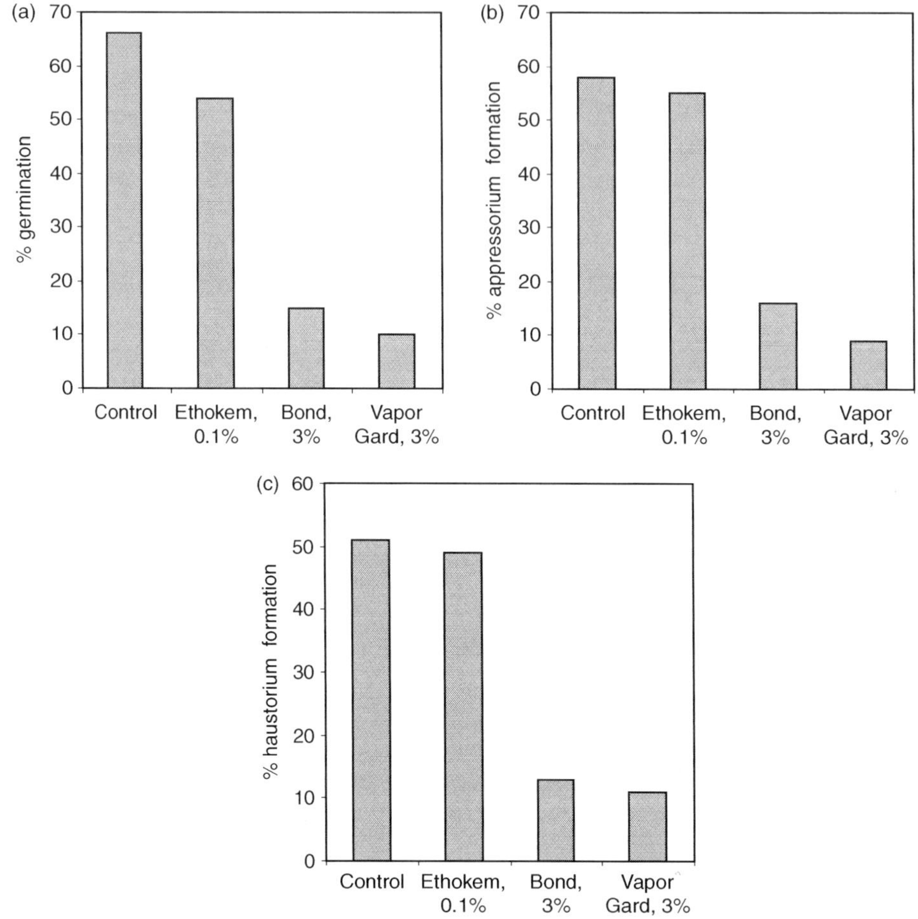

Figure 12.3 Effects of film-forming polymers on (a) spore germination, (b) appressorium formation and (c) haustorium formation. Bond and Vapor Gard treatments significantly different from the control at $P < 0.01$. Data taken from Sutherland & Walters (2002), with kind permission of Springer Science and Business Media.

altered germling development on the leaf surface, reducing germination of conidia, as well as the subsequent formation of appressoria and haustoria (Figure 12.3) (Sutherland & Walters, 2002). However, the film-forming polymers performed less well under field conditions. Thus, Bond used as a 3% spray reduced mildew infection by 63%, in comparison with the commercial fungicide cyproconazole, which reduced mildew infection by 84% (Sutherland & Walters, 2002). In a study to determine whether film-forming polymers exerted a direct effect on fungal pathogens, Sutherland & Walters (2001) found that Ethokem and Bond completely inhibited linear growth of *Pyricularia oryzae* when incorporated into the growth medium at 1% and 2%, and strongly inhibited linear growth of *Pyrenophora avenae* at the same concentrations. Less of an effect on linear growth of these fungi was obtained with Vapor Gard, although all three polymers decreased cell lengths and led to gross changes in hyphal morphology, including swollen, shortened cells, granulation of cytoplasm, increased branching and collapsed empty cells (Sutherland & Walters, 2001). The mechanism(s) by which Bond and Vapor Gard alter hyphal morphology is not known. However, Ethokem is a cationic surfactant and as such may be active at the cell membrane. Wade *et al.* (1993) reported that cationic surfactants such as ethoxylated tallow amines increase plasma membrane permeability in plants. The possibility that Ethokem may exert its effects on hyphal morphology by altering the permeability of fungal membranes is worthy of investigation.

From the above, it is unclear whether the film-forming polymers act either as a physical barrier or to disguise the leaf surface. In fact, two mechanisms have been proposed to account for the effects of film-forming polymers in reducing fungal infection on leaf surfaces (Gale & Poljakoff-Mayber, 1962; Ziv & Frederiksen, 1983; Osswald *et al.*, 1984). First, it has been suggested that leaf surfaces coated with polymers are hydrophobic, leading to low water potential at infection sites (Gale & Hagan, 1966). Second, various workers have suggested that coated leaf surfaces may be impenetrable due to the thickness, hardness, or resistance to enzymic attack of the film-forming polymer (e.g. Ziv & Frederiksen, 1983). Working on the effects of film-forming polymers on development of the rust *Puccinia recondita* f. sp. *tritici* on wheat, Zekaria-Oren & Eyal (1991) found that polymers applied prior to inoculation had a greater effect on rust infection than compounds applied post-inoculation. They found that increasing the concentration of the film-forming polymers led to a progressive reduction in infection intensity, suggesting that efficacy was related to thickness and uniformity of the coat on the leaf surface. Using fluorescence microscopy and scanning electron microscopy, these workers found that surfaces coated with film-forming polymers interfered with fungal penetration of the leaf (Zekaria-Oren & Eyal, 1991). They observed that both the orientation of the germinating uredospore towards the stomata and the formation of appressoria were altered on coated surfaces. The leaf surface is known to provide various stimuli that orient the germinating uredospore towards a stoma (Wynn, 1981; Wynn & Staples, 1981) and induce appressorium formation (Collins *et al.*, 2001). Zekaria-Oren & Eyal (1991) suggested that the significant reduction in appressorium formation and the distribution of appressoria on the coated leaf surface was associated with disruption of mechanisms responsible for orientation of the germinating uredospores towards stomata and triggering appressorium formation. In their work on the effects of film-forming polymers on powdery mildew infection of barley, Sutherland & Walters (2002) suggested that the polymers might have disrupted the ‘first touch’ phenomenon described by Neilsen *et al.* (2000). The ‘first touch’ of conidia of *Blumeria graminis* on the barley leaf surface is associated with conidial uptake of anionic, low-molecular-weight materials before germination. Neilsen *et al.* (2000) suggest that this could be a mechanism for recognition of the host and determination of the direction of growth of the emerging germ tube toward the leaf surface. There is an urgent need for work on the mechanisms by which film-forming polymers reduce fungal infection on leaf surfaces. Since the fungal pathogen on a leaf surface will capitalise on any break or weakness in the film produced by the polymer, research on polymers with increased stretching or ‘self-healing’ properties would also be useful. Interestingly, polymer composites with ‘self-healing’ properties have been reported (White *et al.*, 2001; Brown *et al.*, 2004), although such polymers have been designed for military and medical purposes and are likely to be too costly and impractical for crop protection purposes.

12.3 Particle films as agents for control of plant diseases

The concept of hydrophobic particle film technology for the control of pests and diseases was introduced by Glenn *et al.* in 1999. Hydrophobic particle films are based on the inert mineral, kaolin, which is treated with a water-repelling agent. Kaolin is a white, non-porous, non-swelling, non-abrasive, fine-grained aluminosilicate mineral ($Al_4Si_4O_{10}(OH)_8$) that easily disperses in water and is chemically inert over a wide pH range. Kaolin particles can be made with varying degrees of hydrophobicity by coating them with waterproofing agents such as chrome complexes, stearic acid and organic zirconate. By dusting fruit trees with hydrophobic kaolin particles, Glenn *et al.* (1999) obtained control of arthropod pests and fungal and bacterial pathogens. The authors suggested that disease control was achieved because plants were enveloped in a hydrophobic particle film barrier that prevented pathogen propagules or water from directly contacting the leaf surface. Interestingly however, hydrophilic kaolin particles can also provide plant disease control. For example, Puterka *et al.* (2000) found that hydrophilic particle films controlled fabraea leaf spot on pear, while the hydrophilic kaolin-based product Surround WP® (BASF, NJ, USA) was shown to control *Zygophiala jamaicensis* and *Gloeodes pomigena* on apple fruits and *Phoma* sp. on apple leaves, although it was not consistently effective against the cedar apple rust pathogen *Gymnosporangium juniperi-virginianae* (Thomas *et al.*, 2004). In a recent four-year study, hydrophobic kaolin particle films failed to control *Cladosporium carpophilum* or *Podosphaera leucotricha* on peach, although it did control *Monilinia fructicola* (Lalancette *et al.*, 2005). In contrast, hydrophilic kaolin particle films did not control any of the peach pathogens, leading the authors to suggest that hydrophobicity and deposit density may be important factors for effective disease management (Lalancette *et al.*, 2005).

Whether or not kaolin particle films provide disease control, other beneficial effects of applying such agents have been observed. For example, liquid formulations of both hydrophobic and hydrophilic particle films were shown to double yields of pear in field trials, while delayed fruit maturation, increased fruit size, increased fruit number and increased fruit yield were obtained following use of kaolin particle films (Lalancette *et al.*, 2005). In the latter case, the authors attributed the effects to a reduction in plant stress. In fact, reduced plant stress during extreme temperature conditions was also observed following applications of kaolin particle films to apple trees (Thomas *et al.*, 2004).

12.4 Disrupting spore adhesion to the leaf surface

It is thought that adhesion of fungal spores to the plant surface is the essential first step in the infection process (e.g. Epstein & Nicholson, 1997). In *Magnaporthe grisea*, the causal agent of rice blast, hydration of conidia leads to the release of an adhesive mucilage from the spore tip that forms a viscous pad for attaching the conidium to the plant surface (Hamer *et al.*, 1988). Once a spore has successfully adhered to the leaf surface, germ tube formation is rapid, whereas if conidia fail to adhere to the substrate, viability is rapidly lost (Talbot, 1995). Once formed, germ tubes must also adhere to the plant surface, since this process is important in the perception of signals for further differentiation, for example formation of the appressorium. The importance of spore adhesion to successful infection of plant surfaces suggests that disruption of this process could be useful in disease control. Indeed, some fungicides used for control of rice blast have been shown to interfere with spore adhesion (Inoue *et al.*, 1987). In later work, it was shown that adhesion of spores of phytopathogenic fungi on artificial and plant surfaces could be inhibited using zosteric acid (Stanley *et al.*, 2002). Zosteric acid (*p*-(sulphoxy) cinnamic acid) is a naturally occurring compound found in the eelgrass Zostera marina and which inhibits the attachment of marine bacteria and barnacle larvae (Todd *et al.*, 1993). Stanley *et al.* (2002) showed that zosteric acid not only inhibited attachment of spores of *M. grisea* and *Colletotrichum lindemuthianum*, but also inhibited formation of appressoria, leading to a failure to infect leaves. In fact, on intact plants, zosteric acid reduced lesion development on rice leaves caused by inoculation with *M. grisea* and delayed lesion development on bean leaves following inoculation with *C. lindemuthianum* (Stanley *et al.*, 2002). These workers found that zosteric acid was not toxic to the fungi and that the inhibition of spore adhesion in *M. grisea* was reversible by washing. Although this could lead to transitory protection under field conditions, Stanley *et al.* (2002) point out that reduced adhesion could lower inoculum potential, either because non-attached spores could be more easily detached from leaf surfaces by wind or rain splash, or as in the case of *M. grisea*, the spores quickly lose viability. This work hints at the considerable potential for the design of environmentally benign strategies for plant disease control based on agents which inhibit adhesion (Stanley *et al.*, 2002).

12.5 Conclusions

In order to gain access to the nutrient supplies required for their continued growth and survival, fungal pathogens must breach the outer surfaces of their host plants. Plant surfaces provide chemical and physical cues that are important factors in the development of infection structures of many plant pathogenic fungi. It is no surprise therefore that disrupting these processes, by coating the leaf with polymer films or applying agents that interfere with spore adhesion, can reduce infection and provide disease control.

In many parts of the world, farmers and growers have become accustomed to very high levels of disease control achieved by using fungicides. Similar levels of disease control are unlikely to be achieved with many novel disease control methods, like polymer or particle films, or adhesion inhibitors. However, the problems of fungicide resistance, breakdown in host resistance and increased public concern for the environment, means that the development of new disease control methods cannot be ignored. Further, there are many diseases of agricultural and horticultural crops for which no adequate disease control measures currently exist. New and innovative control measures might be the answer, but the adoption of such measures will not be easy, since, in some cases, it will require changes in crop protection practices. For the approaches described in this chapter, application before the pathogen arrives on the crop is important, since these approaches will act essentially as protectants. This in turn will require robust systems of disease forecasting. Some changes, for example alternating use of a novel method with fungicides in the spray programme, may be easier to integrate into current practice. However, since information in many of these areas is lacking, there is a clear need for research on integrating potential novel methods of disease control into crop protection programmes.

12.6 Acknowledgements

Scottish Agricultural College receives financial support from the Rural Environment, Research and Analysis Directorate of the Scottish Government.

12.7 References

Backman PA, 1978. Fungicide formulation: Relationship to biological activity. *Annual Review of Phytopathology* **16,** 211–237.

Brown EN, White SR, Sottos NR, 2004. Microcapsule induced toughening in a self-healing polymer composite. *Journal of Materials Science* **39,** 1703–1710.

Collins TJ, Moerschbacher BM, Read ND, 2001. Synergistic induction of wheat stem rust appressoria by chemical and topographical signals. *Physiological and Molecular Plant Pathology* **58,** 259–266.

Doss RP, Christian JK, Chastagner GA, 1988. Infection of Easter lily leaves from conidia of *Botrytis elliptica*. *Canadian Journal of Botany* **66,** 1204–1208.

Elad Y, Ziv O, Aynish N, Katan J, 1989. The effect of film-forming polymers on powdery mildew of cucumber. *Phytoparasitica* **17,** 279–288.

Elad Y, Aynish N, Ziv O, Katan J, 1990. Control of grey mould (*Botrytis cinerea*) with film-forming polymers. *Plant Pathology* **39,** 249–254.

Epstein L, Nicholson RL, 1997. Adhesion of spores and hyphae to plant surfaces. In: Carroll G, Tudsynski P, eds. *The Mycota*, Vol. 5. Berlin: Springer-Verlag, 11–25.

Gale J, Hagan RM, 1966. Plant antitranspirants. *Annual Review of Plant Physiology* **17,** 269–279.

Gale J, Poljakoff-Mayber A, 1962. Prophylactic effect of a plant antitranspirant. *Phytopathology* **52,** 715–717.

Glenn DM, Puterka G, Vanderzwet T, Byers RE, Feldhake C, 1999. Hydrophobic film particles: A paradigm for suppression of arthropod pests and plant diseases. *Journal of Economic Entomology* **92,** 759–771.

Hamer JE, Howard RJ, Valent B, Chumley FG, 1988. A mechanism for surface attachment in spores of a plant pathogenic fungus. *Science* **239,** 288–290.

Han J, 1990. Use of antitranspirant epidermal coatings for plant protection in China. *Plant Disease* **74,** 263–266.

Hsieh TF, Huang JW, 1999. Effect of film-forming polymers on control of lily leaf blight caused by *Botrytis elliptica*. *European Journal of Plant Pathology* **105,** 501–508.

Inoue S, Kato T, Jordan VWL, 1987. Inhibition of appressorial adhesion of *Pyricularia oryzae* to barley leaves by fungicides. *Pesticide Science* **19,** 145–152.

Lalancette N, Belding RD, Shearer PW, Frecon JL, Tietjen WH, 2005. Evaluation of hydrophobic and hydrophilic particle films for peach crop arthropod and disease management. *Pest Management Science* **61,** 25–39.

McGrath MT, Shishkoff N, 1999. Evaluation of biocompatible products for managing curcubit powdery mildew. *Crop Protection* **18,** 471–478.

Nielsen KA, Nicholson RL, Carver TLW, Kuhoh H, Oliver RP, 2000. First touch: An immediate response to surface recognition in conidia of *Blumeria graminis*. *Physiological and Molecular Plant Pathology* **56,** 63–70.

Osswald W, Niehuss M, Huber W, Elstner EF, 1984. Support of non-host resistance by artificial leaf coating. *Journal of Plant Diseases and Protection* **91,** 337–344.

Puterka GJ, Glenn DMM, Sekutowski DG, Unruh TR, Jones SK, 2000. Progress toward liquid formulations of particle films for insect and disease control in pear. *Environmental Entomology* **29,** 329–339.

Quarles W, 1991. Antitranspirants show promise as non-toxic fungicides. *The IPM Practitioner* **8 (8),** 1–10.

Royle DJ, 1975. Structural features of resistance to plant diseases. In: Friend J, Threlfall DR, eds. *Biochemical Aspects of Plant–Parasite Relationships*. New York: Academic Press, 161–194.

Smith R, 1996. Evaluation of materials for control of bell pepper powdery mildew. *Fungicide Nematicide Tests* **51,** 120.

Stanley MS, Callow ME, Perry R, Alberte RS, Smith R, Callow JA, 2002. Inhibition of fungal spore adhesion by zosteric acid as the basis for a novel, nontoxic crop protection technology. *Phytopathology* **92,** 378–383.

Sutherland F, Walters DR, 2001. In vitro effects of film-forming polymers on the growth and morphology of *Pyrenophora avenae* and *Pyricularia oryzae*. *Journal of Phytopathology* **149,** 621–624.

Sutherland F, Walters DR, 2002. Effect of film-forming polymers on infection of barley with the powdery mildew fungus, *Blumeria graminis* f. sp. *hordei*. *European Journal of Plant Pathology* **108,** 385–389.

Talbot NJ, 1995. Having a blast: Exploring the pathogenicity of *Magnaporthe grisea*. *Trends in Microbiology* **3,** 9–16.

Thomas AL, Muller ME, Dodson BR, Ellersieck MR, Kaps M, 2004. A kaolin-based particle film suppresses certain insect pests while reducing heat stress in apples. *Journal of the American Pomological Society* **58,** 42–51.

Todd JS, Zimmerman RC, Crews P, Alberte RS, 1993. The antifouling activity of natural and synthetic phenolic acid sulphate esters. *Phytochemistry* **34,** 401–404.

Wade BR, Reichers DE, Leibl RA, Wax LM, 1993. The plasma membrane as a barrier to herbicide penetration and site for adjuvant action. *Pesticide Science* **37,** 195–202.

Walters DR, 1992. The effects of three film-forming polymers, with and without a polyamine biosynthesis inhibitor, on powdery mildew infection of barley seedlings. *Annals of Applied Biology* **120,** 41–46.

White SR, Sottos NR, Geubelle PH, Moore JS, 2001. Autonomic healing of polymer composite. *Nature* **409,** 794–797.

Wilcox WF, Riegel DG, 1996. Evaluation of fungicides and spray timings for control of grape powdery mildew. *Fungicide Nematicide Tests* **51,** 76.

Wilcox WF, Riegel DG, 1997. Evaluation of fungicide programs for control of grapevine powdery mildew. *Fungicide Nematicide Tests* **52,** 86.

Wright AJ, Thomas BJ, Carver TLW, 2002. Early adhesion of *Blumeria graminis* to plant and artificial surfaces demonstrated by centrifugation. *Physiological and Molecular Plant Pathology* **61,** 217–226.

Wynn WK, 1981. Tropic and toxic responses of pathogens to plants. *Annual Review of Phytopathology* **19,** 237–255.

Wynn WK, Staples RC, 1981. Tropisms of fungi in host recognition. In: Staples RC, Toennissen GH, eds. *Plant Disease Control: Resistance and Susceptibility*. New York: John Wiley & Sons, 45–69.

Zekaria-Oren J, Eyal Z, 1991. Effect of film-forming compounds on the development of leaf rust on wheat seedlings. *Plant Disease* **75,** 231–234.

Ziv O, Frederiksen RA, 1983. Control of foliar diseases with epidermal coating materials. *Plant Disease* **67,** 212–214.

Ziv O, Frederiksen RA, 1987. The effect of film-forming antitranspirants on leaf rust and powdery mildew incidence on wheat. *Plant Pathology* **36,** 242–245.

Chapter 13
Bacteriophages as agents for the control of plant pathogenic bacteria

Botond Balogh[1,2], *Timur Momol*[3], *Aleksa Obradovic*[4] *and Jeffrey Jones*[1]

[1]University of Florida, Gainesville, FL 32611, USA
[2]The Connecticut Agricultural Experiment Station, New Haven, CT 06504, USA
[3]Plant Pathology Department, University of Florida, 1052 McCarty Hall, Gainsville, FL 32611, USA
[4]Faculty of Agriculture, University of Belgrade, 11080 Belgrade-Zemun, Serbia

13.1 Introduction – disease control for bacterial diseases

Diseases incited by bacterial plant pathogens are responsible for major economic losses to agricultural production. Disease control is challenging for many diseases incited by bacteria (Civerolo, 1982). The factors that contribute to those challenges are: (a) pathogen variability, (b) high probability of overcoming plant genetic resistance and/or bactericide sensitivity as a result of the high rate of mutation and gene transfer in the pathogen population, (c) the pathogen's ability to reach high populations in a relatively short period of time when conditions are conducive for disease development and (d) lack of effective bactericides. For most plant diseases, including bacterial incited diseases, an integrated management strategy is essential, combining proper cultural practices, biological control, bactericides or plant activators, where applicable, and plant resistance (Obradovic *et al.*, 2004, 2005).

Chemical control has been a significant component of management strategies for controlling plant diseases incited by bacteria. However, control with bactericides has been extremely difficult because few effective bactericides are available. Copper has been used more extensively than any other chemical for control of bacterial plant diseases; however, copper resistance is present in many plant pathogenic bacteria and is associated with plasmid-borne and chromosomal resistance (Bender & Cooksey, 1986; Stall *et al.*, 1986; Bender & Cooksey, 1987; Bender *et al.*, 1990; Lee *et al.*, 1994; Basim *et al.*, 1999; Canteros, 1999). Antibiotics have also been used as part of a management strategy for various bacterial diseases. Since the 1950s, the aminoglycoside antibiotic, streptomycin, has been used (Thayer & Stall, 1961). As a result of extensive use, streptomycin-resistant strains became prevalent, which resulted in reduced disease control efficacy of bacterial spot of tomato and pepper (Thayer & Stall, 1961), fire blight of apple and pear (Manulis *et al.*, 1998), and many other bacterial plant pathogens (Cooksey, 1990). More recently, systemic acquired resistance (SAR) plant inducers have been used and have shown some success against bacterial diseases of tomato and pepper (Louws *et al.*, 2001;

Romero *et al.*, 2001; Obradovic *et al.*, 2004), *Xanthomonas* leaf blight on onion (Gent & Schwartz, 2005) and fire blight on apple (Maxson-Stein *et al.*, 2002) (see also Chapter 4). Although these inducers may reduce disease, they may also have some negative effects on certain plant species, such as by affecting yield (Romero *et al.*, 2001; Gent & Schwartz, 2005). In some pathosystems, plant inducers have been ineffective for disease control (Graham & Leite, 2004; Chapter 4).

13.2 Biological control

Biological control has gained recent interest for controlling bacterial diseases (see Chapter 3). Various strategies for using biological control for bacterial diseases include the use of nonpathogenic or pathogenically attenuated strains of the pathogen (Frey *et al.*, 1994; Liu, 1998; Hert, 2007), saprophytic bacteria (Ji *et al.*, 2006), and plant growth-promoting rhizobacteria (PGPR) (Ji *et al.*, 2006) to suppress pathogen populations or induce a reaction in the plant such that the pathogen is reduced in its ability to colonize the plant and cause disease. Disease control using these approaches has been variable.

13.3 Early use of bacteriophages in agriculture

Bacteriophages were evaluated for controlling a number of human and animal bacterial diseases, soon after their discovery by Twort (1915) and by d'Herelle (1917) at the beginning of the twentieth century (Brunoghe & Maisin, 1921; Beckerich & Hauduroy, 1922; Davison, 1922). Fairly soon after, they were proposed as plant disease control agents (Moore, 1926).

In 1924, Mallman & Hemstreet (1924) isolated the 'cabbage-rot organism,' *Xanthomonas campestris* pv. *campestris*, from rotting cabbage and demonstrated that the filtrate of the liquid collected from the decomposed cabbage inhibited *in vitro* growth of the pathogen. The following year, Kotila & Coons (1925) isolated bacteriophages from soil samples that were active against the causal agent of blackleg disease of potato, *Erwinia carotovora* subsp. *atroseptica*. They demonstrated in growth chamber experiments that co-inoculation of *E. carotovora* subsp. *atroseptica* with phage successfully inhibited the pathogen and prevented rotting of tubers (Kotila & Coons, 1925). These workers also isolated phages against *E. carotovora* subsp. *carotovora* and *Agrobacterium tumefaciens* from various sources, such as soil, rotting carrots and river water (Coons & Kotila, 1925). Thomas (1935) treated corn seeds that were infected with *Pantoea stewartii*, the causal agent of Stewart's wilt of corn, with bacteriophage isolated from diseased plant material. The seed treatment reduced disease incidence from 18% to 1.4%.

Despite promising early results, phage was generally considered an ineffective and unreliable means for controlling plant pathogenic bacteria. Okabe stated in 1963, in a review article on bacteriophages, that the phage in general appears to be ineffective as a control strategy (Okabe & Goto, 1963). Goto concluded in 1992 that practical use of phages for control of bacterial plant disease in the field was unsuccessful (Goto, 1992). Furthermore, it was believed that because of their narrow spectrum of activity, phages were much more likely to fail than antibiotics (Summers, 2005). As a result of these factors, interest in phages waned for controlling bacterial plant diseases and antibiotics and copper compounds became standard.

13.4 Recent approaches for using phages in plant pathology

Studies conducted since the early 1990s have helped to identify strategies for improving phage efficacy. Various approaches were attempted to improve the competitive advantage of phage in the environment in order to improve efficacy.

A major factor limiting the use of phages for control of plant diseases was the probability of developing bacterial strains resistant to the phage. This risk was addressed in 1937 by Katznelson (1937) and later in two reviews on phages (Okabe & Goto, 1963; Vidaver, 1976). Jackson (1989) developed a strategy to prevent occurrence of phage-resistant mutants. This involved preparing mixtures of host range mutant phages (h-mutants) that lyse bacterial strains that are resistant to the parent phage (Adams, 1959), while maintaining the ability to lyse the wild-type bacterium. Using this strategy, a mixture of four phages including wild-type and h-mutant phages were applied twice weekly and provided significantly better disease control and produced greater yield of extra large fruits than the standard copper-mancozeb (Flaherty *et al.*, 2000).

An important and oftentimes neglected aspect of phage-based biological control is pre-screening phages for their biocontrol value before application; that is, identifying specific bacteriophages with particular characteristics that may prove effective in control rather than arbitrarily selecting them based strictly on lytic activity for disease control studies. Saccardi and co-workers (Zaccardelli *et al.*, 1992; Saccardi *et al.*, 1993) selected from a collection of eight phages a lytic phage with the broadest host range to use in studies. The proper assay for phage selection is critical. Although *in vitro* assays are frequently used as a selection process for phages, these may not be good predictors of biological control ability. Balogh (2006) found no correlation between *in vitro* characteristics, such as antibacterial activity or phage multiplication rate, and disease control efficacy (unpublished results). On the other hand, he found that phages which multiplied more efficiently on their host in the phyllosphere, were also better in disease control.

Timing of bacteriophage applications relative to the arrival of the pathogen influenced efficacy of disease control in several instances. Civerolo & Keil (1969) achieved a marked reduction of peach bacterial spot only if phage treatment was applied one hour or one day before inoculation with the pathogen. There was a slight disease reduction when phage was applied one hour after inoculation and no effect if applied one day later. Civerolo (1972) suggested that bacteria were inaccessible to phage in the intercellular spaces, or there were not enough phages reaching the pathogen. Schnabel *et al.* (1999) achieved a significant reduction of fire blight on apple blossoms when the phage mixture was applied at the same time as the pathogen, *Erwinia amylovora*. In contrast, disease reduction was not significant when phages were applied a day before inoculation. Bergamin Filho & Kimati (1981) investigated the effect of timing on the efficacy of phage treatment in greenhouse trials with two pathosystems: black rot of cabbage, caused by *Xanthomonas campestris* pv. *campestris* and bacterial spot of pepper, caused by *Xanthomonas campestris* pv. *vesicatoria*. Phage treatment was applied once varying from seven days before to four days after pathogen inoculation. On cabbage, significant disease reduction was achieved if the phage treatment was applied three days before to one day after inoculation, whereas on pepper from three days before to the day of inoculation. The greatest disease reduction occurred with application of phages the same day as inoculation in both pathosystems.

Maximizing the chances for an interaction between phage and target bacterium is of critical importance. The success of any biocontrol treatment is influenced by agent and target densities (Johnson, 1994). In the case of phage therapy, there is a need for high populations of both phage and bacterium, in order to start the 'chain reaction' of bacterial lysis (Gill & Abedon, 2003). Therefore, a threshold concentration should exist above which phages provide good control regardless of the applied concentration but below that threshold phages will not exert a pronounced effect on the bacterial population. Several findings support this threshold effect hypothesis. Balogh (2002) found that a phage mixture provided similar levels of control of tomato bacterial spot if applied at 10^6 or 10^8 PFU ml^{-1} concentration, but was ineffective at 10^4 PFU ml^{-1}.

Phages encounter a number of adverse factors in the phyllosphere, which substantially reduce their persistence, rapidly diminishing their populations under the threshold level, thus reducing their residual activity. These include sunlight irradiation, especially in the UV A and B regions, desiccation and exposure to certain chemical pesticides, such as copper compounds (Iriarte *et al.*, 2007). Additionally, phage persistence varies depending on the ambient temperature (Iriarte *et al.*, 2007). Under field conditions sunlight irradiation is the single major factor hindering phage persistence (Iriarte *et al.*, 2007). In studies by Iriarte *et al.* (2007), phage population numbers declined sharply during the early afternoon hours but persisted at much higher levels when applied in the early evening, and were highly correlated with the intensity of sunlight UV irradiation (Figure 13.1).

Several strategies have been evaluated for increasing phage persistence, including the use of protective formulations, application scheduling for sunlight avoidance and co-application of bacterial hosts for *in vivo* phage propagation.

Development of solar protectants for increasing biocontrol efficacy has been the focus of considerable research not only in the case of bacteriophages, but also for entomopathogenic viruses and proteinacous biopesticides (Behle *et al.*, 1996; Ignoffo *et al.*, 1997; Balogh *et al.*, 2003; Arthurs *et al.*, 2006). Balogh (2002) identified compounds that, when

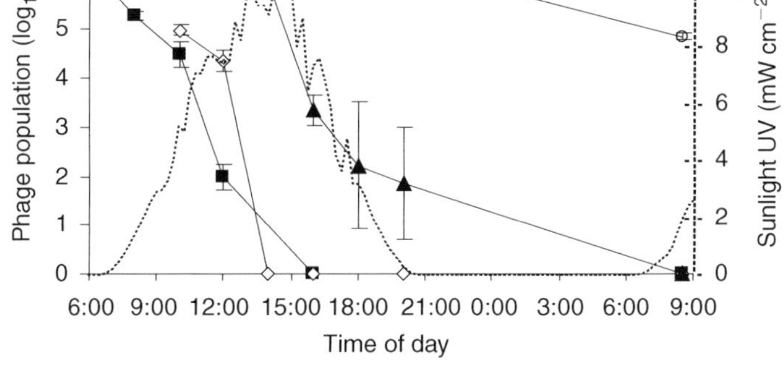

Figure 13.1 Changes in bacteriophage populations applied to field tomatoes at different times of day. Phage was applied at 6:00 (solid squares), at 10:00 (open diamonds), at 14:00 (solid triangles) or at 19:00 (open circles). Dotted line indicates sunlight UV irradiation.

mixed with phage, extended the ability of phage to survive on leaf surfaces. Furthermore, Balogh *et al.* (2003) enhanced the efficacy of phage treatment with protective formulations that increased phage persistence on tomato foliage (Figure 13.2a). These formulations enhanced the phages' ability to persist in the presence of UV and fluorescent light (Iriarte *et al.*, 2007). The use of these formulations led to increased phage residual activity and, consequently, to enhanced disease control efficacy (Balogh *et al.*, 2003) (Figure 13.2b,c).

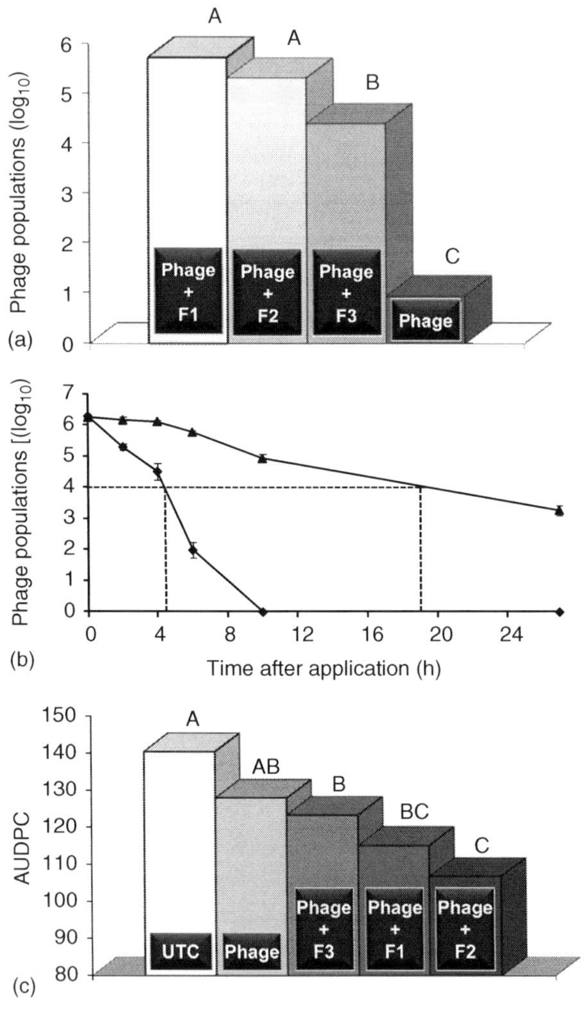

Figure 13.2 Protective formulations increase phage persistence and enhance disease control efficacy. (a) Effect of protective formulations on phage persistence in the greenhouse. Phage suspension was applied on greenhouse tomatoes in water (phage) or in different formulations (F1, F2 and F3) and recovered from the plants 48 h after application. (b) Effect of protective formulation on phage residual activity in the field. Phage suspension was applied to field tomatoes with formulation (solid diamonds) or without it (triangles). Vertical dotted lines indicate the end of the predicted residual activity based on 10^4 PFU cm^{-2} threshold. (c) Effect of protective formulations on phage disease control efficacy. Phage mixture was applied twice-weekly throughout the tomato growing season alone or with different formulations (F1, F2 and F3). Area under the disease progress curve (AUDPC) values indicate the overall disease progress. UTC: untreated control.

Obradovic and co-workers (Obradovic *et al.*, 2004, 2005) determined that application of formulated phages resulted in reduced disease and increased yield.

The issue of sunlight avoidance was illustrated when Flaherty *et al.* (2000) effectively controlled the tomato bacterial spot pathogen in field experiments by applying phages in the early morning hours prior to sunrise and achieved no control if phages were applied during the day (unpublished results). Iriarte *et al.* (2007) showed that phages persist better if applied in the evening rather than in the morning (Figures 13.1 and 13.3a). Balogh *et al.* (2003) more definitively demonstrated that sunlight avoidance during phage application led to increased control by showing that bacteriophages applied to tomato plants in the field in the evening significantly reduced disease compared to morning applications, resulting in 26.9% and 13.1% disease reduction, respectively (Figure 13.3b).

A unique advantage of bacteriophages in comparison with chemical pesticides is the ability to increase their numbers in the target environment by multiplying on a bacterial host. This ability can be exploited if phages are applied into an environment where a phage-sensitive bacterium is present, or alternatively, they are delivered together with the host. In an environment where high host populations are present and conditions are favorable, phages persist much better than without the host (Balogh, 2006) (Figure 13.4).

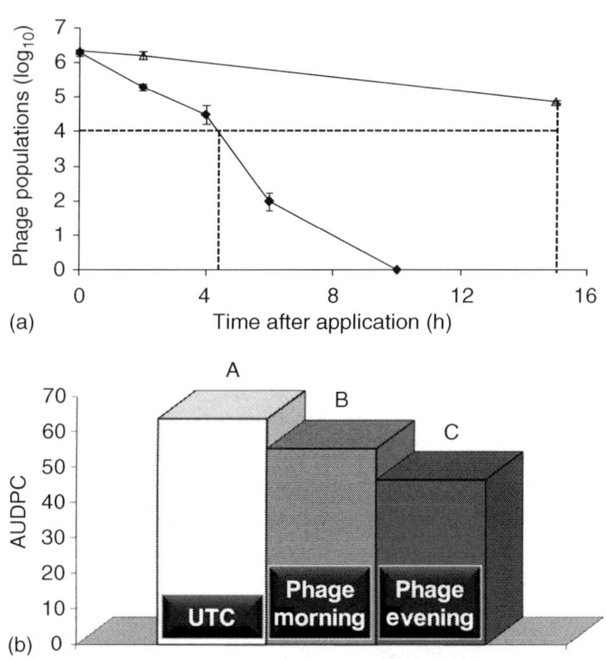

Figure 13.3 Evening phage application results in increased residual activity and better disease control. (a) Effect of application time on phage residual activity in the field. Phage suspension was applied to field tomatoes in the morning at 6:00 (solid diamonds) or in the evening at 19:00 (open triangles). Vertical dotted lines indicate the end of the predicted residual activity based on 10^4 PFU cm^{-2} threshold. (b) Effect of application time on phage disease control efficacy. Phage mixture was applied twice-weekly throughout the tomato growing season in the evening or in the morning. Area under the disease progress curve (AUDPC) values indicate the overall disease progress. UTC: untreated control.

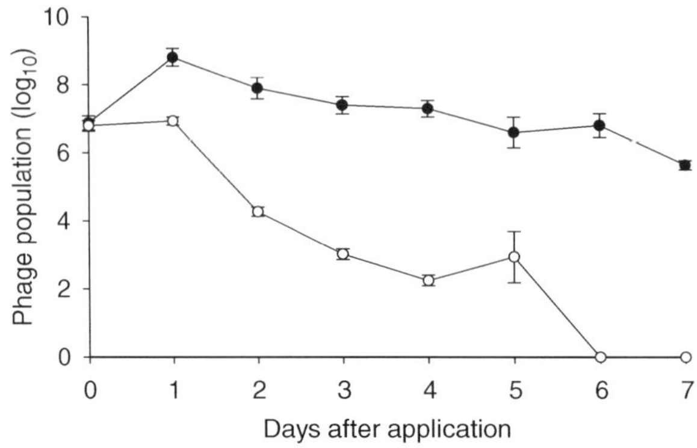

Figure 13.4 The presence of the bacterial host increases phage persistence. Phage suspension was applied to greenhouse tomatoes, which had been pre-inoculated with phage-sensitive bacterium (solid circles) or were uninoculated (open circles).

This approach was investigated for treating both soil-borne and aerial diseases (Tanaka *et al.*, 1990; Svircev *et al.*, 2006). Tanaka *et al.* (1990) used an avirulent strain of *Ralstonia solanacearum* and its phage that was active against both the virulent and avirulent strains to reduce tobacco bacterial wilt incited by *R. solanacearum*. While the application of the avirulent strain alone caused a significant 59% reduction in the number of wilted plants, the co-application of phage with the avirulent strain increased control significantly, to 82%. A similar strategy was employed by Svircev *et al.* (2006) for controlling fire blight of pear. They selected phages based on the ability to lyse both the target organism, the pathogen *Erwinia amylovora*, and also a closely related antagonistic phyllosphere bacterium, *Pantoea agglomerans*. When applied together with the phage, *P. agglomerans* served as a biological control agent, as well as a phage carrier, a vehicle of delivery and medium of propagation on the leaf surface. While *P. agglomerans* alone significantly reduced disease, its combination with phage resulted in significantly better disease control, which was comparable to streptomycin treatment.

13.5 Challenges in using phages for disease control

Biocontrol success with bacteriophages is to a large extent dependent on the population dynamics of both the biological control agent and target bacterium (Johnson, 1994). It is essential that the phage attaches to its host before being destroyed by the various physical factors (Goodridge, 2004). The likelihood for phage–bacterium interactions depends on several key factors: phage concentration at the site of interaction (i.e., phyllosphere or rhizosphere), rates of virion degradation (phage vary in degradative properties), timing of application to optimize efficacy, phage infection and replication ability in the target environment, concentration and accessibility of target bacteria, and the necessary presence of adequate moisture as a medium for phage diffusion (Gill & Abedon, 2003). Additionally, disease control efficacy may be influenced by the relative fitness of phage-resistant bacterial mutants (Gill & Abedon, 2003) and the surrounding environment.

Phages have the potential for controlling plant pathogens in the rhizosphere or phyllosphere. However, Gill & Abedon (2003) identified factors that can hinder success of disease control in the rhizosphere. The relatively low diffusion rate of phages through heterogeneous soil matrices changes as a function of available free water. Biofilms can trap phages (Storey & Ashbolt, 2001), soil clay particles can reversibly adsorb phages (Williams *et al.*, 1987), and low soil pH can inactivate phages (Sykes *et al.*, 1981). In natural environments, as a result of low rates of phage diffusion and high rates of phage inactivation, low numbers of viable phages are available to lyse target bacteria (Gill & Abedon, 2003). One additional factor needed for a high degree of success is that high populations of both phage and bacterium exist in order to initiate a chain reaction of bacterial lysis (Gill & Abedon, 2003).

In the phyllosphere, a harsh environment exists and phages in this location degrade extremely rapidly (Civerolo & Keil, 1969; McNeil *et al.*, 2001; Balogh, 2002; Balogh *et al.*, 2003). This ephemeral existence on plant leaf surfaces is a limiting factor of phage treatment. In field and laboratory studies, viruses were shown to be inactivated by high temperatures, high and low pH and sunlight, and were readily dislodged by rain (Ignoffo *et al.*, 1989; Ignoffo & Garcia, 1992). The environmental factor most destructive to viruses was UV-A and UV-B spectra of sunlight (Ignoffo & Garcia, 1994). Initial studies showed that phage applied in the mid-morning was not effective in controlling bacterial spot of tomato (J. Jones, unpublished results). We speculated that short residual activity of the phage existed as a result of UV degradation and hindered efficacy of phage treatment when applied during daytime. This was confirmed in field experiments, in which phages on tomato foliage exposed to high intensities of sunlight during daytime were eliminated from the phyllosphere within hours after application (Iriarte *et al.*, 2007). In experiments conducted in greenhouses where sunlight UV irradiation is less of a factor because UV cannot penetrate glass, phages can persist up to a week (Balogh, 2006) (Figure 13.4).

13.6 Phages as part of an integrated management strategy

Multifaceted, integrated strategies carry the promise for effective, reliable and sustainable management of bacterial plant diseases. Several approaches have been explored for using phage treatment within an integrated management strategy. Tanaka *et al.* (1990) reduced tobacco bacterial wilt by co-application of an avirulent strain of the pathogen, *R. solanacearum*, with a phage that was active against both virulent and avirulent strains. Using a similar approach, Svircev *et al.* (2006) reduced fire blight of pear with co-application of an antagonistic epiphyte, *Pantoea agglomerans* and a phage that lysed both the antagonist and the pathogen, *Erwinia amylovora*.

Obradovic and co-workers (Obradovic *et al.*, 2004, 2005) used tomato bacterial spot as a model system for developing a comprehensive phage-based integrated management strategy for foliar bacterial diseases. They combined phage treatment with SAR inducers, PGPR and antagonistic bacteria to control bacterial spot of tomato. They achieved better and more reliable disease control when combining phages with SAR inducers; however, integration with bacterial biocontrol agents did not improve control efficacy, as compared to phage alone.

Lang *et al.* (2007) evaluated phage treatment in combination with acibenzolar-*S*-methyl, an SAR inducer, or with copper-mancozeb for the control of *Xanthomonas* leaf blight of onion and found that both combinations resulted in enhanced disease control. On the contrary, Balogh (2006) observed no improvement in control of citrus canker or citrus bacterial spot with the combination of bacteriophages with copper-mancozeb.

13.7 Summary

Bacteriophage-based disease control of plant diseases is a quickly developing field of research. A wide range of strategies has been employed to increase control efficacy. Phages are applied alone as well as part of an integrated disease management approach. Phage treatment is currently used in greenhouses and production fields in Florida as a part of a standard integrated management program for tomato bacterial spot control (Momol *et al.,* 2002). Owing to their increasing efficacy and contribution to sustainable agriculture, phage-based products are likely to gain a bigger share in the bactericide market in the future.

13.8 References

Adams MH, 1959. *Bacteriophages.* New York: Interscience Publishers.

Arthurs SP, Lacey LA, Behle RW, 2006. Evaluation of spray-dried lignin-based formulations and adjuvants as solar protectants for the granulovirus of the codling moth, *Cydia pomonella* (L). *Journal of Invertebrate Pathology* **93,** 88–95.

Balogh B, 2002. Strategies of improving the efficacy of bacteriophages for controlling bacterial spot of tomato. MS Thesis, Gainesville, FL: University of Florida.

Balogh B, 2006. Characterization and use of bacteriophages associated with citrus bacterial pathogens for disease control. PhD Dissertation, Gainesville, FL: University of Florida.

Balogh B, Jones JB, Momol MT, Olson SM, Obradovic A, King P, Jackson LE, 2003. Improved efficacy of newly formulated bacteriophages for management of bacterial spot on tomato. *Plant Disease* **87,** 949–954.

Basim H, Stall RE, Minsavage GV, Jones JB, 1999. Chromosomal gene transfer by conjugation in the plant pathogen *Xanthomonas axonopodis* pv. *vesicatoria. Phytopathology* **89,** 1044–1049.

Beckerich A, Hauduroy P, 1922. Le bactériophage dans le traitment de la fièvre typhoïde. *Comptes Rendus des Séances de la Société de Biologie* **86,** 168.

Behle RW, McGuire MR, Shasha BS, 1996. Extending the residual toxicity of *Bacillus thuringiensis* with casein-based formulations. *Journal of Economic Entomology* **89,** 1399–1405.

Bender CL, Cooksey DA, 1986. Indigenous plasmids in *Pseudomonas syringae* pv. *tomato*: Conjugative transfer and role in copper resistance. *Journal of Bacteriology* **165,** 534–541.

Bender CL, Cooksey DA, 1987. Molecular cloning of copper resistance genes from *Pseudomonas syringae* pv. *tomato. Journal of Bacteriology* **169,** 470–474.

Bender CL, Malvick DK, Conway KE, George S, Cooksey DA, 1990. Characterization of pXv10A, a copper resistance plasmid in *Xanthomonas campestris* pv. *vesicatoria. Applied and Environmental Microbiology* **56,** 170–175.

Bergamin Filho A, Kimati H, 1981. Estudos sobre um bacteriofago isolado de *Xanthomonas campestris.* II. Seu emprego no controle de *X. campestris* e *X. vesicatoria. Summa Phytopathologica* **7,** 35–43.

Brunoghe R, Maisin J, 1921. Essais de therapeutique au moyen du bacteriophage du staphylocoque. *Comptes Rendus des Séances de la Société de Biologie* **85,** 1020–1021.

Canteros BI, 1999. Copper resistance in *Xanthomonas campestris* pv. *citri.* In: Mahadevan A, ed. *Plant Pathogenic Bacteria – Proceedings of the 9th International Conference,* 26–29 August, 1996. Chennai, India: Centre for Advanced Study in Botany, University of Madras, 455–459.

Civerolo EL, 1972. Interaction between bacteria and bacteriophages on plant surfaces and in plant tissues. In: Maas Geesteranus HP, ed. *Proceedings of the Third International Conference of Plant Pathogenic Bacteria,* April 14–21, 1971. Wageningen: Centre for Agricultural Publishing and Documentation, 25–37.

Civerolo EL, 1982. Disease management by cultural practices and environmental control. In: Mount MS, Lacy GH, eds. *Phytopathogenic Prokaryotes.* New York: Academic Press, 343–360.

Civerolo EL, Keil HL, 1969. Inhibition of bacterial spot of peach foliage by *Xanthomonas pruni* bacteriophage. *Phytopathology* **59,** 1966–1967.

Cooksey DA, 1990. Genetics of bactericide resistance in plant pathogenic bacteria. *Annual Review of Phytopathology* **28,** 201–219.

Coons GH, Kotila JE, 1925. The transmissible lytic principle (bacteriophage) in relation to plant pathogens. *Phytopathology* **15,** 357–370.

Davison WC, 1922. The bacteriolysant therapy of bacillary dysentery in children. Therapeutic application of bacteriolysants; d'Herelle's phenomenon. *American Journal of Diseases of Children* **23,** 531.

d'Herelle F, 1917. Sur un microbe invisible antagoniste des bacilles dysentériques. *Comptes Rendus de l'Académie des Sciences Series D*, 373.

Flaherty JE, Jones JB, Harbaugh BK, Somodi GC, Jackson LE, 2000. Control of bacterial spot on tomato in the greenhouse and field with H-mutant bacteriophages. *HortScience* **35,** 882–884.

Frey P, Prior P, Marie C, Kotoujansky A, Trigalet-Demery D, Trigalet A, 1994. Hrp$^-$ mutants of *Pseudomonas solanacearum* as potential biocontrol agents of tomato bacterial wilt. *Applied and Environmental Microbiology* **60,** 3175–3181.

Gent DH, Schwartz HF, 2005. Management of xanthomonas leaf blight of onion with a plant activator, biological control agents, and copper bactericides. *Plant Disease* **89,** 631–639.

Gill JJ, Abedon ST, 2003. Bacteriophage ecology and plants. *http://www.apsnet.org/online/feature/pistachio/print.asp.*

Goodridge LD, 2004. Bacteriophage biocontrol of plant pathogens: Fact or fiction? *Trends in Biotechnology* **22,** 384–385.

Goto M, 1992. *Fundamentals of Bacterial Plant Pathology.* San Diego, CA: Academic Press.

Graham JH, Leite JRP, 2004. Lack of control of citrus canker by induced systemic resistance compounds. *Plant Disease* **88,** 745–750.

Hert AP, 2007. Evaluation of bacteriocins in *Xanthomonas perforans* for use in biological control of *Xanthomonas euvesicatoria.* PhD Dissertation, Gainesville, FL: University of Florida.

Ignoffo CM, Garcia C, 1992. Combinations of environmental factors and simulated sunlight affecting activity of inclusion bodies of the *Heliothis* (*Lepidoptera: Noctuidae*) nucleopolyhedrosis virus. *Environmental Entomology* **21,** 210–213.

Ignoffo CM, Garcia C, 1994. Antioxidant and oxidative enzyme effects on the inactivation of inclusion bodies of the *Heliothis* baculovirus by simulated sunlight- UV. *Environmental Entomology* **23,** 1025–1029.

Ignoffo CM, Garcia C, Saathoff SG, 1997. Sunlight stability and rain-fastness of formulations of *Baculovirus heliothis. Environmental Entomology* **26,** 1470–1474.

Ignoffo CM, Rice WC, McIntosh AH, 1989. Inactivation of nonoccluded and occluded baculoviruses and baculovirus-DNA exposed to simulated sunlight. *Environmental Entomology* **18,** 177–183.

Iriarte FB, Balogh B, Momol MT, Smith LM, Wilson M, Jones JB, 2007. Factors affecting survival of bacteriophage on tomato leaf surfaces. *Applied and Environmental Microbiology* **73,** 1704–1711.

Jackson LE, 1989. Bacteriophage prevention and control of harmful plant bacteria. *US Patent No. 4828999.*

Ji P, Campbell HL, Kloepper JW, Jones JB, Suslow TV, Wilson M, 2006. Integrated biological control of bacterial speck and spot of tomato under field conditions using foliar biological control agents and plant growth-promoting rhizobacteria. *Biological Control* **36,** 358–367.

Johnson KB, 1994. Dose–response relationships and inundative biological control. *Phytopathology* **84,** 780–784.

Katznelson H, 1937. Bacteriophage in relation to plant diseases. *The Botanical Review* **3,** 499–521.

Kotila JE, Coons GH, 1925. Investigations on the blackleg disease of potato. *Michigan Agricultural Experimental Station Technical Bulletin* **67,** 3–29.

Lang JM, Gent DH, Schwartz HF, 2007. Management of Xanthomonas leaf blight of onion with bacteriophages and a plant activator. *Plant Disease* **91,** 871–878.

Lee Y, Hendson M, Ponopoulos NJ, Schroth MN, 1994. Molecular cloning, chromosomal mapping, and sequence analysis of copper resistance genes from *Xanthomonas campestris* pv. *juglandis*: Homology with small blue copper proteins and multicopper oxidase. *Journal of Bacteriology* **176,** 173–188.

Liu T, 1998. Biological control with tomato bacterial spot with hrp$^-$ mutants of *Xanthomonas campestris* pv. *vesicatoria.* MS Thesis, Gainesville, FL: University of Florida.

Louws FJ, Wilson M, Campbell HL, Cuppels DA, Jones JB, Shoemaker PB, Sahin F, Miller SA, 2001. Field control of bacterial spot and bacterial speck of tomato using a plant activator. *Plant Disease* **85,** 481–488.

Mallmann WL, Hemstreet CJ, 1924. Isolation of an inhibitory substance from plants. *Agricultural Research*, **28,** 599–602.

Manulis S, Zutra D, Kleitman F, Dror O, David I, Zilberstaine M, Shabi E, 1998. Distribution of streptomycin-resistant strains of *Erwinia amylovora* in Israel and occurrence of blossom blight in the autumn. *Phytoparasitica* **26,** 223–230.

Maxson-Stein K, He SY, Hammerschmidt R, Jones AL, 2002. Effect of treating apple trees with acibenzolar-*S*-methyl on fire blight and expression of pathogenesis-related protein genes. *Plant Disease* **86,** 785–790.

McNeil DL, Romero S, Kandula J, Stark C, Stewart A, Larsen S, 2001. Bacteriophages: A potential biocontrol agent against walnut blight (*Xanthomonas campestris* pv. *juglandis*). *New Zealand Plant Protection* **54,** 220–224.

Momol MT, Jones JB, Olson SM, Obradovic A, Balogh B, King P, 2002. Integrated management of bacterial spot on tomato in Florida. Online: EDIS, Institute of Food and Agricultural Sciences, University of Florida. http://edis.ifas.ufl.edu/PP110.

Moore ES, 1926. D'Herelle's bacteriophage in relation to plant parasites. *South African Journal of Science* **23,** 306.

Obradovic A, Jones JB, Momol MT, Balogh B, Olson SM, 2004. Management of tomato bacterial spot in the field by foliar applications of bacteriophages and SAR inducers. *Plant Disease* **88,** 736–740.

Obradovic A, Jones JB, Momol MT, Olson SM, Jackson LE, Balogh B, Guven K, Iriarte FB, 2005. Integration of biological control agents and systemic acquired resistance inducers against bacterial spot on tomato. *Plant Disease* **89,** 712–716.

Okabe N, Goto M, 1963. Bacteriophages of plant pathogens. *Annual Review of Phytopathology* **1,** 397–418.

Romero AM, Kousik CS, Ritchie DF, 2001. Resistance to bacterial spot in bell pepper induced by acibenzolar-*S*-methyl. *Plant Disease* **85,** 189–194.

Saccardi A, Gambin E, Zaccardelli M, Barone G, Mazzucchi U, 1993. *Xanthomonas campestris* pv. *pruni* control trials with phage treatments on peaches in the orchard. *Phytopathologia Mediterranea* **32,** 206–210.

Schnabel EL, Fernando WGD, Meyer MP, Jones AL, Jackson LE, 1999. Bacteriophage of *Erwinia amylovora* and their potential for biocontrol. *Acta Horticulturae* **489,** 649–654.

Stall RE, Loschke DC, Jones JB, 1986. Linkage of copper resistance and avirulence loci on a self-transmissible plasmid in *Xanthomonas campestris* pv. *vesicatoria*. *Phytopathology* **76,** 240–243.

Storey MV, Ashbolt NJ, 2001. Persistence of two model enteric viruses (B40–8 and MS- 2 bacteriophages) in water distribution pipe biofilms. *Water Science and Technology* **43,** 133–138.

Summers WC, 2005. Bacteriophage research: Early history. In: Kutter E, Sulakvelidze A, eds. *Bacteriophages: Biology and Applications.* Boca Raton, FL: CRC Press, 5–27.

Svircev AM, Lehman SM, Kim WS, Barszcz E, Schneider KE, Castle AJ, 2006. Control of the fire blight pathogen with bacteriophages. In: Zeller W, Ullrich C, eds. *Proceedings of the 1st International Symposium on Biological Control of Bacterial Plant Diseases*, October 23–26, 2005. Berlin, Germany: Die Deutsche Bibliothek – CIP-Einheitsaufnahme, 259–261.

Sykes IK, Lanning S, Williams ST, 1981. The effect of pH on soil actinophage. *Journal of General Microbiology* **122,** 271–280.

Tanaka H, Negishi H, Maeda H, 1990. Control of tobacco bacterial wilt by an avirulent strain of *Pseudomonas solanacearum* M4S and its bacteriophage. *Annals of the Phytopathological Society of Japan* **56,** 243–246.

Thayer PL, Stall RE, 1961. A survey of *Xanthomonas vesicatoria* resistance to streptomycin. *Proceedings of the Florida State Horticultural Society* **75,** 163–165.

Thomas RC, 1935. A bacteriophage in relation to Stewart's disease of corn. *Phytopathology*, **25,** 371–372.

Twort FW, 1915. An investigation on the nature of the ultra-microscopic viruses. *Lancet*, **189,** 241–243.

Vidaver AK, 1976. Prospects for control of phytopathogenic bacteria by bacteriophages and bacteriocins. *Annual Review of Phytopathology* **14,** 451–465.

Williams ST, Mortimer AM, Manchester L, 1987. Ecology of soil bacteriophages. In: Goyal SM, Gerba CP, Bitton G, eds. *Phage Ecology.* New York: John Wiley and Sons, 157–179.

Zaccardelli M, Saccardi A, Gambin E, Mazzucchi U, 1992. *Xanthomonas campestris* pv. *pruni* bacteriophages on peach trees and their potential use for biological control. *Phytopathologia Mediterranea* **31,** 133–140.

Chapter 14
Controlling plant disease using biological and environmentally friendly approaches: making it work in practice

Dale Walters

Scottish Agricultural College, Edinburgh, UK

14.1 Introduction

Since the 1960s, aggregate world food production has increased by 145%, with production increasing by 280% in Asia, nearly 200% in Latin America and 140% in Africa. Food production started from a higher base in industrialised countries, although it still grew by 68% in Western Europe (FAO, 2005; Pretty, 2008). During this period, world population doubled to more than six billion, although *per capita* agricultural production has exceeded population growth, with the result that for each person there is an additional 25% more food today compared to the 1960s (Hazell & Wood, 2008; Pretty, 2008). Despite these increases in productivity, there are still some 800 million people hungry, with inadequate access to food (Pretty, 2008). In the United States in 2004, 38.2 million people, including 13.9 million children, lived in food-insecure households (Nord *et al.*, 2005), while in India in 2004, many people went hungry despite bumper harvests (Thurow & Solomon, 2004). The problem is of income distribution rather than food shortages – the hungry are too poor to buy the food (Hazell & Wood, 2008).

Increases in food production in the past 50 years have resulted from increasing the intensity of production on agricultural land, with increased use of machinery, fertilisers and pesticides. Indeed, the use of pesticides in agriculture has increased hugely, now amounting to some 2.56 billion kg yr^{-1} (Hazell & Wood, 2008; Pretty, 2008). However, the inefficient use of these inputs has resulted in considerable damage to the environment, with increased agricultural area contributing greatly to the loss of habitats and biodiversity (MEA, 2005; Scherr & McNeely, 2008). Interest in the sustainability of agricultural systems arose out of concern for the damaging effects of agricultural practices on the environment that began to surface in the 1950s–1960s (Pretty, 2008). Current concerns about sustainability revolve around the need to develop agricultural technologies and practices that: (a) have no adverse effects on the environment, (b) are effective and can be easily accessed by farmers, and (c) lead to increased food productivity, while yielding positive effects on environmental goods and services (Pretty, 2008). The key principles for sustainability are to integrate biological and ecological processes (e.g. nutrient cycling, soil regeneration, predation and parasitism) into food production processes, minimise use of non-renewable inputs that harm the environment, make use of the knowledge and skills of farmers in order to substitute human capital for costly external inputs, and to encourage people to work together to solve common agricultural problems (Pretty, 2008). A number of different terms have been used to imply greater sustainability in some agricultural systems than others, including ecoagriculture (McNeely & Scherr, 2003; Scherr & McNeely, 2008), ecological agriculture (Magdoff, 2007), and low input agriculture (Pretty, 2008). Many of these approaches involve minimising or even eliminating the use of pesticides in favour of biologically based approaches to crop protection.

If food production is to increase to feed the ever-rising world population, either the intensity of agricultural production needs to increase or more land is converted to agriculture. At the same time, the environmental consequences of food production need to be tackled, while scientists grapple with the persistent problems of fungicide insensitivity and breakdown of host resistance. This is a tall order and in terms of crop protection, will require a multi-faceted approach to controlling diseases, pests and weeds. The chapters in this book have dealt with a variety of biologically based approaches to controlling plant disease. Some of these approaches are already used in practice, while others will require further research before practical implementation is possible.

14.2 How might biologically based disease control be used in crop protection practice?

As mentioned above, effective disease control requires a multi-faceted approach, using a number of different methods. Control of certain crop diseases will require biologically based methods to be integrated into disease control programmes, along with other approaches. For other diseases, for example, those for which no adequate control exists, biologically based methods might offer the only hope of reducing disease to acceptable levels. The expected scarcity of pesticides in the EU in the future, as a result of the revision of Directive 91/414 EEC, will eventually force farmers and growers to re-think their crop protection strategies and diminish their reliance on pesticides (Kudsk, 2007). This will require more diversified disease control strategies, based on new technologies and including a broad range of approaches. Such a change requires support and with this in mind, an EU Network of Excellence (ENDURE) was established in January 2007 to facilitate this change. The idea behind ENDURE is the establishment of a European network of expertise, which will share knowledge and facilities, and initiate joint research, with the aim of developing innovative crop protection strategies for dissemination to farmers (Kudsk, 2007).

Sustainable approaches to agriculture, including many biologically based methods of disease control, might be particularly appropriate for fragile and low-yielding farming systems located, for example, in dry lands, uplands, near-deserts and hillsides (Hazell & Wood, 2008). In many developing countries, integrated management practices are used to control important pathogens and pests (Phiri *et al.*, 2007). For example, bean common mosaic virus (BCMV) and bean common mosaic necrotic virus (BCMNV) are controlled using virus-free seed, intercropping with non-host crops, and use of resistant varieties, while loose or head smut, caused by *Sphacelotheca reiliana,* is managed through rotation, deep ploughing and destruction of plant debris, and use of resistant varieties (Phiri *et al.*, 2007).

Interestingly, in China, which grows in excess of 28 million hectares of wheat, 'ecological' control of the stripe rust fungus, *Puccinia striiformis,* has been considered as a major strategy for sustainable disease control (Chen *et al.*, 2007). This approach involves: (a) improving cultivar resistance, (b) changing cultural practices, (c) eradicating volunteer wheat seedlings, (d) regulating wheat planting date and (e) returning land to forestry and pastures (Chen *et al.*, 2007).

Irrespective of the system into which biologically based disease control methods are slotted, their use in crop protection programmes will first require a number of issues to be resolved and barriers to be overcome.

14.3 Biologically based disease control: barriers to implementation

14.3.1 Efficacy of disease control

In most developed countries, high crop yields are maintained through the use of improved varieties, together with fertilisers and pesticides. Indeed, farmers and growers in these countries are accustomed to achieving high levels of disease control with fungicides, although as indicated above and in Chapter 1, the development of fungicide resistance can erode fungicide efficacy. In contrast, levels of disease control obtained with many biologically based control methods are lower than those achieved using fungicides. In addition, many biologically based methods tend to provide inconsistent disease control. For example, although induced resistance can provide high levels of disease control on some crops, with many crops, disease control is less impressive. Expression of induced resistance in crop plants can also be variable, depending on a number of factors, including genotype and environment (see Chapter 4). There are also problems of variability and inconsistency of disease control with some biological control agents (BCAs) (Whipps, 2007; see also Chapter 3). Perceived problems with inadequate and inconsistent disease control will not persuade farmers and growers to adopt biologically based approaches. Minimising the effects of these problems requires further research.

14.3.2 Regulatory issues

Despite the considerable effort by researchers to develop novel biologically based solutions for disease control (e.g. BCAs, plant-derived substances, induced resistance agents), few products have reached the marketplace. The high cost of registration, coupled with limited market size for some products, has been identified as a major barrier (Richardson, 2005; Kleeberg, 2007). However, this problem has been recognised by regulatory authorities and in the United Kingdom, for example, the Pesticides Safety Directorate (PSD) launched a pilot scheme for biopesticides in 2004, allowing the requirements for registration to be tailored to the product type and importantly, offering a significant reduction in the application fee (Richardson, 2005; Whittaker, 2007). This pilot scheme has since evolved into a permanent Biopesticides Scheme run by the PSD. However, this experience contrasts with elsewhere in Europe, where the biopesticide industry has failed to engage effectively with the regulatory authorities (Whittaker, 2007). Unless this situation changes, significant problems, getting biopesticides into commercial practice, will continue.

14.3.3 Farmer adoption of biologically based practices

There is evidence that improved agricultural technologies, such as disease-resistant varieties, precision farming and improved water management practices, can increase crop yields while reducing chemical use (Pingali *et al.*, 1997). However, farmers have been slow to adopt these new practices. There are a number of reasons for this reluctance to switch to improved practices. For example, it might be due, in part, to the continuing subsidies on water and agrochemicals provided by many governments, that is by making these inputs less expensive, subsidies encourage farmers and growers to be more wasteful in their use (Hazell & Wood, 2008). In addition, many of these improved practices are more labour and knowledge intensive than the existing practices, which can make it difficult and costly for farmers and growers to adopt them (Pingali *et al.*, 1997; Hazell & Wood, 2008). Further, if farmers can obtain good levels of disease control using the existing methods, why should they switch to a new method, which might not deliver the same level of control? Such an attitude to risk is understandable and has been identified as one of the factors affecting the uptake of new 'no-tillage' technologies in cotton–wheat farming systems in Pakistan's Punjab (Sheikh *et al.*, 2003).

Changing farmer attitudes to the adoption of new agricultural practices is not easy, but ultimately, unless farmers and growers are prepared to use such technologies, most of them will never find their way into crop protection practice.

14.4 Conclusions

It is clear that effective crop protection cannot rely solely on the use of chemical pesticides. Rather, it requires integration of a number of different approaches, including biologically based methods. Indeed, Integrated Crop Management (ICM) is considered by some to be the basis for sustainable agriculture (Wood, 1993). ICM can be defined as the cost-effective production of high-quality crops, with priority given to use of ecologically safe methods of cultivation, and minimising the use of crop protection chemicals (Dehne & Schonbeck, 1994). There is clearly a role for biologically based approaches to disease control in ICM. However, that role will only be realised if the barriers to implementation, discussed briefly above, are overcome.

14.5 Acknowledgements

Scottish Agricultural College receives financial support from the Scottish Government Rural and Environment Research and Analysis Directorate (RERAD).

14.6 References

Chen WQ, Xu SC, Liu TG, Lin RM, Wu LR, Jin SL, Co SQ, 2007. Wheat stripe rust and its prospects for ecological control in China. *Proceedings of the XVI International Plant Protection Congress*, 15–18 October 2007, Glasgow, Scotland, 812–813.

Dehne H-W, Schönbeck F, 1994. Crop protection – past and present. In: Oerke E-C, Dehne H-W, Schönbeck F, eds. *Crop Production and Crop Protection – Estimated Losses in Major Food and Cash Crops*. Amsterdam: Elsevier Science, 45–71.

FAO, 2005. *FAOSTAT Database*. Rome, Italy: FAO. Hazell P, Wood S, 2008. Drivers of change in global agriculture. *Philosophical Transactions of the Royal Society B* **363,** 495–515.

Kleeberg H, 2007. Biological control agents: Requirements and potential in the market. *Proceedings of the XVI International Plant Protection Congress,* 15–18 October 2007, Glasgow, Scotland, 596–597.

Kudsk P, 2007. Crop protection in Europe at the crossroads: Challenges facing European farmers. *Proceedings of the XVI International Plant Protection Congress,* 15–18 October 2007, Glasgow, Scotland, 734–735.

Magdoff F, 2007. Ecological agriculture: Principles, practices and constraints. *Renewable Agriculture and Food Systems* **22,** 109–117.

McNeely JA, Scherr SJ, 2003. *Ecoagriculture: Strategies for Feeding the World and Conserving Wild Biodiversity.* Washington, DC: Island Press.

Millennium Ecosystem Assessment (MEA), 2005. *Ecosystems and Human Well-being: Synthesis.* Washington, DC: World Resources Institute.

Nord M, Andrews M, Carlson S, 2005. Household food security in the United States, 2004. *Economic Research Report Number 11,* Economic Research Service: USDA.

Phiri NA, Karanja D, Kimani M, Spence NJ, 2007. Integrated management practices for the control of important crop diseases in developing countries. *Proceedings of the XVI International Plant Protection Congress,* 15–18 October 2007, Glasgow, Scotland, 734–735.

Pingali PL, Hossain M, Gerpacio RV, 1997. *Asian Rice Bowls: The Returning Crisis.* Wallingford, UK: CAN International.

Pretty J, 2008. Agricultural sustainability: Concepts, principles and evidence. *Philosophical Transactions of the Royal Society B* **363,** 447–465.

Richardson DM, 2005. The registration process, its effect on active substance availability, and initiatives to reduce the impact on minor crops at both UK and EU level. *Proceedings of the BCPC International Congress – Crop Science and Technology 2005,* Vol. 1, 231–238.

Scherr SJ, McNeely JA, 2008. Biodiversity conservation and agricultural sustainability: Towards a new paradigm of 'ecoagriculture' landscapes. *Philosophical Transactions of the Royal Society B* **363,** 477–494.

Sheikh AD, Rehman T, Yates CM, 2003. Logit models for identifying the factors that influence the uptake of new 'no-tillage' technologies by farmers in the rice–wheat and the cotton–wheat farming systems of Pakistan's Punjab. *Agricultural Systems* **75,** 79–95.

Thurow R, Solomon J, 2004. Want amid plenty, an Indian paradox: Bumper harvests and rising hunger. *Wall Street Journal* **June 25**, A1.

Whipps JM, 2007. Complex multitrophic interactions in the plant environment can affect disease biocontrol. *Proceedings of the XVI International Plant Protection Congress,* 15–18 October 2007, Glasgow, Scotland, 432–433.

Whittaker MS, 2007. Regulatory innovation and the biopesticide industry. *Proceedings of the XVI International Plant Protection Congress,* 15–18 October 2007, Glasgow, Scotland, 600–601.

Wood RKS, 1993. Sustainable agriculture: The role of plant pathology. *Abstracts,* 6th *International Congress of Plant Pathology,* July 28–August 6, 1993, Montreal, Canada, 1.

Index

Printed in the USA/Agawam, MA
January 14, 2011
556132.008